# Lecture Notes in Physics

Edited by H. Araki, Kyoto, J. Ehlers, München, K. Hepp, Zürich
R. Kippenhahn, München, D. Ruelle, Bures-sur-Yvette
H. A. Weidenmüller, Heidelberg, J. Wess, Karlsruhe and J. Zittartz, Köln
Managing Editor: W. Beiglböck



372

Flavio Dobran

# Theory of Structured Multiphase Mixtures

Springer-Verlag
Berlin Heidelberg GmbH

**Author**
Flavio Dobran
New York University
Department of Applied Science
New York, NY 10003, USA

ISBN 978-3-662-13852-6 ISBN 978-3-540-46812-7 (eBook)
DOI 10.1007/978-3-540-46812-7


Originally published by Springer-Verlag Berlin Heidelberg New York in 1991
Softcover reprint of the hardcover 1st edition 1991

2153/3140-543210 – Printed on acid-free paper

# Preface

The purpose of this monograph is to present a theory of multiphase mixtures with structure. The development of the theory is guided by physical principles without the sacrifice of mathematical rigor. The starting point in the development of the theory is based on the author's previous work, utilizing the volume averaging approach to motivate the construction of the theory. In the monograph, this basic idea is exploited further to introduce an additional set of transport equations for the purpose of modeling the structural properties of the mixture. As such, much of the material presented in the book is new and it should be useful for further work by graduate students and applied mathematicians and scientists. These studies may involve constitutive equations, wave propagation, numerical solution of partial differential equations, turbulence, and extensions of the theory to include higher order material deformations.

The prerequisite to understanding the mathematics in this book requires an elementary knowledge of algebra, geometry, and calculus. It should be readily accessible to students having backgrounds in fluid mechanics and thermodynamics in which they have been exposed to the cartesian tensor analysis. To make the book selfcontained, an appendix is provided in which the basic mathematical notions of algebra, geometry, and calculus are reviewed. The employed notation is by no means established in the field, but the one chosen has been extensively used in the literature of the single phase multicomponent mixture theories and many workers in this field are familiar with it.

The book is divided into eight chapters. Chapter one summarizes important historical milestones leading to the development of structured models of multiphase mixtures. In particular, the need for reconciliation between different approaches in modeling multiphase mixtures and the inclusion of structural characteristics of mixtures within the theory are discussed and a set of objectives for the book are stated. The development of the theory is carried out in chapter two, whereas in chapter three this development is continued on the basis of a very general principle of the material deformation with respect to the center of mass of each phase. The restriction of the resulting set of field equations by the principle of material frame-indifference is discussed in chapter four. The special cases of the field equations are compared with the existing models in chapter five. Chapter six includes the development of concepts and principles of the constitutive theory which is used in chapters seven and eight to study constitutive equations. The latter chapters include discussions of special results for two-phase mixtures involving compressible and incompressible phases, and concentrated and dilute suspensions with and without negligible inertial effects.

New York City
May, 1989.

Flavio Dobran

# Contents

## CHAPTER 6



## CHAPTER 7

# APPENDIX

CHAPTER 1

# INTRODUCTION

The purpose of this chapter is to summarize significant historical milestones dealing with the formulation of theories of structured multiphase mixtures. After this presentation, the most significant open problems dealing with the formulations are stated so that these may become the guiding principles in the development of the theory. The last section in the chapter summarizes the general scheme of notation and presents a list of symbols used in the monograph.

## 1.1 Historical Milestones

A *multiphase mixture is characterized by the well-defined interfacial areas between the constituents of the mixture*, with *each phase having a smoothly varying chemical composition and steep gradients of properties close to the interface.* The interface between the phases is usually only a few molecular diameters thick, across which mass, momentum and energy transfer processes take place. A *multicomponent mixture* differs from a multiphase mixture in that no steep gradients of properties between the constituents of the mixture can be *macroscopically* discerned. In the mathematical modeling of multiphase mixtures it may be reasonable, therefore, to utilize the continuum field equations in each phase, and provide the boundary conditions between the phases by another set of continuum field equations derived on the assumption that the interface can be modeled as a surface of discontinuity (ERINGEN,1975), or that the three-dimensional region of the interface can be assigned with appropriate density and fluxes (DEEMER & SLATTERY,1978). Due to the complexity of continuum field equations and a large number of interfaces which are commonly present in a real multiphase flow, the modeling approach described above must be abandoned in favor of a more practical one.

Two modeling approaches, both of which ignore the three-dimensional interfacial structure, have been extensively used in the formulation of practical models of multiphase mixtures. These are: (1) theories of mixtures based on the averaging procedure, and (2) theories constructed on the basis of single phase multicomponent mixtures or postulatory theories of multiphase mixtures. The former theories are constructed by averaging the local macroscopic conservation and balance equations of each phase over suitable time and space segments and using the averaged variables for the description of the mixture dynamics and energetics. In the postulatory approach the formulation of multiphase mixture field equations is basically postulated or guided by the theory of single phase multicomponent mixtures or "generalized variational principles."

The result from the space and time averaging approaches is that the resulting averaged equations are all similar to each other, with the difference between them being expressed by the interpretation of the averaged field variables. Beyond this apparent similarity there are other serious difficulties associated with the area, segment, and time averaging procedures that render these averaging approaches undesirable. These difficulties pertain to singularity problems in the resulting equations when interfaces become stationary in the flow and the difficulty associated with using the principle of material frame-indifference in the time-averaged equations (DOBRAN,1985a). For these reasons, DOBRAN (1984a,1985a) advocates the use of the volume averaging approach for the purpose of giving the theory of multiphase mixtures an *existence property*, or motivating the construction of a theory by physical arguments that are well-established. This volume averaging approach is then shown to assign a mapping transformation from the Euclidean space which contains the phases of a multiphase mixture into points of a subspace of the original Euclidean space, where at every point of this subspace coexist all phases of the mixture as superimposed continua. The theory constructed in this way is shown to be consistent with the theory of single phase multicomponent mixtures (BOWEN,1976) when the interfacial structure of phases is disregarded. The apparent drawback of the averaging procedure is that the averaging process destroys the local structural information of the mixture - an information that somehow needs to be suitably accounted in the theory. The structural properties of multiphase mixtures in the averaged theories have been accounted principally in the constitutive equations. BEDFORD & DRUMHELLER (1983) and DOBRAN (1984b) discuss this in greater detail and these works should be referred to for specific examples.

The theories of multiphase mixtures based on various postulates are also superimposed continua models, except that their motivation is clouded by the single phase multicomponent mixtures, generalized variational approaches, and specialized models of the alleged material deformation. The structural characteristics of multiphase mixtures in these theories are accounted through the use of additional transport equations as well as by the constitutive equations. The use of additional transport equations to model the structural characteristics of mixtures has been advocated by GOODMAN & COWIN (1972), PASSMAN (1977), and AHMADI (1985), among others.

The inclusion of additional balance equations within the framework of the classical theory based on the balance of mass, momentum, energy, and entropy inequality to model the mixture's structural characteristics should, of course, have a strong foundational basis. At the present, however, the number and form of these equations is a speculation, although the selections appear to be guided by reasonable physical arguments. The modeling of structural characteristics of mixtures by the additional balance equations is preferable to the modeling by the constitutive equations, since the former equations are required to be satisfied by arbitrary thermokinetic processes within the restrictions imposed by the extrinsic and intrinsic sources.

When the length scale of an external physical effect or disturbance is comparable to the size of an average grain or particle in a mixture, the intrinsic motions of grains should be taken into consideration in the gross description of the material behavior. The early work in this area was concerned with the solid material with internal structural properties or *polar* materials, and dates back to the COSSERAT (1909) brothers who formulated a theory of deformable bars, surfaces and bodies, with each point in the body endowed with vectors (or directors in the language of TRUESDELL & NOLL (1965)), for the purpose of accounting within the generalized theory of elasticity the body couples and couple stresses, since the CAUCHY's second law of motion on the symmetry of stress tensor excludes these notions from the existence. With the realization of the possibility for the existence of nonsymmetric stress tensors in structured materials came the need for the generalization of the concepts of momentum and energy, and the proposals for additional balance equations. Further significant works in this direction to construct theories of structured media may be traced to GRAD (1952), polyatomic gases; ERICKSEN (1960), anisotropic fluids; TOUPIN (1964), hyperelastic materials; ERINGEN & SUHUBI (1964) and ERINGEN (1964,1967), micropolar elasticity and fluids; TWISS & ERINGEN (1971), single phase mixtures of micromorphic and micropolar materials; and BOWEN (1982), porous media.

A continuum theory of granular material with voids was proposed by GOODMAN & COWIN (1972) with the additional balance equations for the equilibrated inertia and equilibrated force where they treated the volumetric fraction of grains as an independent kinematical variable. The evolution of the equilibrated inertia and force is assumed to be governed by the extrinsic and intrinsic supplies in the form of equilibrated stress, body forces, inertia force, and corresponding supplies or interaction fields. Consequently, the energy equation for the granular media contains the additional terms to reflect the effect of the additional kinematical variables and the extrinsic and intrinsic fields present in the theory. Using this theory and linearized constitutive equations which are restricted by an entropy inequality, GOODMAN and COWIN were able to predict the MOHR-COULOMB yield criterion for the flow of granules and the model was shown to exhibit "dilatancy" (GOODMAN & COWIN,1971), or that regions of high shearing are characterized by a low solid volume fraction, whereas the low shearing regions are characterized by a high solid volume fraction. The GOODMAN and COWIN's theory was subsequently extended by PASSMAN (1977) who generalized the theory by generalizing the equilibrated force equation through the addition of equilibrated force and inertia body force. The subsequent developments of this theory and the construction of constitutive equations which are restricted by the entropy inequality for the mixture as a whole may be found in PASSMAN *et al.* (1984).

BEDFORD & DRUMHELLER (1980) developed a theory of multiphase mixtures based on the "Hamilton's extended variational principle," modeled the structural properties of mixtures through the virtual mass and expansion-contraction effects, and utilized the entropy inequality for the mixture as a whole to restrict the constitutive equations. The constitutive assumptions in the works of PASSMAN and coworkers mentioned above involve, among other variables, the volumetric fractions and actual phase densities as independent constitutive variables, whereas BEDFORD and DRUMHELLER involve instead the actual and partial phase densities. This difference in the choice of independent variables does not appear to be significant, but the "derivation" approach of the field equations *is*, since the validity of the proposed variational principle on which the derivation of field equations rests is in doubt due to the difficulty in clearly separating the field equations from the constitutive equations.

In a series of papers AHMADI (1980,1982,1985) used a varied set of additional balance equations to study the fluid-granular media and developed constitutive equations using the entropy inequality for a mixture as a whole. His additional balance equations include equations for the equilibrated inertia, equilibrated force, and micro-inertia which are chosen from the works of GOODMAN & COWIN (1972) and single phase polar media theories of ERICKSEN (1960) and ERINGEN (1964) as noted above.

The balance equations for the equilibrated inertia and force of discrete mass points and extended to continua with affine structure and chemical reactions, have been rigorously treated by CAPRIZ & PODIO-GUIDUGLI (1981). They showed that if the usual cross product of the moment equation of mass points is generalized to a tensor product, then the symmetric part yields an equation for the equilibrated force, whereas the skew-symmetric part yields the usual angular momentum equation. By *extending* the concepts of kinematical and dynamical variables from the mass point mechanics to the continuous bodies and forming a generalized moment of internal forces and a hyperstress field (a third-order tensor field), CAPRIZ and PODIO-GUIDUGLI derived a balance equation for the equilibrated inertia and generalized the energy equation for the additional work terms. Since their generalized angular momentum and energy equations for the continuum are based on the concepts of mass point mechanics, it is not altogether clear that the resulting field equations are unique, but their work does show that GOODMAN & COWIN's (1972) equilibrated inertia equation is very special in the sense that it includes trivial material deformation processes.

## 1.2 Scope of the Monograph

From the above, it is clear that there is no agreement on the proper set of conservation and balance equations of multiphase mixtures. The averaged theories together with the constitutive equations which model the structural characteristics of mixtures have been extensively used to model many technological processes, whereas the theories of multiphase mixtures of the postulatory varieties which include the additional balance equations have demonstrated their prediction capabilities of many physical phenomena such as the MOHR-COULOMB yield criterion in granular media, the REYLEIGH's bubble equation (cit. VAN VIJNGAARDEN, 1972), dilatancy in granular media, wave dispersion, etc. Since both the averaged and postulatory theories of multiphase mixtures predict a wide variety of physical phenomena it is difficult to dismiss either one of them, and one is led to the search for an answer to the following questions:

1. How strong is the evidence for the existence of additional balance equations in the theory of multiphase mixtures?
2. What is the number and proper form of these equations?
3. Is there a direct correspondence between the averaged and postulatory theories of multiphase mixtures?

In this monograph the three questions posed above will be addressed. Starting from the volume averaging approach and certain restrictions of the motion of phases relative to the center of mass it will be shown that it is possible to derive the additional balance equations advocated in the postulatory approach. Moreover, these additional equations will be shown to be of great generality and to reduce to the form advocated in the postulatory theories of mixtures only under very restricted assumptions of the material deformation. After demonstrating the utility of the theory to reduce to special models, it will be shown how it may be used to model the structural properties of mixtures not possible with the present models, and how, by relaxing certain deformation assumptions, a hierarchy of more complex theoretical structures may be developed.

## 1.3 General Scheme of Notation and List of Symbols

In the book, the direct and tensorial notations are used with the symbols defined below in the list of symbols and in the text when they first appear. A more detailed review of the theorems of algebra, geometry, and calculus is presented in the appendix.

The underlying space is the three-dimensional Euclidean space $\mathcal{E}^3$ with the Cartesian coordinate system. In the direct notation, vectors and vector fields are usually denoted by the Latin or Greek bold-faced miniscules: **a**, **b**, $\boldsymbol{\xi}$,..., whereas second order tensors are viewed as linear transformations in $\mathcal{E}^3$ and are usually denoted by Latin or Greek bold-faced majusculus: **A**, **B**, **T**, $\boldsymbol{\Psi}$,... (except **X** and $\boldsymbol{\Sigma}$ which are vectors of the reference position of the material body point). The product of two linear transformations is a linear transformation, $\mathbf{AB} = \mathbf{C}$, and the product of a linear transformation and a vector is a vector, $\mathbf{x} = \mathbf{Tu}$, $\mathbf{T}^T$ is the transpose of **T**, $\mathbf{T}^{-1}$ is the inverse of **T**, $det\mathbf{T}$ is the determinant of **T**, and the contraction of **T** is its trace $tr\mathbf{T}$. The divergence operator is denoted by $\nabla\cdot$, $\nabla a$ is the gradient of $a$, and the gradient of a vector field **b** is $\nabla\mathbf{b}$. **I** is the unit linear transformation. The tensor product of two vectors **a** and **b** is denoted by $\boldsymbol{a} \times \boldsymbol{b}$, and it is identified as the linear transformation.

The tensorial indices are denoted by the italic light-faced minisculus, *i*,*j*,..., and the summation convention always applies to them. A comma with a tensorial index following a tensorial variable $a_i$, $a_{i,j}$, denotes the gradient (or divergence or trace operation if $i$=$j$), and in the Cartesian system it is a spatial partial derivative, whereas in a general coordinate system it must be interpreted as the covariant derivative. The Greek light-faced symbols $\alpha$ and $\beta$ occur as subscripts or superscripts and denote the phases of the multiphase mixture, $\alpha = \beta = 1, ..., \gamma$, where $\gamma$ denotes the total number of phases present in the mixture. The summation of phases is always denoted explicitly by the summation symbol $\sum_\alpha =^\gamma \sum_{\alpha=1}$.

Index of symbols

| Symbol | Description | Section of first occurrence |
|---|---|---|
| $a$ | Coefficient in the intrinsic stress moment and hyperstress | (7.5) |
| $\mathbf{a}$ | Acceleration | (3.1) |
| $A$ | Area, coefficients in the heat flux vector and intrinsic stress moment | (2.2),(8.3) |
| $b$ | Coefficient in the intrinsic stress moment and hyperstress | (7.5) |
| $\mathbf{b}$ | Body force per unit mass | (2.2) |
| $\mathbf{b}^{(\alpha)}$ | Frame-indifferent gyration tensor | (4.3) |
| $B$ | Coefficient in hyperinertia | (8.3) |
| $\mathbf{B}$ | Variable defined in Table 2.1, left CAUCHY-GREEN tensor | (2.2),(6.8 |
| $c$ | Coefficient in the intrinsic stress moment and hyperstress, cohesion | (7.5),(7.8) |
| $\hat{c}$ | Mass source | (2.4) |
| $\mathbf{C}$ | Right CAUCHY-GREEN tensor | (6.8) |
| $d$ | Coefficient in the intrinsic stress moment and hyperinertia | (7.5) |
| $\mathbf{D}$ | Symmetric part of velocity gradient | (4.3) |
| $e, \hat{e}$ | Coefficients in the stress tensor | (7.5),(7.6) |
| $\hat{e}$ | Energy source | (2.4) |
| $\breve{e}$ | Energy variable | (3.4) |
| $E$ | Coefficient in the source inertia | (7.5) |
| $\mathcal{E}$ | Euclidean space | (2.2) |
| $\hat{f}$ | Energy variable | (7.4) |
| $\tilde{\mathbf{f}}$ | Specific surface traction moment | (3.4) |
| $\mathbf{F}$ | Deformation gradient | (2.4) |
| $g_{\kappa}$ | Isotropy group | (6.9) |
| $\hat{\mathbf{g}}$ | Equilibrated moment source | (3.4) |
| $\breve{\mathbf{g}}$ | Variable | (3.4) |
| $G$ | Constitutive response functional | (6.2) |
| $\bar{\mathbf{h}}$ | Variable | (3.4) |
| $H$ | Coefficient in surface traction moment | (7.5) |
| $\mathbf{H}$ | Unimodular tensor | (6.7) |
| $\tilde{\mathbf{i}}$ | Equilibrated inertia | (3.2) |
| $\hat{\mathbf{i}}$ | Equilibrated source inertia | (3.2) |

Index of symbols

| Symbol | Description | Section of first occurrence |
|---|---|---|
| $I$ | Coefficient in the Helmholtz potential | (7.5) |
| $\mathbf{I}$ | Unit tensor | (4.1) |
| $\hat{\mathbf{I}}$ | Hyperinertia | (3.2) |
| $\mathbf{J}$ | Variable defined in Table 2.1 | (2.2) |
| $\hat{\mathbf{k}}$ | Variable | (5.2) |
| $\mathbf{k}, \check{\mathbf{k}}, \mathbf{K}$ | Variables | (3.4) |
| $K$ | Coefficient in surface traction moment | (7.5) |
| $\tilde{\boldsymbol{\ell}}$ | Body force moment | (3.2) |
| $\tilde{\boldsymbol{\ell}}_\epsilon$ | Internal energy density moment | (3.3) |
| $\mathbf{L}$ | Velocity gradient | (2.4) |
| $m$ | Mass transfer rate | (2.2) |
| $\check{\mathbf{m}}$ | Linear momentum source | (3.4) |
| $\hat{\mathbf{m}}, \hat{\mathbf{M}}$ | Angular momentum sources | (2.4) |
| $M$ | Coefficient in the traction force | (7.5) |
| $\mathbf{M}_\alpha$ | Concentration gradient tensor | (8.5) |
| $\mathbf{n}$ | Unit normal vector | (2.2) |
| $N$ | Normal stress, coefficient in the interaction force | (7.8),(8.3) |
| $O$ | Coefficient in the stress tensor | (7.5) |
| $p$ | Hydrostatic pressure | (6.12) |
| $\hat{\mathbf{p}}$ | Linear momentum source | (2.4) |
| $P$ | Interphase pressure | (7.4) |
| $\mathbf{q}$ | Heat flux vector | (2.2) |
| $\bar{q}_s$ | Interphase heat supply rate | (3.3) |
| $\mathbf{Q}$ | Orthogonal tensor | (4.1) |
| $r$ | Heat generation rate per unit volume | (2.2) |
| $R$ | Bubble radius | (7.7) |
| $\mathbf{R}$ | Rotation tensor | (6.8) |
| $s$ | Specific entropy | (2.2) |
| $\hat{s}$ | Entropy source | (2.4) |
| $S$ | Set of ind. variables, shear stress | (7.4),(7.8) |
| $\mathbf{S}$ | Surface velocity | (2.2) |
| $\bar{\mathbf{S}}$ | Surface traction moment | (3.2) |
| $t$ | Time | (2.2) |
| $t_o$ | Reference time | (2.2) |

**Index of symbols**

| Symbol | Description | Section of first occurrence |
|---|---|---|
| $\mathbf{t}$ | Traction force | (2.4) |
| $\bar{\mathbf{t}}$ | Interaction force | (3.3) |
| $T$ | Coefficient in surface traction moment | (7.5) |
| $\mathbf{T}$ | Stress tensor | (2.2) |
| $\hat{\hat{\mathbf{T}}}$ | Variable | (7.2) |
| $\mathbf{u}$ | Diffusion velocity | (2.4) |
| $U$ | Averaging volume | (2.2) |
| $\mathbf{U}$ | Frame-indifferent hyperinertia | (4.3) |
| $\mathbf{U}^t$ | Right stretch tensor | (6.8) |
| $\mathbf{v}$ | Velocity | (2.2) |
| $V$ | Reference volume | (2.2) |
| $\mathbf{V}$ | Left stretch tensor, vector of indep. constitutive variables | (6.8),(7.6) |
| $\mathbf{W}$ | Skew-symmetric part of velocity grad. | (4.3) |
| $\mathbf{x}$ | Spatial position vector | (2.2) |
| $\mathbf{X}$ | Reference position vector | (2.2) |
| $Z$ | Coefficient in the source inertia, bubble density | (7.5),(7.7) |
| Greek | | |
| $\alpha$ | Coefficient in the stress tensor | (8.5) |
| $\beta$ | Configuration pressure, coefficient in the stress tensor | (7.4),(8.5) |
| $\gamma$ | Coefficient in the traction force, coefficient in the stress tensor | (7.5),(8.5) |
| $\Gamma$ | Coefficient in the heat flux vector | (7.5) |
| $\delta$ | Kronecker delta | (3.1) |
| $\Delta$ | Coefficient in the traction force | (7.5) |
| $\boldsymbol{\Delta}$ | Interfacial source defined in Table 2.1 | (2.2) |
| $\boldsymbol{\nabla}$ | Gradient operator | (2.2) |
| $\epsilon$ | Specific internal energy, coefficient in the energy density moment | (2.2) |
| $\epsilon_{ijk}$ | Alternating tensor | (2.4) |
| $\hat{\epsilon}$ | Energy source | (2.4) |
| $\hat{\hat{\epsilon}}$ | Phase change energy flux | (3.3) |
| $\varepsilon$ | Measure of nonlinear effect | (7.5) |

Index of symbols

| Greek Symbol | Description | Section of first occurrence |
|---|---|---|
| $\zeta$ | Coefficient in the heat flux vector | (7.5) |
| $\theta$ | Temperature | (2.2) |
| $\Theta$ | Angle of internal friction | (7.8) |
| $\iota$ | Coefficient in the stress tensor | (7.5) |
| $\kappa$ | Coefficient in the heat flux vector | (7.5) |
| $\boldsymbol{\kappa}$ | Reference configuration | (2.4) |
| $\lambda$ | Coefficient in the stress tensor | (7.5) |
| $\Lambda$ | Phase surface within averaging volume | (2.2) |
| $\bar{\boldsymbol{\lambda}}$ | Intrinsic stress moment | (3.2) |
| $\bar{\boldsymbol{\lambda}}^R$ | Reduced intrinsic hyperstress | (7.4) |
| $\mu, \mu^o$ | Coefficients in the stress tensor | (7.5),(8.3) |
| $\boldsymbol{\mu}$ | Rotation vector | (5.1) |
| $\nu$ | Coefficient in the heat flux vector | (7.5) |
| $\boldsymbol{\nu}$ | Gyration tensor | (3.1) |
| $\hat{\boldsymbol{\nu}}$ | Rotation tensor | (5.1) |
| $\xi$ | Coefficient in the traction force | (7.5) |
| $\boldsymbol{\xi}$ | Position vector relative to the center of mass in the spatial configuration | (2.2) |
| $\boldsymbol{\Xi}$ | Tensor variable | (3.1) |
| $\bar{\bar{\pi}}$ | Thermodynamic pressure | (7.4) |
| $\Pi$ | Coefficient in surface traction moment | (7.5),(8.3) |
| $\boldsymbol{\Pi}$ | Tensor variable | (3.1) |
| $\rho$ | Mass density | (2.2) |
| $\boldsymbol{\Sigma}$ | Position vector relative to the center of mass in the spatial configuration | (2.2) |
| $\tau$ | Coefficient in phase change energy flux | (7.5) |
| $\boldsymbol{\tau}$ | Shear stress tensor | (6.12) |
| $\Upsilon$ | Set of constitutive variables | (6.2) |
| $\phi$ | Volumetric fraction | (5.1) |
| $\Phi, \Phi^o$ | Coefficients in the surface traction moment | (7.5),(8.3) |
| $\boldsymbol{\Phi}$ | Variable defined in Table 2.1 | (2.2) |
| $\chi$ | Coefficient in the source inertia | (7.5) |
| $\boldsymbol{\chi}$ | Configuration | (2.4) |
| $\boldsymbol{\chi}_{\boldsymbol{\kappa}}$ | Deformation function | (2.4) |

Index of symbols

| Greek Symbol | Description | Section of first occurrence |
|---|---|---|
| $\psi$ | Helmholtz potential | (6.12) |
| $\Psi$ | Coefficient in the source inertia | (7.5) |
| $\boldsymbol{\Psi}$ | Variable defined in Table 2.1 | (2.2) |
| $\omega$ | Coefficient in the energy variable | (7.5) |
| $\Omega$ | Coefficient in the source inertia | (7.5) |
| **Subscripts** | **Description** | **Occurrence** |
| $f$ | Fluid phase | (7.9) |
| $m$ | Mixture | (2.4) |
| $s$ | Solid phase | (7.9) |
| $\alpha, \beta, \delta, \sigma$ | Phases of multiphase mixture | (2.2) |
| **Superscripts** | **Description** | **Occurrence** |
| $\alpha, \beta, \delta$ | Phases of multiphase mixture | (2.2) |
| $\gamma$ | Number of phases in the mixture | (2.2) |
| $''$ | Internal constraint | (6.12) |
| $\grave{}$ | Material derivative | (2.4) |
| **Special Sym.** | **Description** | **Occurrence** |
| $< F >$ | Volume-averaged variable | (2.3) |
| $\tilde{F}$ | Density-weighted average variable | (2.3) |
| $\bar{F}$ | Partial average variable | (2.3) |
| $\bar{\bar{F}}$ | Phase average variable | (2.3) |
| $\mathcal{F}$ | Framing | (2.3) |
| $\sum$ | Summation | (2.3) |

CHAPTER 2

# FIELD EQUATIONS OF MULTIPHASE MIXTURES

The construction of a proper theory of multiphase mixtures calls for a strong foundation of the basic principles on which the theory should be built. For this purpose, the utilization of an averaging procedure to construct a set of conservation and balance equations for multiphase mixtures is strongly desirable, provided, of course, that a suitable averaging procedure exists. The suitability of an averaging procedure must be judged on the basis of its capability to yield physical models of sufficient generality such that the resulting field equations can model many known phenomena and be able to predict new ones. Moreover, such an averaging procedure should not produce inconsistencies when combined with the established constitutive principles and it should be able to produce field equations of multiphase mixtures which reduce to the single phase multicomponent mixture theory when the interfacial structure of phases is disregarded. With these requirements, a perfect averaging procedure may not exist, and a theory built upon an incomplete truth must be open to criticism and critical evaluation. Being able to reproduce the simplest physical phenomena first is a necessary attribute of any theory, and a theory which possesses many necessary conditions for its existence is highly desirable. The purpose of this chapter is to present a set of averaged conservation and balance equations of multiphase mixtures obtained by using the *volume averaging* procedure, after discussing other averaging approaches which appear to be much less desirable than the one selected to construct a theory. The choice of the volume averaging approach to motivate the construction of a theory of structured multiphase mixtures is highly desirable, since this choice yields a set of multiphase field equations which are able to account for the structure of multiphase mixtures.

## 2.1 Choice of the Averaging Approach

In formulating theories of multiphase mixtures, two general approaches have been followed: (1) the *averaging approach* where the local macroscopic conservation and balance equations of each phase are integrated over suitable time and space segments to derive an averaged or macro-macroscopic set of field equations for multiphase mixtures, and (2) the

***postulatory approach*** where the multiphase mixture equations are essentially postulated or their derivation rests upon the specialized models of material deformation. Because of very complex thermomechanical processes that may occur in multiphase flows, the latter approach has not so far produced a theory with strong *physical* foundation and the equations may not even be *consistent* with averaging procedures. The consistency of a theory of multiphase mixtures with a suitable averaging procedure is highly desirable since the fundamental macroscopic conservation and balance equations within each phase are well-established. In either of these two approaches, the details associated with the motion of interfaces are explicitly ignored in the theory and the prohibitive problem of tracking all phase interfaces of the mixture is reduced to a much more managable one which only accounts for the gross material description. An averaging approach yielding a set of macro-macroscopic field equations has the tendency to destroy the local structural information of the mixture, leading many advocates of the postulatory approach to dismiss the averaging approach altogether on the grounds that it cannot produce a theory of mixtures with sufficient generality. It will be demonstrated in this monograph that the local structural information of the mixture need not be neglected by the averaging process and that it is possible to construct various approximations of this information and produce many results of the postulatory theory of mixtures as *special cases*.

An averaged set of multiphase field equations may be obtained by utilizing temporal, spatial, or statistical averaging. DREW (1971) carried out time averaging of the global or integral formulation, whereas ISHII (1975) performed time averaging of the local conservation and balance equations of each phase. The spatial averaging may consist of segment, area and volume averaging, and it has been performed to various degrees of completeness by DELHAYE & ACHARD (1977), TRAP (1976), NIGMATULIN (1979), HASSANIZADEH & GRAY (1979), and DOBRAN (1981,1985). Time, segment, and area averaging procedures produce *singularities* in the resulting set of field equations whenever an interface becomes stationary in the flow field, or its unit normal vector is parallel to the surface normal when this surface crosses an averaging area in the flow. The additional drawbacks of the time averaging procedure are that the effect of *multiphase turbulence* is not clearly separable and that it precludes a *consistent* utilization of the *principle of material frame-indifference* on the averaged field equations, which is an important and powerful tool used widely in continuum mechanics to study the constitutive equations. The principle of the material frame-indifference (see chapter 4) involves a time-dependent rigid body rotation invariance, and the equations or physical variables which are invariant before time averaging are not necessarily invariant after the time averaging. A special case of the principle of material frame-indifference contains the Galilean group of transformations which, of course, *is* invariant under time averaging, since this change of frame transformation does not require the time dependent rigid body invariance. Moreover, a time-averaged set of multiphase mixture equations may not be appropriate for applications involving rapidly changing physical processes where the selection of an appropriate time averaging interval may be unknown.

The volume averaging procedure consists of averaging the local field equations of each phase over an arbitrary volume in space that may be fixed or moving. NIGMATULIN (1979) used the fixed volume averaging to derive *two-phase* flow equations for bubbly liquids, neglected surface tension and surface thermal effects at the interfaces between the phases as well as the discussion of the entropy equations for the mixture. HASSANIZADEH & GRAY (1979) also ignored the interfacial sources in their volume averaging approach containing a preselected set of admissible averaged variables. DOBRAN (1985) used the fixed volume averaging to derive an averaged set of balance of mass, linear momentum, angular momentum, and energy equations, and an entropy inequality for each phase as well as for the mixture as a whole. He also accounted for interfacial sources and rigorously *defined the kinematical concepts* in the theory which are indispensible for the proper study of constitutive equations (DOBRAN,1984b). It is also possible for the volume-averaged equations to be discontinuous, even if the local macroscopic variables are differentiable at each point in the flow field whenever an intersection of the surface of averaging volume with the surface of a phase interface has a nonempty intersection. However, this discontinuity is removable through an alternate definition of the averaging volume. The problems associated with the singularity of governing equations or discontinuity of averaged variables may be eliminated through the use of two or more averagings (DREW,1971) or by multiplying the local macroscopic field equations by a suitable weighting function before carrying out the averaging, and then defining the averaged variables in terms of this function (CELMINS,1984). As noted above, when use is made of the volume averaging procedure it is not necessary to employ a weighting function to make the approach legitimate and such will not be used in this book. In view of this and other positive feature of accounting for the *structure* of the mixture (see chapter 3 and subsequently), the volume averaging procedure appears to be the most desirable to *motivate* the construction of a *physical theory* of multiphase mixtures and it will be used in the book to construct *a theory of structured mixtures.* The development of the theory will be based on the prior work by the author (DOBRAN,1982a,1982b,1984a,1985a) where the full potential of the volume averaging approach was not exploited, since the local structural information of the mixture which is apparently destroyed through the averaging procedure was not reintroduced into the global or averaged form of the field equations, except, possibly, through the constitutive equations (DOBRAN,1984b), for some special circumstances of multiphase flows. The extension of this work which involves the inclusion of structural characteristics into the theory of multiphase mixtures based on the volume averaging procedure was first presented in a seminar by the author (DOBRAN,1985b), which in the considerably expanded form forms a basis for this monograph.

## 2.2 Basic Kinematical Concepts and Local Macroscopic Field Equations

A multiphase flow field in an Euclidean space $\mathcal{E}_o^3$ is illustrated in Figure 2.1. It consists of continua or phases $\alpha = 1, ..., \gamma$, where each phase consists of a single (homogeneous) chemical constituent. (It is easy to extend the theory to the situations where one or more phases consist of more than one chemical constitutent.) With each phase $\alpha$ may be identified a subvolume $U^{(\alpha\delta)}$ and observed its motion through space. As a result of the mass, momentum, and energy transfer processes taking place within the mixture, the volume $U^{(\alpha\delta)}$ changes with time, and at some initial or reference time $t_o$ its volume may be identified as $V^{(\alpha\delta)}$. The surface of the volume $U^{(\alpha\delta)}$ will be denoted by $A^{(\alpha\delta)}$ which has a unit normal vector $\mathbf{n}^{(\alpha\delta)}$ and velocity $\mathbf{S}^{(\Lambda\delta)}$. A material point P of phase $\alpha$ is located within $dU^{(\alpha\delta)}$ in the *spatial configuration* and within $dV^{(\alpha\delta)}$ in the *reference configuration.* The reference configuration may be thought as an undeformed and unstressed state of the multiphase mixture such that a material point P of phase $\alpha$ may be located by its position vector $\mathbf{X}^{(\alpha\delta)}$. During the course of time from the initial state $t_o$ to the present state $t$, the material undergoes mechanical and thermal deformations as a result of mass, momentum, and energy transfer processes taking place within the mixture, and the material point P moves to a position $\mathbf{x}^{(\alpha\delta)}$. The position vector $\mathbf{X}^{(\alpha\delta)}$ may be specified by the rectangular coordinates $X_1^{(\alpha\delta)}$, $X_2^{(\alpha\delta)}$ and $X_3^{(\alpha\delta)}$, or $X_K^{(\alpha\delta)}$, $K = 1,2,3$, whereas the vector $\mathbf{x}^{(\alpha\delta)}$ can be specified by its rectangular coordinates $x_1^{(\alpha\delta)}$, $x_2^{(\alpha\delta)}$ and $x_3^{(\alpha\delta)}$, or $x_k^{(\alpha\delta)}$, $k = 1,2,3$. The position vectors $\mathbf{X}^{(\alpha)}$ and $\mathbf{x}^{(\alpha)}$ denote the *center of mass positions of phase* $\alpha$ in the reference and spatial configurations, respectively, and the position vectors $\mathbf{\Sigma}^{(\alpha\delta)}$ and $\boldsymbol{\xi}^{(\alpha\delta)}$ represent the *positions* of the point P *relative to the center of mass* (defined mathematically in section 3.1) as shown in Figure 2.1, *i.e.*

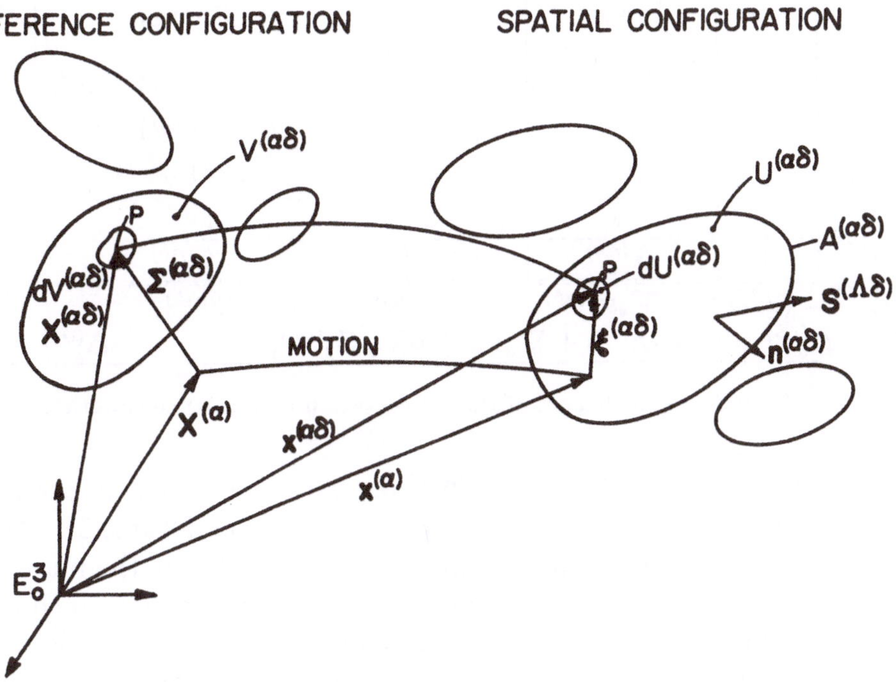


Figure 2.1: Representation of multiphase mixture in reference and spatial configurations

$$\mathbf{X}^{(\alpha\delta)} = \mathbf{X}^{(\alpha)} + \mathbf{\Sigma}^{(\alpha\delta)}, \quad X_K^{(\alpha\delta)} = X_K^{(\alpha)} + \Sigma_K^{(\alpha\delta)}; \quad K = 1,2,3 \tag{2.2.1}$$

$$\mathbf{x}^{(\alpha\delta)} = \mathbf{x}^{(\alpha)} + \boldsymbol{\xi}^{(\alpha\delta)}, \quad x_k^{(\alpha\delta)} = x_k^{(\alpha)} + \xi_k^{(\alpha\delta)}; \quad k = 1,2,3 \tag{2.2.2}$$

From (2.2.1) and (2.2.2) it follows that

$$x_k^{(\alpha)} = x_k^{(\alpha)}(\mathbf{X}^{(\alpha)}, t) \tag{2.2.3}$$

$$\xi_k^{(\alpha\delta)} = \xi_k^{(\alpha\delta)}(\mathbf{X}^{(\alpha)}, \mathbf{\Sigma}^{(\alpha\delta)}, t) \tag{2.2.4}$$

For each phase $\alpha$ and volume $U^{(\alpha\delta)}$, use may be made of the local macroscopic conservation and balance equations of mass, momentum, energy, and entropy for the non-polar media (see, for example, ERINGEN,1975) which may be expressed in the following compact form:

$$\frac{\partial}{\partial t}(\rho^{(\alpha\delta)}\mathbf{\Psi}^{(\alpha\delta)}) + \mathbf{\nabla}^o\cdot(\rho^{(\alpha\delta)}\mathbf{\Psi}^{(\alpha\delta)}\mathbf{v}^{(\alpha\delta)}) + \mathbf{\nabla}^o\cdot\mathbf{J}^{(\alpha\delta)} - \rho^{(\alpha\delta)}\mathbf{\Phi}^{(\alpha\delta)} = \rho^{(\alpha\delta)}\mathbf{B}^{(\alpha\delta)} \tag{2.2.5}$$

| Law | Mass | Linear Momentum | Angular Momentum | Energy | Entropy |
|---|---|---|---|---|---|
| $\mathbf{\Psi}^{(\alpha\delta)}$ | 1 | $\mathbf{v}^{(\alpha\delta)}$ | $\mathbf{r}\times\mathbf{v}^{(\alpha\delta)}$ | $\epsilon^{(\alpha\delta)}+\frac{1}{2}\mathbf{v}^{(\alpha\delta)}\cdot\mathbf{v}^{(\alpha\delta)}$ | $s^{(\alpha\delta)}$ |
| $\mathbf{J}^{(\alpha\delta)}$ | 0 | $-\mathbf{T}^{(\alpha\delta)}$ | $-\mathbf{r}\times\mathbf{T}^{(\alpha\delta)}$ | $\mathbf{q}^{(\alpha\delta)}-\mathbf{T}^{(\alpha\delta)T}\mathbf{v}^{(\alpha\delta)}$ | $\mathbf{h}^{(\alpha\delta)}$ |
| $\mathbf{\Phi}^{(\alpha\delta)}$ | 0 | $\mathbf{b}^{(\alpha\delta)}$ | $\mathbf{r}\times\mathbf{b}^{(\alpha\delta)}$ | $\mathbf{b}^{(\alpha\delta)}.\mathbf{v}^{(\alpha\delta)}$ | $R^{(\alpha\delta)}$ |
| $\mathbf{B}^{(\alpha\delta)}$ | 0 | $\mathbf{0}$ | $\mathbf{0}$ | $r^{(\alpha\delta)}$ | $\zeta^{(\alpha\delta)}$ |
| $\mathbf{\Delta}^{(\alpha\delta)}$ | 0 | $\mathbf{\Delta}_m^{(\alpha\delta)}$ | $\mathbf{r}\times\mathbf{\Delta}_m^{(\alpha\delta)}$ | $\Delta_\epsilon^{(\alpha\delta)}$ | $\Delta_s^{(\alpha\delta)}$ |

$\mathbf{\Delta}_m^{(\alpha\delta)}=(2H\nu\mathbf{n}+\mathbf{\nabla}_s\nu)^{(\alpha\delta)}$, $R^{(\alpha\delta)}=r^{(\alpha\delta)}/\theta^{(\alpha\delta)}$, $\zeta^{(\alpha\delta)}\geq 0$

$\Delta_\epsilon^{(\alpha\delta)}=(2H\nu\mathbf{n}\cdot\mathbf{S}+\mathbf{\nabla}_s\nu\mathbf{S}+\nu\mathbf{\nabla}_s\cdot\mathbf{S})^{(\alpha\delta)}$, $\Delta_s^{(\alpha\delta)}\geq 0$

Table 2.1: Coefficients of the conservation and balance equations

where $\mathbf{\Psi}^{(\alpha\delta)}$, $\mathbf{J}^{(\alpha\delta)}$, $\mathbf{\Phi}^{(\alpha\delta)}$ and $\mathbf{B}^{(\alpha\delta)}$ are variables which depend on the particular conservation or balance law and are given in Table 2.1. $\mathbf{\nabla}^o$ is the gradient operator which operates in the space $\mathcal{E}_o^3$, $\rho^{(\alpha\delta)}$ is the mass density, and $\mathbf{v}^{(\alpha\delta)}$ is the velocity of a material point of phase $\alpha$ contained in $U^{(\alpha\delta)}$. Assuming that the interface can be modeled as a surface of discontinuity (in view of the discussion in chapter 1 this appears to be a very reasonable assumption), it follows that across a phase interface use may be made of the following *jump or boundary conditions* (ERINGEN,1975):

$$(m^{(\alpha\delta)}\mathbf{\Psi}^{(\alpha\delta)}+\mathbf{J}^{(\alpha\delta)}\mathbf{n}^{(\alpha\delta)})+(m^{(\beta\eta)}\mathbf{\Psi}^{(\beta\eta)}+\mathbf{J}^{(\beta\eta)}\mathbf{n}^{(\beta\eta)})=\mathbf{\Delta}^{(\alpha\delta)} \quad (2.2.6)$$

where

$$m^{(\alpha\delta)}=\rho^{(\alpha\delta)}(\mathbf{v}^{(\alpha\delta)}-\mathbf{S}^{(\Lambda\delta)})\cdot\mathbf{n}^{(\alpha\delta)} \quad (2.2.7)$$

is the interphase mass transfer rate and $\mathbf{\Delta}^{(\alpha\delta)}$ is the interfacial source term. This source term can arise from the surface tension and is given in Table 2.1 where H is the mean interface curvature, $\mathbf{\nabla}_s^o$ is the surface gradient operator and $\nu$ is the surface tension coefficient. Notice in (2.2.6) and (2.2.7) that $\alpha\neq\beta$, $\mathbf{n}^{(\alpha\delta)}=-\mathbf{n}^{(\beta\eta)}$ and $\mathbf{S}^{(\Lambda\delta)}=\mathbf{S}^{(B\eta)}$ for a subvolume $\delta$ of phase $\alpha$ in contact with a subvolume $\eta$ of phase $\beta$.

In Table 2.1, $\mathbf{T}^{(\alpha\delta)}$ is the stress tensor, $\mathbf{b}^{(\alpha\delta)}$ is the external body force per unit mass, $\mathbf{v}^{(\alpha\delta)}$ is the velocity, $\rho^{(\alpha\delta)}$ is the mass density, $\epsilon^{(\alpha\delta)}$ is the internal energy per unit mass, $\mathbf{q}^{(\alpha\delta)}$ is the heat flux vector, $s^{(\alpha\delta)}$ is the entropy per unit mass, $\mathbf{h}^{(\alpha\delta)}$ is the entropy flux vector, $\theta^{(\alpha\delta)}$ is the absolute temperature, $r^{(\alpha\delta)}$ is the heat generation rate per unit volume, and $\zeta^{(\alpha\delta)}$ and $\Delta_s$ are the entropy production rates in the continuum and at the interface, respectively. The condition $\zeta^{(\alpha\delta)}\geq 0$, or that the entropy production in each subvolume $\delta$ of phase $\alpha$ is greater than or equal to zero, represents *the local axiom of dissipation* normally adopted in single phase continua (TRUESDELL & NOLL,1965), and the reason for its use at this stage of the analysis is discussed further below in section 2.4.6.

Equations (2.2.6) and (2.2.7) provide the necessary boundary conditions across the interfaces for the solution of the field equations (2.2.5) when supplied by the appropriate constitutive equations and equations of state. As discussed above, such a solution is prohibitive in view of the large number of interfaces present in most practical multiphase mixtures, and the task of the volume averaging procedure is to average (2.2.5) over an arbitrary volume of space in order to reduce the complexity of dealing with the detailed motion of interfaces.

## 2.3 The Volume Averaging Procedure and General Phasic Equation of Balance

The volume averaging procedure involves selecting an *arbitrary fixed volume in space* $U$, in the spatial configuration of the mixture (shown in Figure 2.1) as illustrated in Figure 2.2. At time $t$ the union of subvolumes $U^{(\alpha\delta)}$ of phase $\alpha$ is the volume $U_\alpha = \sum_\delta U^{(\alpha\delta)}$ of phase $\alpha$ which is contained within $U$. The surface of $U^{(\alpha\delta)}$ which is fully contained within $U$ is denoted by $a^{(\Lambda\delta)}$, whereas the surface of intersection of $U^{(\alpha\delta)}$ with $U$ is denoted by $a^{(\alpha\delta)}$. With this notation, the volume averaging of the general balance equation (2.2.5) is performed for each phase over the portion of the volume $U$ which the phase occupies at time $t$, *i.e.* the following operation is carried out:

$$\sum_\delta \int_{U^{(\alpha\delta)}} (equation(2.2.5))\, dU = 0 \tag{2.3.1}$$

The volume averaging procedure has the mathematical property of mapping the entire contents of the flow field at time $t$ in the averaging volume located in the spatial configuration into a point $p(U)$ located at $\mathbf{x}$, such that the mapped configuration space $\mathcal{E}^3$ (to which $\mathbf{x}$ belongs) is the *subspace* of $\mathcal{E}_o^3$ as schematically illustrated in Figure 2.2.

The interchange of the integration and differentiation operators in (2.3.1) is connected with the mapping of $U$ in $\mathcal{E}_o^3$ which consists of continua $\alpha = 1, ..., \gamma$ into a point $p(U)$ in $\mathcal{E}^3$, such that all these continua exist at each point of this space. This mapping is onto and it may also be made one-to-one, since to each point $p(U)$ at the place $\mathbf{x}$ can correspond a unique $U$ at $\mathbf{r}_o = \mathbf{x}$, where $\mathbf{r}_o$ may be identified as the center of $U$. (The center $\mathbf{r}_o$ may be selected as the center of mass of the mixture in $U$ as advocated by DOBRAN(1985a), but this choice is not necessary at this stage of analysis.)

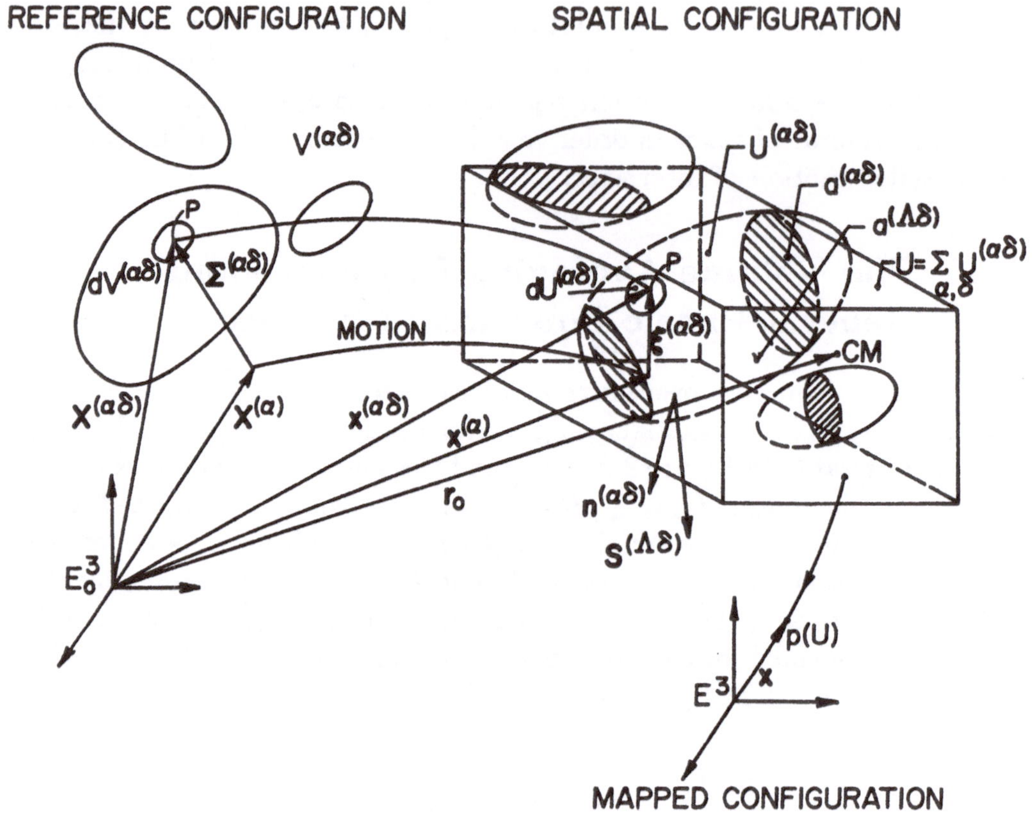


**Figure 2.2: Representation of a multiphase mixture in reference, spatial, and mapped configurations**

Returning to the problem of interchanging time and space integration and differentiation operators in (2.3.1), the following results from calculus may be used (see also, for example, DOBRAN(1985a)):

**Leibnitz's Theorem**

$$\sum_{\delta} \int_{U^{(\alpha\delta)}} \frac{\partial \mathbf{F}^{(\alpha\delta)}}{\partial t}\, dU = \frac{\partial}{\partial t} \sum_{\delta} \int_{U^{(\alpha\delta)}} \mathbf{F}^{(\alpha\delta)}\, dU - \sum_{\delta} \int_{a^{(\Lambda\delta)}} \mathbf{F}^{(\alpha\delta)} \mathbf{S}^{(\Lambda\delta)} . \mathbf{n}^{(\alpha\delta)}\, da \tag{2.3.2}$$

**Divergence Theorem**

$$\sum_{\delta} \int_{U^{(\alpha\delta)}} \boldsymbol{\nabla}^{o} . \mathbf{F}^{(\alpha\delta)}\, dU = \boldsymbol{\nabla} \cdot \sum_{\delta} \int_{U^{(\alpha\delta)}} \mathbf{F}^{(\alpha\delta)}\, dU + \sum_{\delta} \int_{a^{(\Lambda\delta)}} \mathbf{F}^{(\alpha\delta)} . \mathbf{n}^{(\alpha\delta)}\, da \tag{2.3.3}$$

where $\mathbf{F}^{(\alpha\delta)}$ represents a differentiable tensor field in $U^{(\alpha\delta)}$, and on $a^{(\alpha\delta)}$ and $a^{(\Lambda\delta)}$. Notice in (2.3.3) that $\boldsymbol{\nabla}^{o}$ operates in $\mathcal{E}_o^3$, whereas $\boldsymbol{\nabla}$ is an operator associated with the space $\mathcal{E}^3$. The derivation of (2.3.2) is given in any standard work on advanced calculus, whereas (2.3.3) follows by using the GREEN-GAUSS's divergence theorem and the fundamental theorem of calculus. Since the mappings (2.3.2) and (2.3.3) are also linear, it is possible to sum these equations over $\alpha$ and interchange the summation with the operators $\partial/\partial t$ and $\boldsymbol{\nabla}$.

Using the fundamental relations (2.3.2) and (2.3.3) in (2.3.1), the following *general equation of balance for the phase* $\alpha$ is obtained:

$$\begin{aligned} &\frac{\partial}{\partial t} U_\alpha < \rho_\alpha \boldsymbol{\Psi}_\alpha > + \boldsymbol{\nabla} \cdot U_\alpha < \rho_\alpha \boldsymbol{\Psi}_\alpha \mathbf{v}_\alpha > + \boldsymbol{\nabla} \cdot U_\alpha < \mathbf{J}_\alpha > - U_\alpha < \rho_\alpha \boldsymbol{\Phi}_\alpha > \\ &\qquad - U_\alpha < \rho_\alpha \mathbf{B}_\alpha > = - \sum_{\delta} \int_{a^{(\Lambda\delta)}} (m^{(\alpha\delta)} \boldsymbol{\Psi}^{(\alpha\delta)} + \mathbf{J}^{(\alpha\delta)} \mathbf{n}^{(\alpha\delta)})\, da \end{aligned} \tag{2.3.4}$$

where the volume-averaged quantities are defined as follows

$$< F_\alpha > = \frac{1}{U_\alpha} \sum_{\delta} \int_{U^{(\alpha\delta)}} F^{(\alpha\delta)}\, dU \tag{2.3.5}$$

Equation (2.3.4) may be reduced further by defining the *density-weighted average variables*, $\tilde{F}_\alpha$, *partial variables*, $\bar{F}_\alpha$, and *phase average variables*, $\bar{\bar{F}}_\alpha$, *i.e.*

$$\tilde{F}_\alpha = \frac{< \rho_\alpha F_\alpha >}{< \rho_\alpha >} = \frac{1}{\bar{\rho}_\alpha} \frac{U_\alpha}{U} < \rho_\alpha F_\alpha > \tag{2.3.6}$$

$$\bar{F}_\alpha = \frac{U_\alpha}{U} < F_\alpha > \tag{2.3.7}$$

$$\bar{\bar{F}}_\alpha =< F_\alpha > \tag{2.3.8}$$

The *partial density of phase* $\alpha$ is defined by the equation

$$\bar{\rho}_\alpha = \frac{U_\alpha}{U} < \rho_\alpha > \tag{2.3.9}$$

whereas the *mixture density* is obtained by summing up the partial densities of phases, *i.e.*

$$\rho = \sum_\alpha \bar{\rho}_\alpha \tag{2.3.10}$$

The *velocity of phase* $\alpha$, $\tilde{\mathbf{v}}_\alpha$, and the *mixture velocity*, $\mathbf{v}$, are defined by the following relations:

$$\tilde{\mathbf{v}}_\alpha = \frac{1}{\bar{\rho}_\alpha}\frac{U_\alpha}{U} < \rho_\alpha \mathbf{v}_\alpha > \tag{2.3.11}$$

$$\rho\mathbf{v} = \sum_\alpha \bar{\rho}_\alpha \tilde{\mathbf{v}}_\alpha \tag{2.3.12}$$

The above definition of phase velocity as the density weighted velocity can be motivated by specializing (2.3.4) for the case of conservation of mass from Table 2.1, *i.e.*

$$\frac{\partial}{\partial t}U_\alpha < \rho_\alpha > +\nabla\cdot U_\alpha < \rho_\alpha \mathbf{v}_\alpha >= -\sum_\delta \int_{a(\Lambda\delta)} m^{(\alpha\delta)}\, da \tag{2.3.13}$$

which upon using (2.3.11) and noting that $U$ is a constant, it is reduced to an equation that is similar to the form of the balance of mass equation of single phase multicomponent mixtures. Hence

$$\frac{\partial \bar{\rho}_\alpha}{\partial t} + \nabla\cdot\bar{\rho}_\alpha\tilde{\mathbf{v}}_\alpha = -\frac{1}{U}\sum_\delta \int_{a(\Lambda\delta)} m^{(\alpha\delta)}\, da \tag{2.3.14}$$

## 2.4 Basic Conservation and Balance Equations of Multiphase Mixtures

The basic conservation and balance equations of multiphase mixtures consist of the balance of mass, linear momentum, angular momentum and energy equations, and entropy inequality for each phase and for the mixture as a whole. The derivation of these equations from the general equation of balance (2.3.4) also rests upon the fundamental kinematical concepts that should be rigorously defined for the mapped configuration space $\mathcal{E}^3$ at which points all phases of the multiphase mixture exist as superimposed continua.

### 2.4.1 Kinematics of the Superimposed Continua

Similar to the theory of mixtures (BOWEN,1976), a multiphase mixture can be thought to consist of $\gamma$ diffusing bodies (phases) $\mathcal{B}_\alpha$, $\alpha = 1, ..., \gamma$, where each $\mathcal{B}_\alpha$ belongs to a differentiable manifold that is isomorphic to the three-dimensional Euclidean space $\mathcal{E}^3$. Each body $\mathcal{B}_\alpha$ consists of *body points* (particles) $X_\alpha$, and the *configuration* $\chi_\alpha$ of $\mathcal{B}_\alpha$ is a homeomorphism of $\mathcal{B}_\alpha$ into $\mathcal{E}^3$. Corresponding to the body points $X_\alpha$ and the time $t$, a *motion* is assigned to $\mathcal{B}_\alpha$, and the position $\mathbf{x}$ in $\mathcal{E}^3$ of the body points $X_\alpha$ at time $t$ may be represented by

$$\mathbf{x} = \chi_\alpha(X_\alpha, t) \tag{2.4.1}$$

To each $\mathcal{B}_\alpha$ may be assigned a *reference configuration* $\boldsymbol{\kappa}_\alpha$ at time $t_o < t$ of the superimposed continua, where the place of particle $X_\alpha$ in $\boldsymbol{\kappa}_\alpha$ is given by

$$\mathbf{X}_\alpha = \boldsymbol{\kappa}_\alpha(X_\alpha, t_o) \tag{2.4.2}$$

Combining (2.4.1) and (2.4.2) yields

$$\mathbf{x} = \chi_\alpha(\boldsymbol{\kappa}_\alpha^{-1}(\mathbf{X}_\alpha, t_o), t) = \chi_{\alpha\boldsymbol{\kappa}}(\mathbf{X}_\alpha, t) \tag{2.4.3}$$

where $\chi_{\alpha\boldsymbol{\kappa}}$ is the *deformation function* of $\mathcal{B}_\alpha$. Notice that the meanings of reference positions $\mathbf{X}_\alpha$ and $\mathbf{X}^{(\alpha\delta)}$ (in Figure 2.1) are different and must not be confused for the proper appreciation of the theory.

The expression (2.4.3) is assumed to be invertible for each phase $\alpha$ at each time $t$, *i.e.* $\mathbf{X}_\alpha = \chi_{\alpha\boldsymbol{\kappa}}^{-1}(\mathbf{x}, t)$, for the reason that $\mathbf{X}_\alpha$ cannot occupy two different positions in space and two particles $\mathbf{X}_{\alpha 1}$ and $\mathbf{X}_{\alpha 2}$ cannot occupy the same spatial positions. The deformation function $\chi_{\alpha\boldsymbol{\kappa}}$ is further assumed to be twice differentiable so that it is able to represent the velocity and acceleration in $\mathcal{E}^3$. The spatial coordinates will be expressed by the lower case italic indices, $\mathbf{x} = x_i \mathbf{e}_i$, whereas the material coordinates will be expressed by the upper case italic indices, $\mathbf{X}_\alpha = X_{\alpha I} \mathbf{e}_I$, where $\mathbf{e}_i \cdot \mathbf{e}_i = \delta_{ik}$ and $\mathbf{e}_I \cdot \mathbf{e}_K = \delta_{IK}$ are the Kronecker deltas.

With the above definition of the kinematic concepts, the *velocity* and *acceleration* of the body point $X_\alpha$ or phase $\alpha$ at time $t$ and place $\mathbf{x}$ are defined as follows:

$$\tilde{\mathbf{v}}_\alpha = \frac{\partial \boldsymbol{\chi}_{\alpha\kappa}(\mathbf{X}_\alpha, t)}{\partial t}, \quad \grave{\tilde{\mathbf{v}}}_\alpha = \frac{\partial^2 \boldsymbol{\chi}_{\alpha\kappa}(\mathbf{X}_\alpha, t)}{\partial t^2} \tag{2.4.4}$$

where the *backward primes* affixed to the symbols with $\alpha$ always indicate the *material derivative* following the motion of the $\alpha$'th phase. The *deformation gradient* of phase $\alpha$ is a linear transformation defined by

$$\mathbf{F}_\alpha = GRAD\boldsymbol{\chi}_{\alpha\kappa}(\mathbf{X}_\alpha, t), \quad F_{\alpha kJ} = \frac{\partial x_k}{\partial X_{\alpha J}} \tag{2.4.5}$$

$$\mathbf{F}_\alpha^{-1} = \nabla \boldsymbol{\chi}_{\alpha\kappa}^{-1}(\mathbf{x}, t), \quad F_{\alpha Jk}^{-1} = \frac{\partial X_{\alpha J}}{\partial x_k} \tag{2.4.6}$$

and the *velocity gradient* of phase $\alpha$ is given as

$$\mathbf{L}_\alpha = \nabla \tilde{\mathbf{v}}_\alpha(\mathbf{x}, t), \quad L_{\alpha ij} = \frac{\partial \tilde{v}_{\alpha i}}{\partial x_j} \tag{2.4.7}$$

Combining (2.4.4)-(2.4.7) yields

$$\mathbf{L}_\alpha = \grave{\mathbf{F}}_\alpha \mathbf{F}_\alpha^{-1} \tag{2.4.8}$$

Of particular usefulness in the discussion of field equations are two mathematical identities which are discussed by TRUESDELL & TOUPIN (1960) and DOBRAN (1985a), among others. If $\Gamma$ is a differentiable function of $\mathbf{x}$ and $t$, then the time derivatives following the motion of the $\alpha$'th phase and the mixture as a whole are given, respectively, as:

$$\grave{\Gamma} = \frac{\partial \Gamma}{\partial t} + (\nabla \Gamma)\tilde{\mathbf{v}}_\alpha, \quad \dot{\Gamma} = \frac{\partial \Gamma}{\partial t} + (\nabla \Gamma)\mathbf{v} \tag{2.4.9}$$

from which it follows that

$$\grave{\Gamma} - \dot{\Gamma} = (\nabla \Gamma)\mathbf{u}_\alpha \tag{2.4.10}$$

where

$$\mathbf{u}_\alpha = \tilde{\mathbf{v}}_\alpha - \mathbf{v} \tag{2.4.11}$$

is the *diffusion velocity* of phase $\alpha$.

If a mixture property $\Gamma$ is defined by the equation

$$\rho \Gamma = \sum_\alpha \bar{\rho}_\alpha \tilde{\Gamma}_\alpha \tag{2.4.12}$$

then from the above equations it follows that

$$\rho \dot{\Gamma} = \sum_\alpha [\bar{\rho}_\alpha \grave{\tilde{\Gamma}}_\alpha - \nabla\cdot \bar{\rho}_\alpha \tilde{\Gamma}_\alpha \mathbf{u}_\alpha + (\frac{\partial \bar{\rho}_\alpha}{\partial t} + \nabla\cdot \bar{\rho}_\alpha \tilde{\mathbf{v}}_\alpha)\tilde{\Gamma}_\alpha - (\frac{\partial \rho}{\partial t} + \nabla\cdot \rho \mathbf{v})\tilde{\Gamma}_\alpha] \tag{2.4.13}$$

### 2.4.2 Balance of Mass

The balance of mass equation of phase $\alpha$ is obtained from the general equation of balance (2.3.4) and Table 2.1 with the aid of definitions (2.3.9), (2.3.11), and $(2.4.9)_1$. Thus

$$\grave{\bar{\rho}}_\alpha + \bar{\rho}_\alpha \nabla \cdot \tilde{\mathbf{v}}_\alpha = \hat{c}_\alpha \tag{2.4.14}$$

where

$$\hat{c}_\alpha = -\frac{1}{U}\sum_\delta \int_{a(\Lambda\delta)} m^{(\alpha\delta)}\, da = \frac{1}{U}\sum_\delta \int_{a(\Lambda\delta)} \rho^{(\alpha\delta)}(\mathbf{S}^{(\alpha\delta)} - \mathbf{v}^{(\alpha\delta)})\cdot\mathbf{n}^{(\alpha\delta)}\, da \tag{2.4.15}$$

is the *mass supply or source of phase* $\alpha$. This mass source arises due to the phase change or chemical reactions.

To obtain an equation for the conservation of mass for the mixture as a whole, equation (2.4.14) is summed up over $\alpha$ and use made of (2.3.10), (2.3.12), $(2.4.9)_2$ and (2.2.6) with $\Delta^{(\alpha\delta)} = 0$ in Table 2.1. The result is

$$\dot{\rho} + \rho \nabla \cdot \mathbf{v} = 0 \tag{2.4.16}$$

with the condition that the mass sources satisfy

$$\sum_\alpha \hat{c}_\alpha = 0 \tag{2.4.17}$$

Substituting (2.4.14) and (2.4.16) into the identity (2.4.13) yields

$$\rho\dot{\Gamma} = \sum_\alpha [\bar{\rho}_\alpha \grave{\tilde{\Gamma}}_\alpha - \nabla\cdot\bar{\rho}_\alpha\tilde{\Gamma}_\alpha \mathbf{u}_\alpha + \hat{c}_\alpha\tilde{\Gamma}_\alpha] \tag{2.4.18}$$

### 2.4.3 Balance of Linear Momentum

The linear momentum balance of phase $\alpha$ is obtained from (2.3.4) by using the linear momentum coefficients from Table 2.1 and definitions (2.3.6)-(2.3.8), *i.e.*

$$\bar{\rho}_\alpha \grave{\tilde{\mathbf{v}}}_\alpha = \nabla\cdot\bar{\mathbf{T}}_\alpha + \bar{\rho}_\alpha \tilde{\mathbf{b}}_\alpha + \hat{\mathbf{p}}_\alpha \tag{2.4.19}$$

where the *linear momentum supply or source* is expressed by the following equation:

$$\hat{\mathbf{p}}_\alpha = -\hat{c}_\alpha\tilde{\mathbf{v}}_\alpha - \frac{1}{U}\sum_\delta \int_{a(\Lambda\delta)} (m^{(\alpha\delta)}\mathbf{v}^{(\alpha\delta)} - \mathbf{T}^{(\alpha\delta)}\mathbf{n}^{(\alpha\delta)})\, da - \nabla\cdot\mathbf{C}_{1\alpha} \tag{2.4.20}$$

$\mathbf{C}_{1\alpha}$ in the above equation is the covariance coefficient that is defined as

$$\mathbf{C}_{1\alpha} = \frac{U_\alpha}{U} < \rho_\alpha \mathbf{v}_\alpha \times \mathbf{v}_\alpha > -\bar{\rho}_\alpha \tilde{\mathbf{v}}_\alpha \times \tilde{\mathbf{v}}_\alpha \tag{2.4.21}$$

It expresses the effect of the finite interfacial area between the phases or the structural characteristics of the mixture. It will be seen in the next chapter how this quantity is determined when the structural characteristics of the mixture are accounted in the theory.

The *existence* of the stress tensor $\bar{\mathbf{T}}_\alpha$ is tied to the existence of $\mathbf{T}^{(\alpha\delta)}$ which is proved based on the Theorems of NOLL (1973). It may be sufficient to state here that this proof can be established from writing the EULER's first law of motion for a material contained in a tetrahedron with surface forces $\mathbf{t}$ acting on its surfaces. In the limit in the situation when the volume of the tetrahedron is reduced to zero and use made of the CAUCHY's lemma that the forces acting on either side of the surface are equal in magnitude and opposite in direction, $\mathbf{t}(\mathbf{x}, \mathbf{n}) = -\mathbf{t}(\mathbf{x}, -\mathbf{n})$, it is possible to associate $T_{ij}$ as the $i$'th component of the stress vector $\mathbf{t}$ acting on the positive side of the plane $x_j$=constant, thus establishing that $\mathbf{t} = \mathbf{T}\mathbf{n}$ or the existence of $\mathbf{T}$ for a *particular* surface normal vector. An alternate argument for the existence of $\bar{\mathbf{T}}_\alpha$ is presented in section 2.4.7.

The balance of the linear momentum for the mixture as a whole is obtained by summing up $\alpha$ in (2.4.19) and using (2.4.18) with $\tilde{\Gamma}_\alpha = \tilde{\mathbf{v}}_\alpha$. Hence

$$\rho\dot{\mathbf{v}} = \boldsymbol{\nabla}\cdot\mathbf{T} + \rho\mathbf{b} + \hat{\mathbf{p}} \tag{2.4.22}$$

where the stress tensor for the mixture $\mathbf{T}$, the mixture body force $\mathbf{b}$, and the mixture momentum supply or source $\hat{\mathbf{p}}$, are defined as follows:

$$\mathbf{T} = \sum_\alpha (\bar{\mathbf{T}}_\alpha - \bar{\rho}_\alpha \mathbf{u}_\alpha \times \mathbf{u}_\alpha) \tag{2.4.23}$$

$$\rho\mathbf{b} = \sum_\alpha \bar{\rho}_\alpha \tilde{\mathbf{b}}_\alpha \tag{2.4.24}$$

$$\hat{\mathbf{p}} = -\frac{1}{U} \sum_{\alpha,\delta} \int_{a(\wedge\delta)} \boldsymbol{\Delta}_m^{(\alpha\delta)}\, da - \boldsymbol{\nabla}\cdot \sum_\alpha \mathbf{C}_{1\alpha} \tag{2.4.25}$$

Summing up $\alpha$ in (2.4.20) and combining with (2.4.25) results in the condition for the momentum source, *i.e.*

$$\hat{\mathbf{p}} = \sum_\alpha (\hat{c}_\alpha \tilde{\mathbf{v}}_\alpha + \hat{\mathbf{p}}_\alpha) = \sum_\alpha (\hat{c}_\alpha \mathbf{u}_\alpha + \hat{\mathbf{p}}_\alpha) \tag{2.4.26}$$

### 2.4.4 Balance of Angular Momentum

From (2.3.4) and Table 2.1, the balance of the angular momentum for phase $\alpha$ can be expressed by the equation

$$\bar{\rho}_\alpha \overline{\mathbf{x}\times\grave{\mathbf{v}}_\alpha} = \nabla\cdot(\mathbf{x}\times\bar{\mathbf{T}}_\alpha) + \mathbf{x}\times(\bar{\rho}_\alpha\tilde{\mathbf{b}}_\alpha + \hat{\mathbf{p}}_\alpha) + \hat{\mathbf{m}}_\alpha \tag{2.4.27}$$

which upon using the linear momentum equation (2.4.19) is reduced to the form

$$\hat{\mathbf{m}}_\alpha + \mathbf{e}_i\epsilon_{ijk}\bar{T}_{\alpha kj} = 0 \tag{2.4.28}$$

where $\epsilon_{ijk}$ is the alternating symbol and $\hat{\mathbf{m}}_\alpha$ is the *angular momentum source* that is defined by the following equation:

$$\begin{aligned}\hat{\mathbf{m}}_\alpha = -\mathbf{x}\times(\hat{\mathbf{p}}_\alpha + \hat{c}_\alpha\tilde{\mathbf{v}}_\alpha) - \frac{1}{U}\sum_\delta\int_{a(\Lambda\delta)} \mathbf{r}\times\left(m^{(\alpha\delta)}\mathbf{v}^{(\alpha\delta)} - \mathbf{T}^{(\alpha\delta)}\mathbf{n}^{(\alpha\delta)}\right) da \\ -\bar{\rho}_\alpha\overline{(\mathbf{r}-\mathbf{x})\times\grave{\mathbf{v}}_\alpha} - \hat{c}_\alpha\overline{(\mathbf{r}-\mathbf{x})\times\mathbf{v}_\alpha} + \nabla\cdot\overline{(\mathbf{r}-\mathbf{x})\times\mathbf{T}_\alpha} \\ +\bar{\rho}_\alpha\overline{(\mathbf{r}-\mathbf{x})\times\mathbf{b}_\alpha} - \nabla\cdot\bar{\rho}_\alpha\overline{(\mathbf{r}\times\mathbf{v}_\alpha)(\mathbf{v}_\alpha-\tilde{\mathbf{v}}_\alpha)}\end{aligned} \tag{2.4.29}$$

By defining a skew-symmetric tensor $\hat{\mathbf{M}}_\alpha$ with components $\hat{m}_{\alpha k}$, *i.e.*

$$\hat{M}_{\alpha ij} = \epsilon_{ijk}\hat{m}_{\alpha k} \tag{2.4.30}$$

where $\epsilon_{ijk}$ is the alternating tensor, equation (2.4.28) may be written as

$$\hat{\mathbf{M}}_\alpha = \bar{\mathbf{T}}_\alpha - \bar{\mathbf{T}}_\alpha^T \tag{2.4.31}$$

The angular momentum equation for the multiphase mixture as a whole is obtained by summing up $\alpha$ in (2.4.27), *i.e.*

$$\rho\overline{\mathbf{x}\times\dot{\mathbf{v}}} = \nabla\cdot(\mathbf{x}\times\mathbf{T}) + \mathbf{x}\times\rho\mathbf{b} + \hat{\mathbf{m}} + \mathbf{x}\times\hat{\mathbf{p}} \tag{2.4.32}$$

where

$$\begin{aligned}\hat{\mathbf{m}} = \sum_\alpha\hat{\mathbf{m}}_\alpha = \mathbf{x}\times\nabla\cdot\sum_\alpha\mathbf{C}_{1\alpha} - \frac{1}{U}\sum_{\alpha,\delta}\int_{a(\Lambda\delta)}(\mathbf{r}-\mathbf{x})\times\mathbf{\Delta}_m^{(\alpha\delta)}\,da \\ -\sum_\alpha\bar{\rho}_\alpha\overline{(\mathbf{r}-\mathbf{x})\times\grave{\mathbf{v}}_\alpha} - \sum_\alpha\hat{c}_\alpha\overline{(\mathbf{r}-\mathbf{x})\times\mathbf{v}_\alpha} + \nabla\cdot\sum_\alpha\overline{(\mathbf{r}-\mathbf{x})\times\mathbf{T}_\alpha} \\ +\sum_\alpha\bar{\rho}_\alpha\overline{(\mathbf{r}-\mathbf{x})\times\mathbf{b}_\alpha} - \nabla\cdot\sum_\alpha\bar{\rho}_\alpha\overline{(\mathbf{r}\times\mathbf{v}_\alpha)(\mathbf{v}_\alpha-\tilde{\mathbf{v}}_\alpha)}\end{aligned} \tag{2.4.33}$$

or upon using (2.4.31) and (2.4.23) it follows that

$$\hat{\mathbf{M}} = \sum_\alpha\hat{\mathbf{M}}_\alpha = \mathbf{T} - \mathbf{T}^T \tag{2.4.34}$$

The skew-symmetric tensor $\hat{\mathbf{M}}_\alpha$ expresses the nonsymmetry of the stress tensor of phase $\alpha$. From (2.4.29) and (2.4.31), this nonsymmetry can be associated with the structural characteristics of the multiphase mixture that also affects the mixture stress tensor as may be seen from (2.4.33). The angular momentum sources may be affected by the internal particle spins, couple stresses, and body moments.

### 2.4.5 Balance of Energy

The internal energy equation of phase $\alpha$ is obtained from Table 2.1, general balance equation (2.3.4), and by the use of balance of mass and linear momentum equations (2.4.14) and (2.4.19). After some algebra, the result is

$$\bar{\rho}_\alpha \grave{\bar{\epsilon}}_\alpha = tr(\bar{\mathbf{T}}_\alpha^T \nabla \tilde{\mathbf{v}}_\alpha) - \nabla \cdot \bar{\mathbf{q}}_\alpha + \bar{\rho}_\alpha \tilde{r}_\alpha + \hat{\epsilon}_\alpha \tag{2.4.35}$$

where the *energy source* is defined by

$$\begin{aligned} \hat{\epsilon}_\alpha = &-\hat{\mathbf{p}}_\alpha \cdot \tilde{\mathbf{v}}_\alpha - \hat{c}_\alpha(\tilde{\epsilon}_\alpha + \frac{1}{2}\tilde{\mathbf{v}}_\alpha \cdot \tilde{\mathbf{v}}_\alpha) + \nabla \cdot (\mathbf{c}_{3\alpha} - \mathbf{c}_{2\alpha}) + c_{4\alpha} \\ &-\frac{1}{U}\sum_\delta \int_{a(\Lambda\delta)} [m^{(\alpha\delta)}(\epsilon^{(\alpha\delta)} + \frac{1}{2}\mathbf{v}^{(\alpha\delta)} . \mathbf{v}^{(\alpha\delta)}) + (\mathbf{q}^{(\alpha\delta)} - \mathbf{T}^{(\alpha\delta)T}\mathbf{v}^{(\alpha\delta)}) . \mathbf{n}^{(\alpha\delta)}]\, da \\ &+\frac{1}{2}\bar{\rho}_\alpha(\overline{\tilde{\mathbf{v}}_\alpha \cdot \tilde{\mathbf{v}}_\alpha - \overline{\mathbf{v}_\alpha \cdot \mathbf{v}_\alpha}}) + \frac{1}{2}\hat{c}_\alpha(\tilde{\mathbf{v}}_\alpha \cdot \tilde{\mathbf{v}}_\alpha - \widetilde{\mathbf{v}_\alpha \cdot \mathbf{v}_\alpha}) \end{aligned} \tag{2.4.36}$$

and the covariance coefficients $\mathbf{c}_{2\alpha}$, $\mathbf{c}_{3\alpha}$ and $c_{4\alpha}$ are given by

$$\mathbf{c}_{2\alpha} = \frac{U_\alpha}{U} < \rho_\alpha(\epsilon_\alpha + \frac{1}{2}\mathbf{v}_\alpha \cdot \mathbf{v}_\alpha)\mathbf{v}_\alpha > -\bar{\rho}_\alpha(\tilde{\epsilon}_\alpha + \frac{1}{2}\widetilde{\mathbf{v}_\alpha \cdot \mathbf{v}_\alpha})\tilde{\mathbf{v}}_\alpha \tag{2.4.37}$$

$$\mathbf{c}_{3\alpha} = \frac{U_\alpha}{U} < \mathbf{T}_\alpha^T \mathbf{v}_\alpha > -\bar{\mathbf{T}}_\alpha^T \tilde{\mathbf{v}}_\alpha \tag{2.4.38}$$

$$c_{4\alpha} = \frac{U_\alpha}{U} < \rho_\alpha \mathbf{v}_\alpha \cdot \mathbf{b}_\alpha > -\bar{\rho}_\alpha \tilde{\mathbf{v}}_\alpha \cdot \tilde{\mathbf{b}}_\alpha \tag{2.4.39}$$

These coefficients express the structural characteristics of the mixture and do not appear in a theory of mixtures where the interfacial area is neglected.

An equation for the energy balance for the mixture as a whole is obtained by summing up $\alpha$ in (2.4.35). The result is

$$\rho\dot{\epsilon} = tr(\mathbf{T}^T \nabla \mathbf{v}) - \nabla \cdot \mathbf{q} + \rho r + (\hat{e} - \mathbf{v} \cdot \hat{\mathbf{p}}) \tag{2.4.40}$$

where the mixture properties $\epsilon$, $\mathbf{q}$, $r$ and $\hat{e}$ are defined by the following equations:

$$\rho\epsilon = \sum_\alpha (\bar{\rho}_\alpha \bar{\epsilon}_\alpha + \frac{1}{2}\bar{\rho}_\alpha \mathbf{u}_\alpha \cdot \mathbf{u}_\alpha) \tag{2.4.41}$$

$$\mathbf{q} = \sum_\alpha (\bar{\mathbf{q}}_\alpha - \bar{\mathbf{T}}_\alpha^T \mathbf{u}_\alpha + \bar{\rho}_\alpha \bar{\epsilon}_\alpha \mathbf{u}_\alpha + \frac{1}{2}\bar{\rho}_\alpha (\mathbf{u}_\alpha \cdot \mathbf{u}_\alpha)\mathbf{u}_\alpha) \tag{2.4.42}$$

$$\rho r = \sum_\alpha (\bar{\rho}_\alpha \bar{r}_\alpha + \bar{\rho}_\alpha \tilde{\mathbf{b}}_\alpha \cdot \mathbf{u}_\alpha) \tag{2.4.43}$$

$$\begin{aligned} \hat{e} = & \sum_\alpha [c_{4\alpha} + \nabla \cdot (\mathbf{c}_{3\alpha} - \mathbf{c}_{2\alpha}) + \frac{1}{2}\bar{\rho}_\alpha \overline{(\tilde{\mathbf{v}}_\alpha \cdot \tilde{\mathbf{v}}_\alpha - \overline{\mathbf{v}_\alpha \cdot \mathbf{v}_\alpha})} \\ & + \frac{1}{2}\hat{c}_\alpha (\tilde{\mathbf{v}}_\alpha \cdot \tilde{\mathbf{v}}_\alpha - \overline{\mathbf{v}_\alpha \cdot \mathbf{v}_\alpha})] - \frac{1}{U}\sum_{\alpha,\delta} \int_{a(\Lambda\delta)} \Delta_\epsilon^{(\alpha\delta)} \, da \\ & = \mathbf{v} \cdot \hat{\mathbf{p}} + \sum_\alpha [\hat{\epsilon}_\alpha + \mathbf{u}_\alpha \cdot \hat{\mathbf{p}}_\alpha + \hat{c}_\alpha (\bar{\epsilon}_\alpha + \frac{1}{2}\mathbf{u}_\alpha \cdot \mathbf{u}_\alpha)] \end{aligned} \tag{2.4.44}$$

The second equality in (2.4.44) follows from (2.4.26) and (2.4.36).

The inclusion of structural characteristics of the mixture into the theory in the next chapter will require the redefinition of the mixture characteristics expressed by (2.4.41)-(2.4.44). Without accounting for such a structure these properties are formally similar to the ones defined in the single phase multicomponent mixture theory where it is assumed that $\hat{e} = 0$ and $\hat{\mathbf{p}} = \mathbf{o}$.

### 2.4.6 Entropy Inequality

The role of entropy in continuum mechanics is a controversial issue. Classical thermodynamics requires only that the entropy of the system and its environment be nondecreasing, which leaves open the question whether or not it is legitimate to state that $\zeta^{(\alpha\delta)} \geq 0$. Because of this, the statement that $\zeta^{(\alpha\delta)} \geq 0$ as stated in Table 2.1 is referred to as the *local axiom of dissipation* and it is advocated by some that $\zeta$ can only be properly determined by the constitutive theory. While it is possible to allow for such a generalization in a completely postulated theory of mixtures, it should be clear that in a theory of mixtures which is *consistent* with volume averaging of local macroscopic field equations where the local axiom of dissipation has proven to provide very general and useful results, it should be legitimate to state that each sub-body $\delta$ of phase $\alpha$ produces $\zeta^{(\alpha\delta)} \geq 0$ as stated in Table 2.1. The adequacy of employing the local axiom of dissipation in a theory of mixtures can only be properly assessed by comparing the results of the theory with experiment, and it may indeed be necessary in the future revisions of the theory to employ a more general condition on $\zeta^{(\alpha\delta)}$.

In the formulation of an entropy inequality there is also some question whether the entropy flux $\mathbf{h}$ is equal to the heat flux $\mathbf{q}$ divided by the absolute temperature $\theta$. Although the kinetic theory of gases is a distinct theory from the one considered here, we may nevertheless employ it to provide us with a possible resolution of this problem. In the kinetic theory, the entropy flux is *not* equal in general to the heat flux divided by absolute temperature (MÜLLER,1968,1983). Moreover, this theory *motivates* the single phase theory of continua and therefore may be used to establish the result that $\mathbf{h} \neq \mathbf{q}/\theta$. There *does not appear to be* a kinetic theory-independent proof, however, that in a local macroscopic continuum this condition is satisfied, and it is common to assume that $\mathbf{h}=\mathbf{q}/\theta$. This being the case, and since an averaging procedure is used to construct a theory of multiphase mixtures starting from equations of single phase continua, it is also appropriate to assume herein the condition that $\bar{\mathbf{h}}=\overline{(\mathbf{q}/\theta)}$. With this in view, the entropy inequality for phase $\alpha$ is obtained from (2.3.4) as:

$$\bar{\rho}_\alpha \grave{\tilde{s}}_\alpha + \nabla\cdot\left(\frac{\bar{\mathbf{q}}_\alpha}{\bar{\bar{\theta}}_\alpha}\right) - \frac{\bar{\rho}_\alpha \tilde{r}_\alpha}{\bar{\bar{\theta}}_\alpha} + \hat{c}_\alpha \tilde{s}_\alpha + \hat{s}_\alpha \geq 0 \tag{2.4.45}$$

where

$$\hat{s}_\alpha = \nabla\cdot(\mathbf{c}_{5\alpha} + \mathbf{c}_{6\alpha}) - c_{7\alpha} + \frac{1}{U}\sum_\delta \int_{a(\Lambda\delta)} \left(m^{(\alpha\delta)} s^{(\alpha\delta)} + \frac{\mathbf{q}^{(\alpha\delta)}\cdot\mathbf{n}^{(\alpha\delta)}}{\theta^{(\alpha\delta)}}\right) da \tag{2.4.46}$$

is the *entropy source of phase* $\alpha$, whereas the covariance coefficients $\mathbf{c}_{5\alpha}$, $\mathbf{c}_{6\alpha}$ and $c_{7\alpha}$ are defined as follows:

$$\mathbf{c}_{5\alpha} = \frac{U_\alpha}{U} < \rho_\alpha s_\alpha \mathbf{v}_\alpha > - \bar{\rho}_\alpha \tilde{s}_\alpha \tilde{\mathbf{v}}_\alpha \tag{2.4.47}$$

$$\mathbf{c}_{6\alpha} = \frac{U_\alpha}{U} < \frac{\mathbf{q}_\alpha}{\theta_\alpha} > - \frac{\bar{\mathbf{q}}_\alpha}{\bar{\bar{\theta}}_\alpha} \tag{2.4.48}$$

$$c_{7\alpha} = \frac{U_\alpha}{U} < \frac{\rho_\alpha r_\alpha}{\theta_\alpha} > - \frac{\bar{\rho}_\alpha \tilde{r}_\alpha}{\bar{\bar{\theta}}_\alpha} \tag{2.4.49}$$

The entropy inequality for the mixture as a whole is obtained from (2.4.45) by summing over $\alpha$ and using (2.4.18) with $\tilde{\Gamma}_\alpha = \tilde{s}_\alpha$, *i.e.*

$$\rho\dot{s} + \nabla\cdot\sum_\alpha\left(\frac{\bar{\mathbf{q}}_\alpha}{\bar{\bar{\theta}}_\alpha} + \bar{\rho}_\alpha \tilde{s}_\alpha \mathbf{u}_\alpha\right) - \sum_\alpha \frac{\bar{\rho}_\alpha \tilde{r}_\alpha}{\bar{\bar{\theta}}_\alpha} + \hat{s} \geq 0 \tag{2.4.50}$$

where

$$\rho s = \sum_{\alpha} \bar{\rho}_\alpha \tilde{s}_\alpha \tag{2.4.51}$$

$$\hat{s} = \sum_{\alpha} \hat{s}_\alpha = \sum_{\alpha} [\nabla \cdot (\mathbf{c}_{5\alpha} + \mathbf{c}_{6\alpha}) - c_{7\alpha}] + \frac{1}{U} \sum_{\alpha,\delta} \int_{a(\Lambda\delta)} \Delta_s^{(\alpha\delta)}\, da \tag{2.4.52}$$

$\Delta_s^{(\alpha\delta)}$ is the entropy production rate at the interface and it is positive semidefinite (Table 2.1) within the assumption of the validity of the local axiom of dissipation as discussed above. The entropy source $\hat{s}$ does not necessarily assume a positive definite value, since it also includes the divergence of covariance coefficients (the effects of structure of the mixture) that may have negative values.

### 2.4.7 Proof of the Existence of Phasic Stress Tensor

The *existence* of $\bar{\mathbf{T}}_\alpha$ *independent of the volume averaging approach* can be established by accepting the EULER's first law of motion of phase $\alpha$ (DOBRAN, 1985b) where it is assumed that the body $\mathcal{B}_{m\alpha}$ of phase $\alpha$ in $\mathcal{E}^3$ is a *homogeneous* continuum body, *i.e.*

$$\frac{d}{dt} \int_{V_m} \bar{\rho}_\alpha \tilde{\mathbf{v}}_\alpha \, dV = \int_{A_m} \mathbf{t}_\alpha \, dA + \int_{V_m} (\bar{\rho}_\alpha \tilde{\mathbf{b}}_\alpha + \hat{\mathbf{p}}_\alpha + \hat{c}_\alpha \tilde{\mathbf{v}}_\alpha) \, dV \tag{2.4.53}$$

where $V_m$ is the volume containing the mass of phase $\alpha$ and $A_m$ is its surface at time $t$. As noted earlier in section (2.4.3) the existence of stress tensor can be proved rigorously by employing the Theorems of NOLL (1973). Less formally, however, equation (2.4.53) is now applied to the stress tetrahedron shown in Figure 2.3 where $\mathbf{t}_\alpha$ and $\mathbf{n}$ are the stress vector and unit normal vector, respectively, acting on the slanted face with area A. The stress $\mathbf{t}_1$ is acting on the face $A_1$, $\mathbf{t}_2$ is acting on $A_2$, and $\mathbf{t}_3$ is acting on $A_3$. Applying the Transport Theorem[1] to the term on the left, using the mean value theorem, and taking the limit as the height $h$ of the tetrahedron is reduced to zero, yields

---

[1]The Transport Theorem for phase $\alpha$ is

$$\frac{d}{dt} \int_{V_m} \bar{\rho}_\alpha \tilde{\mathbf{v}}_\alpha \, dV = \int_{V_m} \frac{\partial \bar{\rho}_\alpha \tilde{\mathbf{v}}_\alpha}{\partial t} dV + \int_{A_m} \bar{\rho}_\alpha \tilde{\mathbf{v}}_\alpha \tilde{\mathbf{v}}_\alpha \cdot \mathbf{n} \, dA = \int_{V_m} (\bar{\rho}_\alpha \dot{\tilde{\mathbf{v}}}_\alpha + \hat{c}_\alpha \tilde{\mathbf{v}}_\alpha) \, dV$$

where use is made of the balance of mass equation (2.4.14) and Green-Gauss *Divergence Theorem*

$$\int_A \mathbf{F}\mathbf{n} \, dA = \int_V \nabla \cdot \mathbf{F} \, dV$$

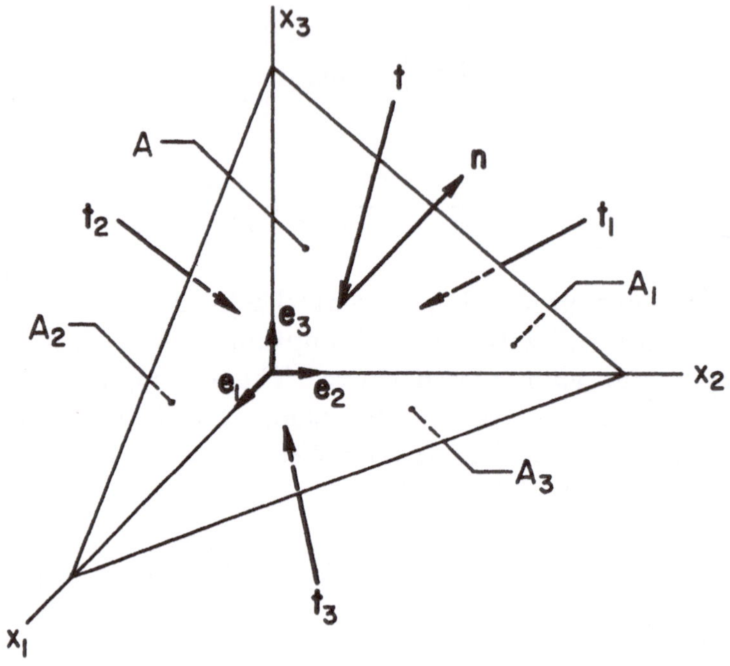


Figure 2.3: The stress tetrahedron

$$lim_{h\to 0}(\bar{\rho}_\alpha \dot{\bar{\mathbf{v}}}_\alpha - \bar{\rho}_\alpha \tilde{\mathbf{b}}_\alpha - \hat{\mathbf{p}}_\alpha)\frac{hA}{3} = A(\mathbf{t} + n_1\mathbf{t}_1 + n_2\mathbf{t}_2 + n_3\mathbf{t}_3) = \mathbf{0} \tag{2.4.54}$$

or

$$\mathbf{t}_\alpha = -n_1\mathbf{t}_1 - n_2\mathbf{t}_2 - n_3\mathbf{t}_3 \tag{2.4.55}$$

But, by the CAUCHY's lemma

$$\mathbf{t}(\mathbf{x},\mathbf{n}) = -\mathbf{t}(\mathbf{x},-\mathbf{n}) \tag{2.4.56}$$

so that

$$-\mathbf{t}_1 = \bar{T}_{i1}\mathbf{e}_i, \quad -\mathbf{t}_2 = \bar{T}_{i2}\mathbf{e}_i, \quad -\mathbf{t}_3 = \bar{T}_{i3}\mathbf{e}_i \tag{2.4.57}$$

where $\bar{T}_{ij}$ is *defined* to be the $i$'th component of the stress vector acting on the *positive* side of the plane $x_j$=constant, it follows that

$$\mathbf{t} = \bar{T}_{ij}\mathbf{e}_i n_j = \bar{\mathbf{T}}\mathbf{n} \tag{2.4.58}$$

which establishes the *existence* of $\bar{\mathbf{T}}_\alpha$. If this result is now substituted into (2.4.53), and the transport and GREEN-GAUSS divergence theorems employed for the term on the left and area integral for the term on the right side of the equation, respectively, gives

$$\int_{V_m} (\bar{\rho}_\alpha \dot{\bar{\mathbf{v}}}_\alpha - \nabla \cdot \bar{\mathbf{T}}_\alpha - \bar{\rho}_\alpha \tilde{\mathbf{b}}_\alpha - \hat{\mathbf{p}}_\alpha)\, dV = 0 \qquad (2.4.59)$$

where the multiphase variables are assumed to be differentiable in $V_m$ and on $A_m$. Requiring that the integrand in this equation vanishes for all arbitrary volumes $V_m$ produces the balance of linear momentum (2.4.19).

## 2.5 Properties of the Averaged Equations

The phasic equations for the balance of mass (2.4.14), balance of linear momentum (2.4.19), balance of angular momentum (2.4.31), and balance of energy (2.4.35), are similar to the balance equations of constituents in the single phase multicomponent mixture theory (BOWEN,1976) and would reduce to these equations in the absence of the interfacial structure of the mixture. This reduction would not, however, be valid for the case of the entropy inequality (2.4.45), since an entropy inequality for each constituent of a single phase multicomponent mixture would yield results that are inconsistent with the classical thermochemistry in the sense that an equilibrium chemical potential would depend only on the properties of that constituent. This result is, of course, too special and it would describe only ideal mixtures (BOWEN,1976). For this reason, in the theory of single phase multicomponent mixtures use is made only of the global or entropy inequality for the mixture as a whole, since then the theory does not produce any inconsistency with the classical results and it is moreover consistent with the results of the kinetic theory of gases (TRUESDELL,1984). The derivation of the multiphase mixture field equations using the averaging approach leads naturally to an entropy inequality for *each phase*, as long as it is assumed that the local axiom of dissipation or entropy inequality within each sub-body $\delta$ of phase $\alpha$ is assumed valid, as common in modeling the single phase continua (see also the discussion earlier in section 2.4.6). All models of multiphase mixtures obtained from averaging the local macroscopic field equations with the assumed validity of the local axiom of dissipation generally require a second law of thermodynamics for each phase, whereas the *postulated* theories of mixtures discussed in chapter 1 do not. These latter models only assume an entropy inequality for the mixture as a whole (DRUMHELLER & BEDFORD (1980), PASSMAN *et al.* (1984)), similarly as in the single phase multicomponent mixture theory.

In the postulated theories of multiphase mixtures, the source terms of the mixture $\hat{\mathbf{p}}$, $\hat{\mathbf{M}}$, $\hat{\epsilon}$ and $\hat{s}$ are assumed to be equal to zero. The motivation for this assumption comes from a principle of TRUESDELL (1984, page 221): "The motion of the mixture is governed by the same equations as is a single body." The premise of this postulate is that mechanical and energetic processes taking place within a mixture come from reactions and transfers between the constituents of the mixture, rather than from the true processes of "creation or destruction." The *mixture source terms* (2.4.25), (2.4.34), (2.4.44) and (2.4.52) *arise precisely from reactions and transfers between the phases*, and in order for the above principle to apply in the theory of mixtures the interfacial effects arising in these sources would have to be ignored or incorporated within the definition of other mixture variables. For example, the right side of the momentum source equation (2.4.25) could be incorporated within the definition of $\mathbf{T}$ in (2.4.23) producing $\hat{\mathbf{p}}=\mathbf{0}$. This procedure does not, however, appear to be completely successful in reformulating other balance equations to the form which is required by the principle, even if all mixture source terms are set equal to zero, due to the complications arising in the entropy inequality (2.4.50). In the latter case with $\hat{s} = 0$, we may define an entropy flux $\mathbf{H}$ and an entropy supply $\sigma$, *i.e.*

$$\mathbf{H} = \sum_\alpha \left(\frac{\bar{\mathbf{q}}_\alpha}{\bar{\bar{\theta}}_\alpha} + \bar{\rho}_\alpha \tilde{s}_\alpha \mathbf{u}_\alpha\right) \tag{2.5.1}$$

$$\rho\sigma = \sum_\alpha \frac{\bar{\rho}_\alpha \tilde{s}_\alpha}{\bar{\bar{\theta}}_\alpha} \tag{2.5.2}$$

thereby transforming (2.4.50) into the form

$$\rho\dot{s} + \nabla\cdot\mathbf{H} - \rho\sigma \geq 0 \tag{2.5.3}$$

But as discussed earlier, in single phase continuum theories of matter it is common to assume that $\mathbf{H} = \mathbf{q}/\theta$, which in theories of mixtures is not true unless diffusion and temperature differences between the phases are neglected. To circumvent this difficulty PASSMAN *et al.* (1984) suggest the use of two different temperatures: one for *conduction* and the other for *radiation* defined as follows:

$$\frac{\mathbf{q}}{\theta_c} = \sum_\alpha \left(\frac{\bar{\mathbf{q}}_\alpha}{\bar{\bar{\theta}}_\alpha} + \bar{\rho}_\alpha \tilde{s}_\alpha \mathbf{u}_\alpha\right) \tag{2.5.4}$$

$$\frac{\rho r}{\theta_r} = \sum_\alpha \frac{\bar{\rho}_\alpha \tilde{r}_\alpha}{\bar{\bar{\theta}}_\alpha} \tag{2.5.5}$$

so that (2.4.50) could be written as

$$\rho\dot{s} + \nabla\cdot\left(\frac{\mathbf{q}}{\theta_c}\right) - \frac{\rho r}{\theta_r} \geq 0 \tag{2.5.6}$$

This procedure is, however, highly speculative as no such temperatures have been observed by experiments in multiphase flows and the heat generation rate $r$ may exist due to the sources other than by radiation. The theory of multiphase mixtures obtained from an averaging approach provides a clearer definition of the mixture source terms as arising from the interphase interactions. Their justification for the presence in the mixture field equations comes from averaging, which is simply a method whereby the local properties of the mixture at a point in space and time is replaced by an average over the neighborhood of this point. In this sense, the *source terms represent the nonlocal or structural characteristics of the mixture* and need not vanish, as it is assumed to be the case in a local or structureless theory.

## 2.6 Concluding Remarks

In this chapter, the concepts and methods of the volume averaging approach were presented and used to derive a basic set of multiphase field equations. This basic set of equations consists of the balance of mass, momentum and energy, and entropy inequality for each phase in the mixture. The averaged equations contain the structural properties of the mixture and can be adjoined with the additional balance equations that can model these characteristics as discussed in the following chapter. An alternative approach in establishing the phasic and mixture balance laws is to *postulate* the *integral balance laws* as in DOBRAN (1985b), and require the differentiability of all multiphase fields within an arbitrary material volume and on its surface. The application of the transport theorems then establishes the phasic and mixture balance laws discussed in this chapter. Moreover, if a region of discontinuity is assumed to exist within the material volume, then the integral balance laws produce differential balance equations in the regions removed from the discontinuity and jump or boundary conditions or balance laws across the surface of discontinuity. The jump conditions can be utilized to study the acceleration and shock waves in multiphase flows DOBRAN (1983).

CHAPTER 3

# THE MATERIAL DEFORMATION POSTULATE AND FIELD EQUATIONS OF MULTIPHASE MIXTURES WITH STRUCTURE

In the development of the basic conservation and balance equations of multiphase mixtures by the use of volume averaging procedure in chapter 2, no method was proposed within the theory to account for the structural characteristics of mixtures. The averaging approach has also left us with two problems. The first problem pertains to the proper selection of the position $\mathbf{x}$ in $\mathcal{E}^3$ onto which the continua in the averaging volume $U$ are mapped, whereas the second problem pertains to the selection of a proper averaging volume $U$ in the averaged field equations. Although DOBRAN (1984b,1985a) alluded to the possibility of choosing $\mathbf{x}$ as the center of mass of the mixture in $U$ and accounting for the size of $U$ in the averaged field equations through a *nonlocal theory of constitutive equations*, no such program was carried out. In this chapter, a method will be presented whereby the structural characteristics of the mixture can be incorporated explicitly in the theory to various degrees of approximation. The starting point of the analysis is a *basic material deformation postulate* which is used to derive *additional transport equations* for multiphase mixtures. This will be followed by showing how the structural characteristics of mixtures can be modeled in a more explicit manner.

## 3.1 The Basic Material Deformation Postulate

The basic material deformation postulate pertains to an assumption of the deformation of material particles of each phase when they undergo a motion from the reference to the spatial configuration as illustrated in Figure 2.1.

**The basic material deformation postulate.** The motion of material particles of each phase $\alpha$ relative to the center of mass undergo a homogeneous deformation expressed by the following equations:

$$\xi_k^{(\alpha\delta)} = \Pi_{kK}^{(\alpha)} \Sigma_K^{(\alpha\delta)} \tag{3.1.1}$$

$$\Sigma_K^{(\alpha\delta)} = \Xi_{Kk}^{(\alpha)} \xi_k^{(\alpha\delta)} \tag{3.1.2}$$

This asumption of the material deformation relative to the center of mass of each phase is a special case of the more general material motion which can be constructed from (2.2.4). With the assumption that this equation is differentiable with respect to $\mathbf{\Sigma}^{(\alpha\delta)}$ it can be expanded in series about $\mathbf{\Sigma}^{(\alpha\delta)}=0$, yielding

$$\xi_k^{(\alpha\delta)}(\mathbf{X}^{(\alpha)},\mathbf{\Sigma}^{(\alpha\delta)},t)=\xi_k(\mathbf{X}^{(\alpha)},\mathbf{\Sigma}^{(\alpha\delta)}=0,t)+\left(\frac{\partial\xi_k^{(\alpha\delta)}}{\partial\Sigma_K^{(\alpha\delta)}}\right)_{\Sigma_K^{(\alpha\delta)}=0}\Sigma_K^{(\alpha\delta)}$$
$$+\left(\frac{\partial^2\xi_k^{(\alpha\delta)}}{\partial\Sigma_K^{(\alpha\delta)}\partial\Sigma_L^{(\alpha\delta)}}\right)_{\Sigma_K^{(\alpha\delta)}=0,\Sigma_L^{(\alpha\delta)}=0}\Sigma_K^{(\alpha\delta)}\Sigma_L^{(\alpha\delta)}+\cdots \tag{3.1.3}$$

Referring to Figure 2.1 and the motion of a material particle $P$ from the reference to the spatial configuration, it is clear that physically we expect distinct material particles in the reference configuration at time $t_o$ to be located at different places in the spatial configuration at time $t$, and vice versa if these particles were to retrace the motion to their original reference states. This physical requirement of inpenetrability of matter implies that the mapping between two configurations must be assumed mathematically to be *bijective* and we may, therefore, set in (3.1.3)

$$\xi_k^{(\alpha\delta)}(\mathbf{X}^{(\alpha)},\mathbf{\Sigma}^{(\alpha\delta)}=0,t)=0 \tag{3.1.4}$$

Moreover, by defining

$$\Pi_{kK}^{(\alpha)}=\left(\frac{\partial\xi_k^{(\alpha\delta)}}{\partial\Sigma_K^{(\alpha\delta)}}\right)_{\Sigma_K^{(\alpha\delta)}=0} \tag{3.1.5}$$

it may be seen that (3.1.3) is reduced to (3.1.1) if second and higher order terms in the series expansion are neglected. The basic material deformation postulate (3.1.1) or (3.1.2) accounts, therefore, only for the *first order* deformation of the material relative to the center of mass, and although it is possible to relax this assumption by including higher order terms in the series expansion, a theory based on higher order approximations would loose its simplicity and possibly also its practical usefulness.

The basic material deformation postulate implies that the motion of material of phase $\alpha$ is comprised of:

1. An arbitrary deformation of the center of mass.

2. A rotation of the mass points about the center of mass.

3. An affine deformation of material particles relative to the center of mass carrying ellipsoids into ellipsoids.

As will be seen below, this material deformation assumption allows for the inclusion of considerable structure into the theory of multiphase mixtures. An assumption similar to (3.1.1) is also central in the development of the micromorphic theory of ERINGEN (1964) and TWISS & ERINGEN (1971), who developed structural theories for *single phase/single component and single phase/multicomponent* mixtures. The adoption of the deformation principle (3.1.1) in the theory of *multiphase mixtures* is, however, reflected in the definitions of mixtures' structural properties which are *different* from the ones in the micromorphic theory.

Equations (3.1.1) and (3.1.2) imply that

$$\Pi_{kL}^{(\alpha)}\,\Xi_{L\ell}^{(\alpha)} = \delta_{k\ell} \tag{3.1.6}$$

$$\Xi_{Kk}^{(\alpha)}\,\Pi_{kL}^{(\alpha)} = \delta_{KL} \tag{3.1.7}$$

where $\delta_{k\ell}$ and $\delta_{KL}$ are the Kronecker's deltas defined in chapter 1.

The velocity of the material point $P$ (Figure 2.1) and of the center of mass of phase $\alpha$ are defined by the usual relations,

$$\mathbf{v}^{(\alpha\delta)} = \left(\frac{\partial \mathbf{x}^{(\alpha\delta)}}{\partial t}\right)_{\mathbf{X}^{(\alpha\delta)}}, \quad \mathbf{v}^{(\alpha)} = \left(\frac{\partial \mathbf{x}^{(\alpha)}}{\partial t}\right)_{\mathbf{X}^{(\alpha)}} \tag{3.1.8}$$

whereas the material derivative of $\boldsymbol{\xi}^{(\alpha\delta)}$ is the time derivative following the motion of particle $P$ with $\mathbf{X}^{(\alpha)}$ remaining constant. Thus,

$$\grave{\boldsymbol{\xi}}^{(\alpha\delta)} = \left(\frac{\partial \boldsymbol{\xi}^{(\alpha\delta)}}{\partial t}\right)_{\mathbf{X}^{(\alpha\delta)},\mathbf{X}^{(\alpha)}} \tag{3.1.9}$$

and upon using (3.1.1) and (3.1.2) it may be written as

$$\grave{\xi}_k^{(\alpha\delta)} = \grave{\Pi}_{kK}^{(\alpha)}\,\Sigma_K^{(\alpha\delta)} = \grave{\Pi}_{kK}^{(\alpha)}\,\Xi_{K\ell}^{(\alpha)}\xi_\ell^{(\alpha\delta)} = \nu_{k\ell}^{(\alpha)}\xi_\ell^{(\alpha\delta)} \tag{3.1.10}$$

where $\nu_{k\ell}^{(\alpha)}$ is, in general, a function of time and position in the spatial configuration. In a special situation when $\nu_{k\ell}^{(\alpha)}$ is a skew-symmetric tensor, (3.1.10) represents a *rotation* of material particles relative to the center of mass (as will be discussed below in more detail), and for this reason $\nu_{k\ell}^{(\alpha)}$ in single phase structured theories has acquired the name of the **gyration tensor**. Using $(3.1.8)_1$ and (3.1.10) it follows from (2.2.2) that the local macroscopic velocity of a particle of phase $\alpha$, $\mathbf{v}^{(\alpha\delta)}$, and of the center of mass of the same phase, $\mathbf{v}^{(\alpha)}$, are related by

$$v_k^{(\alpha\delta)} = v_k^{(\alpha)} + \nu_{k\ell}^{(\alpha\delta)}\xi_\ell^{(\alpha\delta)} \tag{3.1.11}$$

Multiplying (3.1.11) by $\rho^{(\alpha\delta)}$ and integrating over the volume of phase $\alpha$ contained in the averaging volume shown in Figure 2.2 gives:

$$\sum_{\delta}\int_{U(\alpha\delta)}\rho^{(\alpha\delta)}v_k^{(\alpha\delta)}\,dU = v_k^{(\alpha)}\sum_{\delta}\int_{U(\alpha\delta)}\rho^{(\alpha\delta)}\,dU + \nu_{k\ell}^{(\alpha)}\sum_{\delta}\int_{U(\alpha\delta)}\rho^{(\alpha\delta)}\xi_\ell^{(\alpha\delta)}\,dU \tag{3.1.12}$$

where use was made of the fact that $v_k^{(\alpha)}$ and $\nu_{k\ell}^{(\alpha)}$ can be taken out of the integral. But *from the definition of the center of mass of phase* $\alpha$,

$$\sum_{\delta}\int_{U(\alpha\delta)}\rho^{(\alpha\delta)}\xi_\ell^{(\alpha\delta)}\,dU = 0 \tag{3.1.13}$$

and noting that

$$\sum_{\delta}\int_{U(\alpha\delta)}\rho^{(\alpha\delta)}\,dU = \bar{\rho}_\alpha U \tag{3.1.14}$$

which follows from (2.3.5) and (2.3.7), (3.1.12) is reduced to the following form:

$$\bar{\rho}_\alpha v_k^{(\alpha)}U = \sum_{\delta}\int_{U(\alpha\delta)}\rho^{(\alpha\delta)}v_k^{(\alpha\delta)}\,dU \tag{3.1.15}$$

Utilizing (2.3.5) and (2.3.6) in the above equation gives

$$\mathbf{v}^{(\alpha)} = \tilde{\mathbf{v}}_\alpha \tag{3.1.16}$$

or that the density-weighted velocity of phase $\alpha$ is equal to the center of mass velocity of the same phase. Similarly, the mixture velocity $\mathbf{v}$, defined by (2.3.12), is equal to the center of mass velocity of the mixture, *i.e.*

$$\rho\mathbf{v} = \sum_{\alpha,\delta}\int_{U(\alpha\delta)}\rho^{(\alpha\delta)}\mathbf{v}^{(\alpha\delta)}\,dU = \sum_{\alpha}\bar{\rho}_\alpha\mathbf{v}^{(\alpha)} \tag{3.1.17}$$

This result follows from (3.1.15) when it is summed up over $\alpha$ and use made of (3.1.16). From (3.1.10) and (3.1.11), the accelerations of a material particle of phase $\alpha$ and of the center of mass are related by

$$a_k^{(\alpha\delta)} = a_k^{(\alpha)} + (\grave{\nu}_{k\ell}^{(\alpha)} + \nu_{km}^{(\alpha)}\nu_{m\ell}^{(\alpha)})\xi_\ell^{(\alpha\delta)} \tag{3.1.18}$$

Again, multiplying (3.1.18) by $\rho^{(\alpha\delta)}$ and integrating over $U_\alpha$ produces the results

$$\mathbf{a}^{(\alpha)} = \grave{\tilde{\mathbf{v}}}_\alpha \tag{3.1.19}$$

$$\rho \mathbf{a} = \rho \dot{\mathbf{v}} = \sum_{\alpha} \bar{\rho}_\alpha \dot{\mathbf{v}}_\alpha \tag{3.1.20}$$

where the last equation is obtained by summing up $\alpha$ in (3.1.19). As in the case of velocities, the density-weighted accelerations of each phase and of the mixture are equal to the corresponding accelerations of the center of mass. These results are clearly the consequence of the basic material deformation postulate and do not follow directly from the volume averaging procedure discussed in the last chapter. The association of the position $\mathbf{x}$ in the mapped configuration space in Figure 2.2 with the center of mass of the mixture in the averaging volume $U$ is now clear.

The introduction of the above material deformation postulate is useful for resolving a problem stated at the beginning of the chapter on a choice of the mapping position vector $\mathbf{x}$ in $\mathcal{E}^3$. More importantly, this postulate plays a much more significant role, for it reveals a rich theoretical structure on which foundations may be built a theory of structured multiphase mixtures. In the next section, it will be shown how (3.1.1) and (3.1.2) allow for the derivation of additional transport equations, and in later sections how the basic balance equations discussed in chapter 2 may be recast into forms which reveal in more detail the structural characteristics of the mixture.

Before closing this section, however, it is useful for later developments to summarize two results which follow from (3.1.1), (3.1.2), (3.1.6), (3.1.7) and (3.1.10). These results are

$$\nu_{i\ell}^{(\alpha)} = \dot{\Pi}_{iK}^{(\alpha)} \Xi_{K\ell}^{(\alpha)} \tag{3.1.21}$$

$$\dot{\Pi}_{iL}^{(\alpha)} = \nu_{i\ell}^{(\alpha)} \Pi_{\ell L}^{(\alpha)} \tag{3.1.22}$$

## 3.2 Additional Transport Equations in the Theory of Multiphase Mixtures

The inclusion of structural characteristics into the theory of multiphase mixtures may be performed through the additional transport or balance equations which model these characteristics. In this section, equations for equilibrated inertia and equilibrated moments will be derived using the volume averaging procedure and the basic material deformation postulate from the previous section.

### 3.2.1 Balance Equation for the Equilibrated Inertia

The balance equation for the equilibrated inertia is derived by taking a moment about the center of mass of the conservation of mass equation

(2.2.5) and performing volume averaging of the resulting expression, *i.e.*

$$\sum_{\delta}\int_{U^{(\alpha\delta)}} \xi_k^{(\alpha\delta)}\xi_\ell^{(\alpha\delta)}\left(\frac{\partial}{\partial t}\rho^{(\alpha\delta)} + \frac{\partial}{\partial x_m^{(\alpha\delta)}}\rho^{(\alpha\delta)}v_m^{(\alpha\delta)}\right) dU = 0 \tag{3.2.1}$$

After some algebraic manipulations using (2.2.7), (2.3.2), (2.3.3), (2.4.15), (3.1.10), (3.1.11) and (3.1.14) it may be shown that (3.2.1) can be reduced to an *equation for the equilibrated inertia of phase* $\alpha$, *i.e.*

$$\bar{\rho}_\alpha(\dot{\bar{i}}_{\alpha k\ell} - \nu_{km}^{(\alpha)}\bar{i}_{\alpha m\ell} - \nu_{\ell m}^{(\alpha)}\bar{i}_{\alpha km}) = -\hat{c}_\alpha(\bar{i}_{\alpha k\ell} - \hat{i}_{\alpha k\ell}) - (\bar{\rho}_\alpha \hat{I}_{\alpha mk\ell})_{,m} \tag{3.2.2}$$

where $\bar{i}_{\alpha k\ell}$ is the *equilibrated inertia* of phase $\alpha$ defined by

$$\bar{i}_{\alpha k\ell} = \frac{1}{\bar{\rho}_\alpha}\frac{1}{U}\sum_{\delta}\int_{U^{(\alpha\delta)}} \rho^{(\alpha\delta)}\xi_k^{(\alpha\delta)}\xi_\ell^{(\alpha\delta)}\, dU \tag{3.2.3}$$

and $\hat{i}_{\alpha k\ell}$ and $\hat{I}_{\alpha mk\ell}$ are the *equilibrated source inertia* and *hyperinertia tensor*, respectively, which are defined by the following expressions:

$$\hat{c}_\alpha \hat{i}_{\alpha k\ell} = \frac{1}{U}\sum_{\delta}\int_{a^{(\Lambda\delta)}} \rho^{(\alpha\delta)}\xi_k^{(\alpha\delta)}\xi_\ell^{(\alpha\delta)}(\mathbf{S}^{(\Lambda\delta)} - \mathbf{v}^{(\alpha\delta)})\cdot\mathbf{n}^{(\alpha\delta)}\, da \tag{3.2.4}$$

$$\bar{\rho}_\alpha \hat{I}_{\alpha mk\ell} = \nu_{mn}^{(\alpha)}\frac{1}{U}\sum_{\delta}\int_{U^{(\alpha\delta)}} \rho^{(\alpha\delta)}\xi_k^{(\alpha\delta)}\xi_\ell^{(\alpha\delta)}\xi_n^{(\alpha\delta)}\, dU \tag{3.2.5}$$

The equilibrated source inertia $\hat{i}_{\alpha k\ell}$, arises from the phase change within the mixture, whereas the hyperinertia tensor $\hat{I}_{\alpha mk\ell}$ represents a third order moment effect. Using (3.1.11), (3.2.3) and definition (2.3.5), (3.2.5) may be expressed as

$$\bar{\rho}_\alpha \hat{I}_{\alpha mk\ell} = \frac{U_\alpha}{U} < \rho_\alpha v_{\alpha m}\xi_{\alpha k}\xi_{\alpha\ell} > - \bar{\rho}_\alpha \tilde{v}_{\alpha m}\bar{i}_{\alpha k\ell} \tag{3.2.6}$$

The hyperinertia tensor $\hat{I}_{\alpha mk\ell}$ represents, therefore, a *covariance effect* that vanishes if the local velocity in each phase is equal to the center of mass velocity.

### 3.2.2 Balance Equation for the Equilibrated Moments

The local momentum equation (2.2.5) may be used to form a balance equation for the equilibrated moment. This is obtained by taking a moment of (2.2.5) and integrating the resulting expression over the volume $U_\alpha$, *i.e.* the following procedure is carried out:

$$\sum_{\delta}\int_{U^{(\alpha\delta)}} x_j^{(\alpha\delta)}\Big[\frac{\partial}{\partial t}\rho^{(\alpha\delta)}v_k^{(\alpha\delta)} + \frac{\partial}{\partial x_m^{(\alpha\delta)}}\rho^{(\alpha\delta)}v_k^{(\alpha\delta)}v_m^{(\alpha\delta)} - T_{k\ell,\ell}^{(\alpha\delta)} - \rho^{(\alpha\delta)}b_k^{(\alpha\delta)}\Big]\, dU = 0 \tag{3.2.7}$$

After rearranging and using (2.3.2), (2.3.3), (3.1.10), (3.1.11) and (3.2.2)-(3.2.5) it may be shown that the above equation can be reduced to

$$
\begin{aligned}
&x_j^{(\alpha)}[\bar{\rho}_\alpha \dot{\tilde{v}}_{\alpha k} - \hat{p}_{\alpha k} - \bar{T}_{\alpha km,m} - \bar{\rho}_\alpha \tilde{b}_{\alpha k}] + \bar{\rho}_\alpha \tilde{i}_{\alpha jn}[\dot{\nu}_{kn}^{(\alpha)} + \nu_{km}^{(\alpha)}\nu_{mn}^{(\alpha)}] \\
&+\bar{\rho}_\alpha \nu_{mq}^{(\alpha)} \tilde{v}_{\alpha k,m} \tilde{i}_{\alpha jq} = -\nu_{kn,m}^{(\alpha)} \nu_{mq}^{(\alpha)} \frac{1}{U} \sum_\delta \int_{U(\alpha\delta)} \rho^{(\alpha\delta)} \xi_j^{(\alpha\delta)} \xi_n^{(\alpha\delta)} \xi_q^{(\alpha\delta)} \, dU \\
&+\Big(\frac{1}{U} \sum_\delta \int_{U(\alpha\delta)} \xi_j^{(\alpha\delta)} T_{km}^{(\alpha\delta)} \, dU\Big)_{,m} + \frac{1}{U} \sum_\delta \int_{a(\Lambda\delta)} \xi_j^{(\alpha\delta)} T_{km}^{(\alpha\delta)} n_m^{(\alpha\delta)} \, da \\
&+\frac{1}{U} \sum_\delta \int_{U(\alpha\delta)} \rho^{(\alpha\delta)} \xi_j^{(\alpha\delta)} b_k^{(\alpha\delta)} \, dU
\end{aligned}
\tag{3.2.8}
$$

A close examination of this equation reveals that the right side contains various moments of the *intrinsic* and *extrinsic* forces of phase $\alpha$, whereas the left side contains the inertial or convective effects of the material particles with respect to the center of mass. These extrinsic and intrinsic stresses and moments will be defined as follows:

**Stress average, $\bar{\mathbf{T}}_\alpha$:**

$$\bar{T}_{\alpha kj} = \frac{1}{U} \sum_\delta \int_{U(\alpha\delta)} T_{kj}^{(\alpha\delta)} \, dU \tag{3.2.9}$$

**Body force moment, $\tilde{\boldsymbol{\ell}}_\alpha$:**

$$\tilde{\ell}_{\alpha jk} = \frac{1}{\bar{\rho}_\alpha} \frac{1}{U} \sum_\delta \int_{U(\alpha\delta)} \rho^{(\alpha\delta)} \xi_j^{(\alpha\delta)} b_k^{(\alpha\delta)} \, dU \tag{3.2.10}$$

**Intrinsic stress moment, $\bar{\boldsymbol{\lambda}}_\alpha$:**

$$\bar{\lambda}_{\alpha jkm} = \frac{1}{U} \sum_\delta \int_{U(\alpha\delta)} \xi_j^{(\alpha\delta)} T_{km}^{(\alpha\delta)} \, dU \tag{3.2.11}$$

**Surface traction moment, $\bar{\mathbf{S}}_\alpha$:**

$$\bar{S}_{\alpha jk} = \frac{1}{U} \sum_\delta \int_{a(\Lambda\delta)} \xi_j^{(\alpha\delta)} T_{km}^{(\alpha\delta)} n_m^{(\alpha\delta)} \, da \tag{3.2.12}$$

Substituting definitions (3.2.9)-(3.2.12) into (3.2.8) and requiring that the terms multiplied by $x_j^{(\alpha)}$ vanish (*cit.* 2.4.19) results in a *balance equation for equilibrated moments of phase $\alpha$*:

$$
\begin{aligned}
\bar{\rho}_\alpha \tilde{i}_{\alpha jn}(\dot{\nu}_{kn}^{(\alpha)} + \nu_{km}^{(\alpha)}\nu_{mn}^{(\alpha)}) &= \bar{S}_{\alpha jk} + \bar{\rho}_\alpha \tilde{\ell}_{\alpha jk} + \bar{\lambda}_{\alpha jkm,m} \\
&-\bar{\rho}_\alpha \nu_{kn,m}^{(\alpha)} \hat{I}_{\alpha mjn} - \bar{\rho}_\alpha \nu_{mq}^{(\alpha)} \tilde{v}_{\alpha k,m} \tilde{i}_{\alpha jq}
\end{aligned}
\tag{3.2.13}
$$

The stress average defined by (3.2.9) is not a new definition as it occurs in the basic balance equations (2.4.19) and (2.4.35). The body force moment $\tilde{\boldsymbol{l}}_\alpha$ results from the nonuniformity of the external body force fields and body force moments acting on phase $\alpha$, whereas the intrinsic stress moment $\bar{\boldsymbol{\lambda}}_\alpha$ arises from internal stresses acting on material particles of phase $\alpha$. The intrinsic stress moment is a third order tensor and it may be called a *hyperstress.* The surface traction moment $\bar{\mathbf{S}}_\alpha$ comes from the forces acting on the surface of $U_\alpha$.

The above balance equations for the equilibrated inertia and moments reveal the structural characteristics of the mixture. They show that the replacement of the local force balance equation in each phase by an averaged equation has the effect of replacing the continuous distribution of forces by resultant forces and couples. This result is, of course, also consistent with the particle mechanics where the forces acting on a collection of particles may be replaced by a resultant force and a resultant couple. The extrinsic forces, which may be of the gravitational or electromagnetic origin, arise from the external effects and they act on the entire set of particles. Their local distribution in $U_\alpha$ is equivalent to a resultant force $\tilde{\mathbf{b}}_\alpha$, and a resultant couple $\tilde{\boldsymbol{l}}_\alpha$. A similar situation arises for the forces acting on the surface of $U_\alpha$. When averaged over the surface of $U_\alpha$, they may be replaced by a surface traction force $\bar{\mathbf{T}}_\alpha \mathbf{n}_\alpha$, and a couple $\bar{\mathbf{S}}_\alpha$. In a continuum of each phase, the interparticle forces cancel out by the NEWTON's third law of action and reaction and there is no net resultant force. The internal forces produce, however, the stress and couple stress when a particle of small volume is isolated from the rest of the body and the effect of the body on the surface of the particle is considered in terms of resultant forces and couples. The result of this, in the averaged field equations, is the existence of the stress field $\bar{\mathbf{T}}_\alpha$ and of a hyperstress field $\bar{\boldsymbol{\lambda}}_\alpha$. The reason for possible nonsymmetry of the stress tensor $\bar{\mathbf{T}}_\alpha$ is also contained in (3.2.13) and it will reveal itself more succinctly when we consider a special case of this equation in chapter 5.

## 3.3 Momentum and Energy Sources

The momentum and energy sources are defined by (2.4.20) and (2.4.36), and in this section their form will be examined in more detail with the aid of the basic deformation postulate. Using these sources, the energy balance equation (2.4.35) will then be transformed into an alternate form so that the structural characteristics of the mixture can reveal themselves more explicitly. The results in this section form a preliminary analysis for the next section where the basic field equations of chapter 2 and additional balance equations of the previous section are manipulated into alternate forms for further analysis in later chapters.

The covariance coefficient $\mathbf{C}_{1\alpha}$ appears in the momentum source $\hat{\mathbf{p}}_\alpha$ in (2.4.20) and it is defined by (2.4.21). Using (2.4.20), (3.1.1), (3.1.11), (3.1.13) and (3.2.3), it is reduced to the following form:

$$C_{1\alpha ij} = \bar{\rho}_\alpha \nu_{im}^{(\alpha)} \nu_{jn}^{(\alpha)} \bar{i}_{\alpha mn} \tag{3.3.1}$$

from where it can be noticed that the velocity correlation is expressed directly in terms of the inertia and gyration tensors. Equations (3.3.1), (3.1.11) and the definition of the *interaction force*

$$\bar{t}_{\alpha i} = \frac{1}{U} \sum_\delta \int_{a(\Lambda\delta)} T_{ij}^{(\alpha\delta)} n_j^{(\alpha\delta)}\, da \tag{3.3.2}$$

reduce (2.4.20) to

$$\hat{p}_{\alpha i} = \bar{t}_{\alpha i} - \bar{\rho}_\alpha \nu_{iq,j}^{(\alpha)} \nu_{jn}^{(\alpha)} \bar{i}_{\alpha qn} \tag{3.3.3}$$

This equation shows that the momentum source arises from reaction forces acting on the surface of $U_\alpha$ and nonuniform material motion about the center of mass as it is subjected to mechanical and thermal loadings. Note that the interaction force $\bar{\mathbf{t}}_\alpha$ is different from the definition of the traction force $\mathbf{t} = \bar{\mathbf{T}}\mathbf{n}$ in section 2.4.7.

The covariance coefficients $\mathbf{c}_{2\alpha}$, $\mathbf{c}_{3\alpha}$ and $c_{4\alpha}$ in the energy source (2.4.36) and defined by (2.4.37)-(2.4.39) may also be reduced with the aid of (2.3.7), (3.1.11), (3.1.13), (3.2.3), (3.2.5), (3.2.10) and (3.2.11), *i.e.*

$$\hat{c}_{2\alpha j} = \bar{\rho}_\alpha \nu_{jm}^{(\alpha)} \bar{\ell\epsilon}_{\alpha m} + \bar{\rho}_\alpha \bar{v}_{\alpha i} \nu_{im}^{(\alpha)} \nu_{jq}^{(\alpha)} \bar{i}_{\alpha mq} + \frac{1}{2} \bar{\rho}_\alpha \nu_{im}^{(\alpha)} \nu_{in}^{(\alpha)} \hat{I}_{\alpha jmn} \tag{3.3.4}$$

$$\hat{c}_{3\alpha j} = \nu_{im}^{(\alpha)} \bar{\lambda}_{\alpha mij} \tag{3.3.5}$$

$$\hat{c}_{4\alpha} = \bar{\rho}_\alpha \nu_{im}^{(\alpha)} \bar{\ell}_{\alpha mi} \tag{3.3.6}$$

where the *internal energy density moment* $\bar{\ell\epsilon}_\alpha$, is defined by

$$\bar{\ell\epsilon}_{\alpha m} = \frac{1}{\bar{\rho}_\alpha} \frac{1}{U} \sum_\delta \int_{U(\alpha\delta)} \rho^{(\alpha\delta)} \epsilon^{(\alpha\delta)} \xi_m^{(\alpha\delta)}\, dU \tag{3.3.7}$$

Defining further the *phase change energy flux* $\hat{c}_\alpha \hat{\bar{\epsilon}}_\alpha$, and the *interphase heat supply rate* $\bar{q}_{s\alpha}$, by equations

$$\hat{c}_\alpha \hat{\bar{\epsilon}}_\alpha = \frac{1}{U} \sum_\delta \int_{a(\Lambda\delta)} \rho^{(\alpha\delta)} \epsilon^{(\alpha\delta)} (\mathbf{S}^{(\Lambda\delta)} - \mathbf{v}^{(\alpha\delta)}) \cdot \mathbf{n}^{(\alpha\delta)}\, da \tag{3.3.8}$$

$$\bar{q}_{s\alpha} = \frac{1}{U}\sum_{\delta}\int_{a(\Lambda\delta)} \mathbf{q}^{(\alpha\delta)}\cdot\mathbf{n}^{(\alpha\delta)}\,da \tag{3.3.9}$$

the energy source (2.4.36) can be written as:

$$\begin{aligned}\hat{\epsilon}_\alpha = &-\hat{p}_{\alpha i}\tilde{v}_{\alpha i} - \hat{c}_\alpha(\tilde{\epsilon}_\alpha - \hat{\hat{\epsilon}}_\alpha) + (\bar{\rho}_\alpha \nu^{(\alpha)}_{jq}\tilde{i}_{\alpha mq})_{,j}\tilde{v}_{\alpha i}\nu^{(\alpha)}_{im} - \bar{q}_{s\alpha} + \tilde{v}_{\alpha i}\bar{t}_{\alpha i}\\ &+\bar{S}_{\alpha mj}\nu^{(\alpha)}_{jm} + \bar{\rho}_\alpha\tilde{\ell}_{\alpha mi}\nu^{(\alpha)}_{im} - \frac{1}{2}\bar{\rho}_\alpha\grave{\overline{(\nu^{(\alpha)}_{im}\nu^{(\alpha)}_{iq}\tilde{i}_{\alpha mq})}}\\ &+(\nu^{(\alpha)}_{jm}\bar{\lambda}_{\alpha mji} - \nu^{(\alpha)}_{im}\bar{\rho}_\alpha\tilde{\ell}\epsilon_{\alpha m} - \bar{\rho}_\alpha\tilde{v}_{\alpha j}\nu^{(\alpha)}_{jm}\tilde{i}_{\alpha mq}\nu^{(\alpha)}_{iq}\\ &-\frac{1}{2}\bar{\rho}_\alpha\nu^{(\alpha)}_{jm}\nu^{(\alpha)}_{jn}\hat{I}_{\alpha imn})_{,i} + \frac{1}{2}\hat{c}_\alpha\nu^{(\alpha)}_{im}\nu^{(\alpha)}_{iq}(\hat{i}_{\alpha mq} - \tilde{i}_{\alpha mq})\end{aligned} \tag{3.3.10}$$

or upon using (3.3.3) it becomes:

$$\begin{aligned}\hat{\epsilon}_\alpha = &-\hat{c}_\alpha(\tilde{\epsilon}_\alpha - \hat{\hat{\epsilon}}_\alpha) - \bar{q}_{s\alpha} + \bar{S}_{\alpha mj}\nu^{(\alpha)}_{jm} + \bar{\rho}_\alpha\tilde{\ell}_{\alpha mi}\nu^{(\alpha)}_{im} - \frac{1}{2}\bar{\rho}_\alpha\grave{\overline{(\nu^{(\alpha)}_{im}\nu^{(\alpha)}_{iq}\tilde{i}_{\alpha mq})}}\\ &-\tilde{v}_{\alpha j,i}(\bar{\rho}_\alpha\nu^{(\alpha)}_{jm}\nu^{(\alpha)}_{iq}\tilde{i}_{\alpha mq}) + \frac{1}{2}\hat{c}_\alpha\nu^{(\alpha)}_{im}\nu^{(\alpha)}_{iq}(\hat{i}_{\alpha mq} - \tilde{i}_{\alpha mq})\\ &+(\nu^{(\alpha)}_{jm}\bar{\lambda}_{\alpha mji} - \nu^{(\alpha)}_{im}\bar{\rho}_\alpha\tilde{\ell}\epsilon_{\alpha m} - \frac{1}{2}\bar{\rho}_\alpha\nu^{(\alpha)}_{jm}\nu^{(\alpha)}_{jn}\hat{I}_{\alpha imn})_{,i}\end{aligned} \tag{3.3.11}$$

The momentum source (3.3.3) and energy source (3.3.11) can be used in the momentum equation (2.4.19) and energy equation (2.4.35) to form the *total energy balance equation of phase α*. This is accomplished by first forming a scalar product of the momentum equation (2.4.19) with the velocity $\tilde{\mathbf{v}}_\alpha$ and adding the result to (2.4.35). After some algebra, the result is:

$$\begin{aligned}\bar{\rho}_\alpha(\grave{\tilde{\epsilon}}_\alpha + \frac{1}{2}\grave{\overline{\tilde{v}_{\alpha i}\tilde{v}_{\alpha i}}} + \frac{1}{2}\grave{\overline{\nu^{(\alpha)}_{im}\nu^{(\alpha)}_{iq}\tilde{i}_{\alpha mq}}}) = &(\bar{T}_{\alpha ij}\tilde{v}_{\alpha i})_{,j} - \bar{q}_{\alpha i,i}\\ +\bar{\rho}_\alpha\tilde{b}_{\alpha i}\tilde{v}_{\alpha i} + \bar{\rho}_\alpha\tilde{r}_\alpha - \hat{c}_\alpha(\tilde{\epsilon}_\alpha - \hat{\hat{\epsilon}}_\alpha) &+ \tilde{v}_{\alpha i}\nu^{(\alpha)}_{im}(\bar{\rho}_\alpha\nu^{(\alpha)}_{jq}\tilde{i}_{\alpha mq})_{,j}\\ &-\bar{q}_{s\alpha} + \tilde{v}_{\alpha i}\bar{t}_{\alpha i} + \bar{S}_{\alpha mj}\nu^{(\alpha)}_{jm} + \bar{\rho}_\alpha\tilde{\ell}_{\alpha mi}\nu^{(\alpha)}_{im}\\ &+(\nu^{(\alpha)}_{jm}\bar{\lambda}_{\alpha mji} - \nu^{(\alpha)}_{im}\bar{\rho}_\alpha\tilde{\ell}\epsilon_{\alpha m} - \bar{\rho}_\alpha\tilde{v}_{\alpha j}\nu^{(\alpha)}_{jm}\tilde{i}_{\alpha mq}\nu^{(\alpha)}_{iq}\\ &-\frac{1}{2}\bar{\rho}_\alpha\nu^{(\alpha)}_{jm}\nu^{(\alpha)}_{jn}\hat{I}_{\alpha imn})_{,i} + \frac{1}{2}\hat{c}_\alpha\nu^{(\alpha)}_{im}\nu^{(\alpha)}_{iq}(\hat{i}_{\alpha mq} - \tilde{i}_{\alpha mq})\end{aligned} \tag{3.3.12}$$

This equation illustrates that the total energy of phase $\alpha$ consists of the thermal energy, the kinetic energy of the center of mass, and the kinetic energy relative to the center of mass. The rate of change of this energy is balanced by: (1) the power produced by stress, body force, heat transfer and heat generation within the mixture; (2) the energy transfer due to the phase change; (3) the heat transport and rate of work done by forces on the boundary of $U_\alpha$; and (4) the power associated with the spatial nonuniformity of the material motion relative to the center of mass.

## 3.4 Alternate Forms of Balance Equations

The equilibrated inertia equation (3.2.2), equilibrated moments equation (3.2.13), and the energy balance equation (3.3.12) can be transformed into alternate forms depending on the different definitions of source terms or the combination of one or more equations. In this section these equations will be manipulated into the expressions that will be useful for analysis in the next section and for comparison with special models of structured multiphase mixtures in chapter 5. The reinterpretation of source terms in various balance equations also affects the definition of variables for the mixture as a whole, and it will be discussed in the later part of the section.

### 3.4.1 Phasic Balance Equations

The momentum balance equation (2.4.19) can be transformed into an alternate form by redefining the momentum source, *i.e.*

$$\check{\mathbf{m}}_\alpha = \hat{\mathbf{p}}_\alpha + \hat{c}_\alpha \tilde{\mathbf{v}}_\alpha \tag{3.4.1}$$

so that (2.4.19) can be written as

$$\bar{\rho}_\alpha \grave{\tilde{\mathbf{v}}}_\alpha = \nabla\cdot\bar{\mathbf{T}}_\alpha + \bar{\rho}_\alpha \tilde{\mathbf{b}}_\alpha + \check{\mathbf{m}}_\alpha - \hat{c}_\alpha \tilde{\mathbf{v}}_\alpha \tag{3.4.2}$$

In the equilibrated inertia equation (3.2.2) the equilibrated source inertia $\hat{i}_{\alpha k\ell}$, and the hyperinertia tensor $\hat{I}_{\alpha mk\ell}$, may be grouped into one term defined by the equation

$$\hat{k}_{\alpha jn} = \hat{c}_\alpha \hat{i}_{\alpha jn} - (\bar{\rho}_\alpha \hat{I}_{\alpha mjn})_{,m} \tag{3.4.3}$$

such that (3.2.2) reduces to the following form:

$$\bar{\rho}_\alpha(\grave{\tilde{i}}_{\alpha jn} - \nu^{(\alpha)}_{jm}\tilde{i}_{\alpha mn} - \nu^{(\alpha)}_{nm}\tilde{i}_{\alpha jm}) = -\hat{c}_\alpha \hat{i}_{\alpha jn} + \hat{k}_{\alpha jn} \tag{3.4.4}$$

Alternatively, we may define

$$k_{\alpha jni} = -\bar{\rho}_\alpha \hat{I}_{\alpha ijn} \tag{3.4.5}$$

$$K_{\alpha jn} = \bar{\rho}_\alpha \nu^{(\alpha)}_{jm}\tilde{i}_{\alpha mn} + \bar{\rho}_\alpha \nu^{(\alpha)}_{nm}\tilde{i}_{\alpha jm} \tag{3.4.6}$$

$$\check{k}_{\alpha jn} = \hat{c}_\alpha \hat{i}_{\alpha jn} \tag{3.4.7}$$

so that (3.2.2) can also be written as

$$\bar{\rho}_\alpha \grave{\tilde{i}}_{\alpha jn} = -\hat{c}_\alpha \tilde{i}_{\alpha jn} + K_{\alpha jn} + k_{\alpha jni,i} + \check{k}_{\alpha jn} \tag{3.4.8}$$

Multiplying (3.4.4) by $\nu^{(\alpha)}_{kn}$ and adding the result to the equilibrated moments equation (3.2.13) yields:

$$\bar{\rho}_\alpha(\overset{\backprime}{\overline{\tilde{\imath}_{\alpha jn}\nu^{(\alpha)}_{kn}}} - \nu^{(\alpha)}_{jn}\tilde{\imath}_{\alpha nm}\nu^{(\alpha)}_{km}) = \bar{S}_{\alpha jk} + \bar{\rho}_\alpha\tilde{\ell}_{\alpha jk} + \bar{\lambda}_{\alpha jkm,m}$$
$$-\bar{\rho}_\alpha\nu^{(\alpha)}_{mq}\tilde{v}_{\alpha k,m}\tilde{\imath}_{\alpha jq} - \hat{c}_\alpha\tilde{\imath}_{\alpha jn}\nu^{(\alpha)}_{kn} + \hat{g}_{\alpha jk} \qquad (3.4.9)$$

where the equilibrated moment source $\hat{g}_{\alpha jk}$ is defined by

$$\hat{g}_{\alpha jk} = \hat{c}_\alpha\hat{\imath}_{\alpha jn}\nu^{(\alpha)}_{kn} - (\bar{\rho}_\alpha\nu^{(\alpha)}_{kn}\hat{I}_{\alpha mjn})_{,m} \qquad (3.4.10)$$

Defining a new set of variables in (3.2.13) by

$$\bar{h}_{\alpha jkm} = \bar{\lambda}_{\alpha jkm} - \bar{\rho}_\alpha\tilde{v}_{\alpha k}\tilde{\imath}_{\alpha jq}\nu^{(\alpha)}_{mq} \qquad (3.4.11)$$

$$\bar{\rho}_\alpha\tilde{f}_{\alpha jk} = \bar{S}_{\alpha jk} \qquad (3.4.12)$$

$$\breve{g}_{\alpha jk} = \bar{\rho}_\alpha\nu^{(\alpha)}_{jn}\tilde{\imath}_{\alpha nm}\nu^{(\alpha)}_{km} + \hat{g}_{\alpha jk} + \tilde{v}_{\alpha k}(\bar{\rho}_\alpha\tilde{\imath}_{\alpha jq}\nu^{(\alpha)}_{mq})_{,m} \qquad (3.4.13)$$

equation (3.4.9) is reduced to an alternate form of the equilibrated moments equation. Hence

$$\bar{\rho}_\alpha\overset{\backprime}{\overline{\tilde{\imath}_{\alpha jn}\nu^{(\alpha)}_{kn}}} = \bar{h}_{\alpha jkm,m} + \bar{\rho}_\alpha(\tilde{f}_{\alpha jk} + \tilde{\ell}_{\alpha jk}) - \hat{c}_\alpha\tilde{\imath}_{\alpha jn}\nu^{(\alpha)}_{kn} + \breve{g}_{\alpha jk} \qquad (3.4.14)$$

The energy balance equation (3.3.12) may also be transformed into alternate forms. Towards this end two new energy source terms, $\hat{e}_\alpha$ and $\breve{e}_\alpha$, will be defined, *i.e.*

$$\hat{e}_\alpha = \hat{c}_\alpha\hat{\tilde{\varepsilon}}_\alpha + (\bar{\rho}_\alpha\nu^{(\alpha)}_{jq}\tilde{\imath}_{\alpha mq})_{,j}\tilde{v}_{\alpha i}\nu^{(\alpha)}_{im} - \bar{q}_{s\alpha} + \tilde{v}_{\alpha i}\bar{t}_{\alpha i}$$
$$-(\bar{\rho}_\alpha\nu^{(\alpha)}_{im}\tilde{\ell}\epsilon_{\alpha m} + \frac{1}{2}\bar{\rho}_\alpha\nu^{(\alpha)}_{jm}\nu^{(\alpha)}_{jn}\hat{I}_{\alpha imn})_{,i} + \frac{1}{2}\hat{c}_\alpha\nu^{(\alpha)}_{im}\nu^{(\alpha)}_{iq}(\hat{\imath}_{\alpha mq} - \tilde{\imath}_{\alpha mq}) \qquad (3.4.15)$$

$$\breve{e}_\alpha = \hat{c}_\alpha(\frac{1}{2}\tilde{v}_{\alpha i}\tilde{v}_{\alpha i} + \frac{1}{2}\nu^{(\alpha)}_{im}\nu^{(\alpha)}_{in}\tilde{\imath}_{\alpha mn}) + \nu^{(\alpha)}_{jm}\bar{S}_{\alpha mj}$$
$$+\frac{1}{2}\bar{\rho}_\alpha\tilde{\imath}_{\alpha mq}(\nu^{(\alpha)}_{im}\grave{\nu}^{(\alpha)}_{iq} - \grave{\nu}^{(\alpha)}_{im}\nu^{(\alpha)}_{iq}) + \hat{e}_\alpha \qquad (3.4.16)$$

and (3.3.10) and (3.3.12) may be written, respectively, as

$$\hat{\varepsilon}_\alpha = -\hat{p}_{\alpha i}\tilde{v}_{\alpha i} - \hat{c}_\alpha\tilde{\varepsilon}_\alpha + \bar{S}_{\alpha mj}\nu^{(\alpha)}_{jm} + \bar{\rho}_\alpha\tilde{\ell}_{\alpha mi}\nu^{(\alpha)}_{im} - \frac{1}{2}\bar{\rho}_\alpha(\overset{\backprime}{\overline{\nu^{(\alpha)}_{im}\nu^{(\alpha)}_{iq}\tilde{\imath}_{\alpha mq}}})$$
$$+(\nu^{(\alpha)}_{jm}\bar{\lambda}_{\alpha mji} - \bar{\rho}_\alpha\tilde{v}_{\alpha j}\nu^{(\alpha)}_{jm}\tilde{\imath}_{\alpha mq}\nu^{(\alpha)}_{iq})_{,i} + \hat{e}_\alpha \qquad (3.4.17)$$

$$
\begin{aligned}
\bar{\rho}_\alpha\Big(\grave{\tilde{\epsilon}}_\alpha + \frac{1}{2}\overline{\tilde{v}_{\alpha i}\tilde{v}_{\alpha i}}^{\,\grave{}} + \frac{1}{2}\overline{\nu^{(\alpha)}_{im}\nu^{(\alpha)}_{iq}\tilde{i}_{\alpha mq}}^{\,\grave{}}\Big) &= (\bar{T}_{\alpha ij}\tilde{v}_{\alpha i})_{,j} - \bar{q}_{\alpha i,i} \\
&+ \bar{\rho}_\alpha\tilde{b}_{\alpha i}\tilde{v}_{\alpha i} + \bar{\rho}_\alpha\tilde{r}_\alpha - \hat{c}_\alpha\tilde{\epsilon}_\alpha + \bar{S}_{\alpha mj}\nu^{(\alpha)}_{jm} + \bar{\rho}_\alpha\tilde{l}_{\alpha mi}\nu^{(\alpha)}_{im} \\
&+ (\nu^{(\alpha)}_{jm}\bar{\lambda}_{\alpha mji} - \bar{\rho}_\alpha\tilde{v}_{\alpha j}\nu^{(\alpha)}_{jm}\tilde{i}_{\alpha mq}\nu^{(\alpha)}_{iq})_{,i} + \hat{e}_\alpha
\end{aligned}
\tag{3.4.18}
$$

Using (3.4.1), (3.4.11)-(3.4.16) and the balance of kinetic energy equation obtained by taking a scalar product of the momentum equation (3.4.2) with the velocity $\tilde{\mathbf{v}}_\alpha$, the above equation is reduced to

$$
\begin{aligned}
\bar{\rho}_\alpha\grave{\tilde{\epsilon}}_\alpha &= \bar{T}_{\alpha ij}\tilde{v}_{\alpha i,j} - \bar{q}_{\alpha i,i} + \bar{\rho}_\alpha\tilde{r}_\alpha + \nu^{(\alpha)}_{im,q}\bar{h}_{\alpha miq} - \check{m}_{\alpha i}\tilde{v}_{\alpha i} - \nu^{(\alpha)}_{im}\check{g}_{\alpha mi} \\
&- \bar{\rho}_\alpha\nu^{(\alpha)}_{im}\tilde{f}_{\alpha mi} - \hat{c}_\alpha\Big(\tilde{\epsilon}_\alpha - \frac{1}{2}\tilde{v}_{\alpha i}\tilde{v}_{\alpha i} - \frac{1}{2}\nu^{(\alpha)}_{im}\tilde{i}_{\alpha mn}\nu^{(\alpha)}_{in}\Big) \\
&+ \frac{1}{2}\bar{\rho}_\alpha\nu^{(\alpha)}_{im}\nu^{(\alpha)}_{iq}\grave{\tilde{i}}_{\alpha mq} + \check{e}_\alpha
\end{aligned}
\tag{3.4.19}
$$

### 3.4.2 Mixture Balance Equations

The conservation and balance equations for the multiphase mixture as a whole are obtained by summing up the phasic field equations. This task was performed in chapter 2 for the basic set of field equations involving the balance of mass, linear momentum, angular momentum, energy, and entropy inequality. The introduction of the structural characteristics of the mixture into the theory through the additional balance equations expressing the balance of equilibrated inertia and moments necesitates not only the derivation of these balance equations for the mixture as a whole, but also the redefinition of the mixture properties in the basic field equations. Although some of these mixture variables will not change (notably, the mixture stress tensor (2.4.23), mixture body force (2.4.24), and mixture entropy (2.4.51)), presented below is, however, a complete list of these variables for a ready reference.

In the theory of structured multiphase mixtures discussed above, the *mixture variables will be defined as* follows:

$$
\rho = \sum_\alpha \bar{\rho}_\alpha \tag{2.3.10}
$$

$$
\rho v_i = \sum_\alpha \bar{\rho}_\alpha\tilde{v}_{\alpha i} \tag{2.3.12}
$$

$$
T_{ij} = \sum_\alpha (\bar{T}_{\alpha ij} - \bar{\rho}_\alpha u_{\alpha i}u_{\alpha i}) \tag{2.4.23}
$$

$$\rho b_i = \sum_\alpha \bar{\rho}_\alpha \tilde{b}_{\alpha i} \tag{2.4.24}$$

$$\rho i_{ij} = \sum_\alpha \bar{\rho}_\alpha \tilde{i}_{\alpha ij} \tag{3.4.20}$$

$$\rho i_{ij}\nu_{mn} = \sum_\alpha \bar{\rho}_\alpha \tilde{i}_{\alpha ij}\nu^{(\alpha)}_{mn} \tag{3.4.21}$$

$$\rho\epsilon = \sum_\alpha [\bar{\rho}_\alpha\tilde{\epsilon}_\alpha + \frac{1}{2}\bar{\rho}_\alpha u_{\alpha i}u_{\alpha i} + \frac{1}{2}\bar{\rho}_\alpha\tilde{i}_{\alpha mn}(\nu^{(\alpha)}_{im} - \nu_{im})(\nu^{(\alpha)}_{in} - \nu_{in})] \tag{3.4.22}$$

$$\begin{aligned} q_i = \sum_\alpha [&\bar{q}_{\alpha i} - \bar{T}_{\alpha ji}u_{\alpha j} + \bar{\rho}_\alpha\tilde{\epsilon}_\alpha u_{\alpha i} + \frac{1}{2}\bar{\rho}_\alpha u_{\alpha j}u_{\alpha j}u_{\alpha i} \\ &+(\nu_{jk} - \nu^{(\alpha)}_{jk})\bar{h}_{\alpha kji} - \bar{\rho}_\alpha(\nu_{jk} - \frac{1}{2}\nu^{(\alpha)}_{jk})\nu^{(\alpha)}_{jq}\tilde{i}_{\alpha kq}u_{\alpha i}] \end{aligned} \tag{3.4.23}$$

$$\begin{aligned} \rho r = \sum_\alpha [&\bar{\rho}_\alpha\tilde{r}_\alpha + \bar{\rho}_\alpha\tilde{b}_{\alpha i}u_{\alpha i} + \bar{\rho}_\alpha\tilde{\ell}_{\alpha ij}(\nu^{(\alpha)}_{ji} - \nu_{ji}) \\ &-\frac{1}{2}\bar{\rho}_\alpha\tilde{i}_{\alpha mn}(\nu^{(\alpha)}_{im}\grave{\nu}^{(\alpha)}_{in} - \grave{\nu}^{(\alpha)}_{im}\nu^{(\alpha)}_{in})] \end{aligned} \tag{3.4.24}$$

$$\rho s = \sum_\alpha \bar{\rho}_\alpha\tilde{s}_\alpha \tag{2.4.51}$$

$$k_{jni} = \sum_\alpha [k_{\alpha jni} - \bar{\rho}_\alpha\tilde{i}_{\alpha jn}u_{\alpha i}] \tag{3.4.25}$$

$$K_{jn} = \sum_\alpha K_{\alpha jn} \tag{3.4.26}$$

$$\rho \ell_{ij} = \sum_\alpha \bar{\rho}_\alpha\tilde{\ell}_{\alpha ij} \tag{3.4.27}$$

$$\rho f_{ij} = \sum_\alpha \bar{\rho}_\alpha\tilde{f}_{\alpha ij} \tag{3.4.28}$$

$$h_{jik} = \sum_\alpha [\bar{h}_{\alpha jik} - \bar{\rho}_\alpha\nu^{(\alpha)}_{in}\tilde{i}_{\alpha jn}u_{\alpha k}] \tag{3.4.29}$$

**This definition of mixture variables allows for the conservation and balance equations for the mixture as a whole to be expressed in terms of these variables. The conservation of mass and balance of linear momentum, angular momentum, energy, equilibrated inertia, and equilibrated moments, together with the entropy inequality for the mixture, can thus be written as:**

$$\dot{\rho} + \rho v_{i,i} = 0 \tag{2.4.16}$$

$$\rho \dot{v}_i = T_{ij,j} + \rho b_i + \hat{p}_i \tag{2.4.22}$$

$$\hat{M}_{ij} = T_{ij} - T_{ji} \tag{2.4.34}$$

$$\begin{aligned}\rho \dot{\epsilon} = T_{ji} v_{j,i} - q_{i,i} + \rho r + \frac{1}{2}\rho[\nu_{im}\nu_{in}\dot{i}_{mn} + i_{mn}(\nu_{im}\dot{\nu}_{in} - \dot{\nu}_{im}\nu_{in})] \\ + h_{kji}\nu_{jk,i} - \rho \nu_{jk} f_{kj} + \check{e}\end{aligned} \tag{3.4.30}$$

$$\rho \dot{s} + \sum_\alpha \Big(\frac{\bar{q}_{\alpha i}}{\bar{\bar{\theta}}_\alpha} + \bar{\rho}_\alpha \tilde{s}_\alpha u_{\alpha i}\Big)_{,i} - \sum_\alpha \frac{\bar{\rho}_\alpha \tilde{r}_\alpha}{\bar{\bar{\theta}}_\alpha} + \hat{s} \geq 0 \tag{2.4.50}$$

$$\rho \dot{i}_{jn} = k_{jni,i} + K_{jn} + \check{k}_{jn} \tag{3.4.31}$$

$$\rho \overline{\dot{i}_{jn}\nu_{kn}} = h_{jkm,m} + \rho \ell_{jk} + \rho f_{jk} + \breve{g}_{jk} \tag{3.4.32}$$

**where the source terms satisfy the following conditions:**

$$\sum_\alpha \hat{c}_\alpha = 0 \tag{2.4.17}$$

$$\hat{p}_i = \sum_\alpha \check{m}_{\alpha i} \tag{2.4.26}$$

$$\hat{M}_{ij} = \sum_\alpha \hat{M}_{\alpha ij} \tag{2.4.34}$$

$$\check{e} = \sum_\alpha \check{e}_\alpha - v_j \hat{p}_j - \nu_{jk} \breve{g}_{kj} \tag{3.4.33}$$

$$\hat{s} = \sum_\alpha \hat{s}_\alpha \tag{2.4.52}$$

$$\check{k}_{jn} = \sum_{\alpha} \check{k}_{\alpha jn} \tag{3.4.34}$$

$$\check{g}_{jk} = \sum_{\alpha} \check{g}_{\alpha jk} \tag{3.4.35}$$

The reasons for the existence of source terms in the theory of multiphase mixtures was discussed in section 2.5.

The conservation and balance equations of structured multiphase mixtures presented above reduce to the basic set of equations discussed in chapter 2 in the absence of the structural characteristics of the mixture. Thus, setting the phase inertias and gyration tensors equal to zero reduces (3.4.22) to (2.4.41), (3.4.23) to (2.4.42), (3.4.24) to (2.4.43) and $\check{e}$ in (3.4.33) to $(\hat{e} - \mathbf{v}\cdot\hat{\mathbf{p}})$ appearing in (2.4.40).

## 3.5 Concluding Remarks

The basic task of the chapter, to derive a set of field equations for multiphase mixtures with structure, is complete. By invoking a basic postulate of the material deformation with respect to the center of mass has allowed not only the derivation of additional balance equations expressing the evolution of inertia, particles rotation and particle dilatation of each phase, but also the embodiment of these characteristics into the basic field equations expressing the balance of mass, momentum, energy, and entropy inequality. The affine deformation postulate permits *modeling of the center of mass motions with the basic equations*, whereas the additional *balance equations expressing the equilibrated inertia and equilibrated moments allow for the modeling of material deformation relative to the center of mass.* In this sense, these latter equations may be viewed as constitutive, since they are a consequence of an assumption of the material deformation. While a consideration was given only to the first order material deformation relative to the center of mass of each phase, it should be clear that it may be possible to construct higher order theories involving second and higher order approximations in (3.1.3). Since the additional balance equations contain the intrinsic and extrinsic source terms that must be specified by (probably much more restrictive) constitutive equations, it is approprite to view these equations as the *balance equations*, as they have been referred to in this chapter.

In the following chapter, the development of the theory will involve an examination of the kinematic, dynamic, and energetic field variables for the invariance properties under the change of the observer's frame of reference. The reason for this is to properly study the constitutive equations in subsequent chapters and to compare the results of the theory with existing special models of multiphase mixtures.

CHAPTER 4

# RESTRICTIONS IMPOSED ON FIELD EQUATIONS BY THE PRINCIPLE OF MATERIAL FRAME-INDIFFERENCE

**After a brief review of the concepts of frame, change of frame, and frame-indifference, the principle of material frame-indifference for multiphase mixtures will be stated and used to study the restrictions imposed on the kinematic, dynamic and energetic variables appearing in the multiphase field equations developed in the previous chapters. The analysis in this chapter forms a basis for the subsequent chapters involved with the development of constitutive equations and more detailed discussion of special models of structured multiphase mixtures.**

## 4.1 Events, Frames, Change of Frame, and Frame-Indifference

The distance and time intervals are fundamental quantities which are measured in kinematics. They can only be determined with respect to an observer or frame of reference. A *frame of reference* of an observer is, thus, fundamental to the observation of a physical phenomenon and may be the fixed stars, walls of a laboratory, or a set of objects whose mutual distances remain unchanging during the period of observation. When this set of objects is chosen as three mutually perpendicular unit vectors, then a *coordinate system* is specified, which is obviously different from a frame of reference. The time of occurrence of some *event* may be specified only with respect to the time of occurrence of some other event, with such a reference being a part of the specification of a frame of reference. The frame of reference may be viewed, therefore, as a space of events formed by the places and instants which themselves are the elements of Euclidean spaces. This homeomorphic mapping of events into places and instants may be called a *framing* $\mathcal{F}$ (TRUESDELL,1977), where the instants are assumed to be oriented and a tensor is an element in the space of places. A *change of frame* in classical mechanics has the meaning that two observers who have chosen the same units of length and time, set their clocks differently and are in arbitrary rigid motion with respect to one another are equally qualified to represent physical phenomena. If the same event is observed in frames $\mathcal{F}$ and $\mathcal{F}^*$, with corresponding places $\mathbf{x}$ and $\mathbf{x}^*$, then a change of frame from $\mathcal{F}$ to $\mathcal{F}^*$ of an observer is represented by

$$t^* = t + a \tag{4.1.1}$$

$$\mathbf{x}^* = \mathbf{x}_0^*(t) + \mathbf{Q}(t)(\mathbf{x} - \mathbf{x}_0) \tag{4.1.2}$$

where $a$ is a reference time. A place $\mathbf{x}_0$ and time $t$ assigned by an event in $\mathcal{F}$ is assigned a place $\mathbf{x}_0^*$ and time $t^*$ by the same event in $\mathcal{F}^*$, and $\mathbf{Q}(t)$ is a rotation of a vector $(\mathbf{x} - \mathbf{x}_0)$ in $\mathcal{F}$ into a vector $(\mathbf{x}^* - \mathbf{x}_0^*)$ in $\mathcal{F}^*$ with the properties of preserving the inner product and sense of orientation. The inner product is preserved if $\mathbf{Q}$ is an *orthogonal tensor*, *i.e.*

$$\mathbf{Q}(t)\mathbf{Q}(t)^T = \mathbf{I}, \quad Q_{ij}Q_{kj} = Q_{ji}Q_{jk} = \delta_{ik} \tag{4.1.3}$$

where $\mathbf{I}$ is the unit tensor, whereas the sense of orientation is preserved by setting

$$det\mathbf{Q} = +1 \tag{4.1.4}$$

Both $\mathbf{x}_0^*(t)$ and $\mathbf{Q}(t)$ are also assumed to be smooth functions of $t$. Thus, the condition (4.1.3) is necessary and sufficient for $\mathbf{Q}$ to preserve the inner product and distances in vector spaces. [1]

Although a physical phenomenon is assumed to be independent of a frame of reference, its description in one frame is generally different from a description of it in another frame. The physical variables which are functions defined only on the event world in each frame and do not depend for their description on the frame itself are said to be *frame-indifferent*. A frame-indifferent scalar-valued function $f(\mathbf{x}, t)$ has the same value in any frame of reference and it is required to transform under the change of frame according to

$$f(\mathbf{x}^*, t^*) = f(\mathbf{x}, t) \tag{4.1.5}$$

A vector field $\mathbf{y}$, a second order tensor $\mathbf{M}$, and a third order tensor $\mathbf{A}$ to be frame-indifferent are required to transform under the change of frame (4.1.2) as follows:

$$\mathbf{y}^*(\mathbf{x}^*, t^*) = \mathbf{Q}(t)\mathbf{y}(\mathbf{x}, t), \quad y_i^* = Q_{ij}y_j \tag{4.1.6}$$

$$\mathbf{M}^*(\mathbf{x}^*, t^*) = \mathbf{Q}(t)\mathbf{M}(\mathbf{x}, t)\mathbf{Q}(t)^T, \quad M_{ij}^* = Q_{ik}M_{kl}Q_{jl} \tag{4.1.7}$$

[1] This is proved by using (4.1.2) and (4.1.3), *i.e.*

$$(\mathbf{x}^* - \mathbf{x}_0^*)\cdot(\mathbf{y}^* - \mathbf{y}_0^*) = Q_{ij}(x_j - x_{0j})Q_{ik}(y_k - y_{0k}) = (\mathbf{x} - \mathbf{x}_0)\cdot(\mathbf{y} - \mathbf{y}_0)$$

$$\mathbf{A}^*(\mathbf{x}^*, t^*) = \mathbf{Q}(t)\mathbf{A}(\mathbf{x}, t)\mathbf{Q}(t)^T\mathbf{Q}(t)^T, \quad A^*_{ijk} = Q_{i\ell}Q_{jm}Q_{kn}A_{\ell mn} \tag{4.1.8}$$

The result (4.1.6) follows from (4.1.2) directly. To prove (4.1.7) we can take in (4.1.2) $\mathbf{u} = \mathbf{x} - \mathbf{x}_0$ and $\mathbf{u}^* = \mathbf{x}^* - \mathbf{x}_0^* = \mathbf{Q}\mathbf{u}$ and consider the tensor transformation $\mathbf{v} = \mathbf{M}\mathbf{u}$. But

$$\mathbf{v}^* = \mathbf{Q}\mathbf{v} = \mathbf{Q}\mathbf{M}\mathbf{u} = \mathbf{Q}\mathbf{M}\mathbf{Q}^{-1}\mathbf{u}^* = \mathbf{Q}\mathbf{M}\mathbf{Q}^T\mathbf{u}^* = \mathbf{M}^*\mathbf{u}^*$$

and, therefore, $\mathbf{M}^* = \mathbf{Q}\mathbf{M}\mathbf{Q}^T$ as stated. Similarly, we may also consider the tensor transformation $\mathbf{V} = \mathbf{A}\mathbf{v}$, where $\mathbf{V}$ and $\mathbf{v}$ are second order and first order frame-indifferent tensors, respectively. Then

$$\mathbf{V}^* = \mathbf{Q}\mathbf{V}\mathbf{Q}^T = \mathbf{Q}\mathbf{A}\mathbf{v}\mathbf{Q}^T = \mathbf{Q}\mathbf{A}\mathbf{Q}^{-1}\mathbf{v}^*\mathbf{Q}^T = \mathbf{Q}\mathbf{A}\mathbf{Q}^T\mathbf{v}^*\mathbf{Q}^T = \mathbf{A}^*\mathbf{v}^*$$

from where it follows that $\mathbf{A}^* = \mathbf{Q}\mathbf{A}\mathbf{Q}^T\mathbf{Q}^T$, as claimed.

## 4.2 The Principle of Material Frame-Indifference for Multiphase Mixtures

As discussed in chapter 2, it is possible to consider the motion of a particle of phase $\alpha$, $X^{(\alpha\delta)}$, from the reference configuration to the spatial configuration as shown in Figure 2.1, as well as the motion of a particle of phase $\alpha$, $X_\alpha$, from the reference configuration of the superimposed continua to the mapped configuration of the superimposed continua. In both situations one can define respective particle deformation functions such that the change of frame from $\mathcal{F}$ to $\mathcal{F}^*$ of an observer induces also a change in the motion of the material particle of phase $\alpha$, namely:

$$\chi_{\boldsymbol{\kappa}}^{*(\alpha\delta)}(\mathbf{X}^{(\alpha\delta)}, t^*) = \mathbf{Q}(t)\chi_{\boldsymbol{\kappa}}^{(\alpha\delta)} + \mathbf{c}(t) \tag{4.2.1}$$

$$\chi^*_{\alpha\boldsymbol{\kappa}}(\mathbf{X}_\alpha, t^*) = \mathbf{Q}(t)\chi_{\alpha\boldsymbol{\kappa}} + \mathbf{c}(t) \tag{4.2.2}$$

where $\chi_{\boldsymbol{\kappa}}^{(\alpha\delta)}$ is the deformation function associated with the motion of particle $X^{(\alpha\delta)}$ from the reference configuration $\mathbf{X}^{(\alpha\delta)}$ to the spatial configuration $\mathbf{x}^{(\alpha\delta)}$,

$$\mathbf{x}^{(\alpha\delta)} = \chi_{\boldsymbol{\kappa}}^{(\alpha\delta)}(\mathbf{X}^{(\alpha\delta)}, t) \tag{4.2.3}$$

whereas $\chi_{\alpha\boldsymbol{\kappa}}$ denotes the deformation function associated with the motion of particle $X_\alpha$ of the superimposed continua from the reference configuration $\mathbf{X}_\alpha$ to the spatial configuration $\mathbf{x}$, *i.e.*

$$\mathbf{x} = \chi_{\alpha\boldsymbol{\kappa}}(\mathbf{X}_\alpha, t) \tag{4.2.4}$$

The latter deformation function, it may be recalled, was considered in section 2.4.1. $\mathbf{Q}(t)$ in (4.2.1) and (4.2.2) is an orthogonal tensor satisfying (4.1.3) and (4.1.4), $t^*$ is related to $t$ by the shift in time expressed by (4.1.1), and $\mathbf{c}(t)$ is a time-dependent translation vector.

Most of the physical variables that appear in the conservation and balance equations of multiphase mixtures are not, however, frame-indifferent, as may be readily shown for the velocities and accelerations. Differentiating (4.2.1) yields

$$\tilde{\mathbf{v}}_\alpha^* = \dot{\mathbf{Q}}(t)\mathbf{x} + \mathbf{Q}(t)\tilde{\mathbf{v}}_\alpha + \dot{\mathbf{c}}(t) \tag{4.2.5}$$

$$\dot{\tilde{\mathbf{v}}}_\alpha^* = \ddot{\mathbf{Q}}(t)\mathbf{x} + \mathbf{Q}(t)\dot{\tilde{\mathbf{v}}}_\alpha + 2\dot{\mathbf{Q}}(t)\tilde{\mathbf{v}}_\alpha + \ddot{\mathbf{c}}(t) \tag{4.2.6}$$

which clearly illustrates that the velocity and acceleration are *not* frame-indifferent, since they do not transform according to (4.1.6). If, however, $d\mathbf{Q}/dt = \mathbf{0}$, $t^* = t$ and $\mathbf{c} = \mathbf{w}t$, with $\mathbf{w}$ a constant vector, then (4.2.5) and (4.2.6) give $\tilde{\mathbf{v}}_\alpha^* = \mathbf{Q}\tilde{\mathbf{v}}_\alpha + \mathbf{w}$ and $\dot{\tilde{\mathbf{v}}}_\alpha^* = \mathbf{Q}\dot{\tilde{\mathbf{v}}}_\alpha$. This change of frame transformation, where the acceleration is frame-indifferent, is called the group of *Galilean transformations* which relate to observers moving at uniform velocities with respect to one another and with no change of relative orientation in time (TRUESDELL,1977).

Mass, force, torgue, temperature, internal energy, and entropy are all assumed to be frame-indifferent. They are intrinsic or primitive material properties that all observers decide *a priori* to agree upon for any given material.

The requirement of certain thermo-mechanical variables to be material frame-indifferent or form invariant under arbitrary time-dependent rotations and translations of the frame of reference has created some controversy. This controversy has basically resulted from the stress and heat flux computations in the kinetic theory (CHAPMAN & COWLING,1970) using various order iteration procedures for an ideal gas (the CHAPMAN-ENSKOG or MAXWELLIAN iterations). Starting with the third order or BURNET approximation, MÜLLER (1972) and EDELEN & MCLENNAN (1973) demonstrated that the stress and heat flux explicitly depend on the spin tensor or the skew-symmetric part of the velocity gradient which, as will be discussed in the next section, is not frame-indifferent. WANG (1975) and TRUESDELL (1976) argued against using iterative procedures or kinetic theory to prove or disprove the principle of material frame-indifference, since the iteration schemes provide only approximate results, whereas the kinetic theory in general does not provide the constitutive equations determined on the basis of *independent* variables. More recently, however, a consensus appears to have been reached that although the kinetic theory does not provide frame-indifferent constitutive equations for stress and heat flux according to the principle of material frame-indifference, it does provide and excellent approximation to these values for gases which are not highly rarefied, or more precisely, for gases with the ratios of mean free molecular paths to the characteristic time scales of the problem of much less than one (HICKL & MÜLLER,1983; SPEZIALE,1986). From the latest studies this much is clear: the use of the principle of material frame-indifference can be physically justified in many practical applications and should be fully exploited in analyses. The principle of the material frame-indifference is a very powerful mathematical tool for studying the constitutive equations and will be used in this book without further questioning its validity.

## 4.3 Transformation Properties of Multiphase Variables

The density, internal energy, heat generation rate, entropy, and temperature are all scalar functions and are required to be invariant under the change of the observer's frame of reference. Thus,

$$\bar{\rho}^*_\alpha = \bar{\rho}_\alpha, \quad \tilde{\epsilon}^*_\alpha = \tilde{\epsilon}_\alpha, \quad \tilde{r}^*_\alpha = \tilde{r}_\alpha, \quad \tilde{s}^*_\alpha = \tilde{s}_\alpha, \quad \bar{\bar{\theta}}^*_\alpha = \bar{\bar{\theta}}_\alpha \tag{4.3.1}$$

The material derivative of a frame-indifferent scalar function is frame-indifferent. For example, the frame invariance of the material derivative of density $\bar{\rho}_\alpha$ is proved as follows. By definition we have

$$\begin{aligned} \grave{\bar{\rho}}^*_\alpha(\mathbf{x}^*,t^*) &= \left(\frac{\partial \bar{\rho}^*_\alpha(\mathbf{x}^*,t^*)}{\partial t^*}\right)_{\mathbf{x}^*} + \tilde{v}^*_{\alpha i}\left(\frac{\partial \bar{\rho}^*_\alpha(\mathbf{x}^*,t^*)}{\partial x^*_i}\right)_{t^*} \\ &= \left(\frac{\bar{\rho}_\alpha(\mathbf{x},t)}{\partial t^*}\right)_{\mathbf{x}^*} + \tilde{v}^*_{\alpha j}\left(\frac{\partial \bar{\rho}_\alpha(\mathbf{x},t)}{\partial x^*_j}\right)_{t^*} \\ &= \left(\frac{\partial \bar{\rho}_\alpha}{\partial t}\right)_{\mathbf{x}} + \frac{\partial \bar{\rho}_\alpha}{\partial x_i}\left(\frac{\partial x_i}{\partial t}\right)_{\mathbf{x}^*} + \tilde{v}^*_{\alpha j}\frac{\partial \bar{\rho}_\alpha}{\partial x_i}\left(\frac{\partial x_i}{\partial x^*_j}\right)_t \end{aligned} \tag{4.3.2}$$

But from (4.2.2) and (4.2.4)

$$x^*_i = Q_{ij}x_j + c_i \tag{4.3.3}$$

which upon multiplying both sides by $Q_{ik}$ and using (4.1.3) yields

$$x_i = Q_{ji}x^*_j - Q_{ji}c_j \tag{4.3.4}$$

and upon differentiating gives

$$\frac{\partial x_i}{\partial x^*_k} = Q_{ki}, \quad \frac{\partial x_i}{\partial t} = \dot{Q}_{ji}x^*_j - \dot{Q}_{ji}c_j - Q_{ji}\dot{c}_j \tag{4.3.5}$$

Equations (4.3.4) and (4.3.5) reduce therefore (4.3.2) to the form:

$$\grave{\bar{\rho}}_\alpha^*(\mathbf{x}^*, t^*) = \left(\frac{\partial \bar{\rho}_\alpha}{\partial t}\right)_{\mathbf{x}} + \left(\frac{\partial \bar{\rho}_\alpha}{\partial x_i}\right)_t (\dot{Q}_{ji} x_j^* - \dot{Q}_{ji} c_j - Q_{ji} \dot{c}_j + \tilde{v}_{\alpha j}^* Q_{ji}) \quad (4.3.6)$$

From (4.3.3), $x_j^* = Q_{jk} x_k + c_j$, time differentiating (4.1.3), *i.e.*

$$\dot{Q}_{ji} Q_{jk} = -Q_{ji} \dot{Q}_{jk} \quad (4.3.7)$$

$$\dot{Q}_{ji} Q_{ji} = 0 \quad (4.3.8)$$

and using (4.2.5), $\tilde{v}_{\alpha j}^* = \dot{Q}_{jk} x_k + Q_{jk} \tilde{v}_{\alpha k} + \dot{c}_j$, reduces (4.3.6) to the claimed form

$$\grave{\bar{\rho}}_\alpha^*(\mathbf{x}^*, t^*) = \frac{\partial \bar{\rho}_\alpha}{\partial t} + \tilde{v}_{\alpha i} \frac{\partial \bar{\rho}_\alpha}{\partial x_i} = \grave{\bar{\rho}}_\alpha(\mathbf{x}, t) \quad (4.3.9)$$

Thus, we also have

$$\grave{\bar{\epsilon}}_\alpha^* = \grave{\bar{\epsilon}}_\alpha, \quad \grave{\bar{s}}_\alpha^* = \grave{\bar{s}}_\alpha, \quad \grave{\bar{\bar{\theta}}}_\alpha^* = \grave{\bar{\bar{\theta}}}_\alpha \quad (4.3.10)$$

The divergence of the velocity field is a frame-indifferent scalar. This is readily proved as follows

$$\begin{aligned} \nabla^* \cdot \tilde{\mathbf{v}}_\alpha^* = \frac{\partial \tilde{v}_{\alpha i}^*}{\partial x_i^*} &= \frac{\partial}{\partial x_i^*}(\dot{Q}_{ij} x_j + Q_{ij} \tilde{v}_{\alpha ij} + \dot{c}_i) = \dot{Q}_{ij} \frac{\partial x_j}{\partial x_i^*} + Q_{ij} \frac{\partial \tilde{v}_{\alpha j}}{\partial x_k} \frac{\partial x_k}{\partial x_i^*} \\ &= \dot{Q}_{ij} Q_{ij} + Q_{ij} \frac{\partial \tilde{v}_{\alpha j}}{\partial x_k} Q_{ik} = \frac{\partial \tilde{v}_{\alpha i}}{\partial x_i} = \nabla \cdot \tilde{\mathbf{v}}_\alpha \end{aligned} \quad (4.3.11)$$

where use was made of (4.3.8) and (4.1.3). Equations (4.3.1), (4.3.9), and (4.3.11) yield from (2.4.14) the frame-indifference of mass supplies, *i.e.*

$$\hat{c}_\alpha^* = \hat{c}_\alpha \quad (4.3.12)$$

The velocity gradient is not frame-indifferent, since it transforms according to

$$\tilde{v}_{\alpha i,j}^* = Q_{im} \tilde{v}_{\alpha m,n} Q_{jn} + \dot{Q}_{im} Q_{jm} \quad (4.3.13)$$

The velocity gradient can be uniquely decomposed into a symmetric part $D_{\alpha ij}$, and a skew-symmetric part $W_{\alpha ij}$. This is defined by

$$D_{\alpha ij} = \frac{1}{2}(\tilde{v}_{\alpha i,j} + \tilde{v}_{\alpha j,i}) \quad (4.3.14)$$

$$W_{\alpha ij} = \frac{1}{2}(\tilde{v}_{\alpha i,j} - \tilde{v}_{\alpha j,i}) \quad (4.3.15)$$

such that

$$\tilde{v}_{\alpha i,j} = D_{\alpha ij} + W_{\alpha ij} \tag{4.3.16}$$

Thus, $D_{\alpha ij}$ is frame-indifferent, whereas $W_{\alpha ij}$ is not, since

$$D^*_{\alpha ij} = Q_{im} D_{\alpha mn} Q_{jn} \tag{4.3.17}$$

$$W^*_{\alpha ij} = Q_{im} W_{\alpha mn} Q_{jn} + \dot{Q}_{im} Q_{jm} \tag{4.3.18}$$

In the frame $\mathcal{F}^*$, equations (2.2.2) and (3.1.1) yield

$$x_k^{*(\alpha\delta)} = x_k^{*(\alpha)} + \xi_k^{*(\alpha\delta)} = x_k^{*(\alpha)} + \Pi_{kK}^{*(\alpha)} \Sigma_K^{*(\alpha\delta)} \tag{4.3.19}$$

Using (4.2.1) and the change of frame transformation of the center of mass of phase $\alpha$

$$x_k^{*(\alpha)} = Q_{kj} x_j^{(\alpha)} + c_k \tag{4.3.20}$$

gives the results

$$\xi_k^{*(\alpha\delta)} = Q_{kj} \xi_j^{(\alpha\delta)} \tag{4.3.21}$$

$$\Pi_{kK}^{*(\alpha)} = Q_{kj} \Pi_{jK}^{(\alpha)} \tag{4.3.22}$$

which shows that $\boldsymbol{\xi}^{(\alpha\delta)}$ is frame-indifferent, whereas $\boldsymbol{\Pi}^{(\alpha)}$ is not. Taking the material derivative of (4.3.22) in the frame $\mathcal{F}^*$ in the center of mass of phase $\alpha$ and using (4.3.20) gives

$$\dot{\Pi}_{kK}^{*(\alpha)} = \dot{Q}_{kj} \Pi_{jK}^{(\alpha)} + Q_{kj} \dot{\Pi}_{jK}^{(\alpha)} \tag{4.3.23}$$

where $\dot{\Pi}_{jK}^{(\alpha)}$ denotes the material derivative of $\Pi_{jK}^{(\alpha)}$ in the frame $\mathcal{F}$. Starting from (3.1.6) with $\Pi_{kL}^{*(\alpha)} \Xi_{L\ell}^{*(\alpha)} = \delta_{k\ell}$, using (4.3.22), multiplying the result by $Q_{ki} \Xi_{Mi}^{(\alpha)}$, and using (3.1.7) produces the result

$$\Xi_{M\ell}^{*(\alpha)} = Q_{\ell i} \Xi_{Mi}^{(\alpha)} \tag{4.3.24}$$

Moreover, upon using (4.3.23) and (4.3.24) in (3.1.21), $\nu_{i\ell}^{*(\alpha)} = \dot{\Pi}_{iK}^{*(\alpha)} \Xi_{K\ell}^{*(\alpha)}$, (3.1.6), (3.1.7), and (3.1.22), the change of frame transformation property for the gyration tensor is obtained, *i.e.*

$$\nu_{i\ell}^{*(\alpha)} = Q_{ij} \nu_{jm}^{(\alpha)} Q_{\ell m} + \dot{Q}_{im} Q_{\ell m} \tag{4.3.25}$$

Setting $j = \ell$ in (4.3.13) and subtracting the resulting expression from (4.3.25) gives the following frame-indifferent tensor

$$b_{i\ell}^{*(\alpha)} = \nu_{i\ell}^{*(\alpha)} - \bar{v}_{\alpha i,\ell}^{*} = Q_{ij}(\nu_{jm}^{(\alpha)} - \bar{v}_{\alpha j,m})Q_{\ell m} = Q_{ij}b_{jm}^{(\alpha)}Q_{\ell m} \tag{4.3.26}$$

The gradient of the gyration tensor is frame-indifferent. This easily follows from (4.3.25). Thus,

$$\nu_{i\ell,k}^{*(\alpha)} = Q_{ij}Q_{\ell m}Q_{kn}\nu_{jm,n}^{(\alpha)} \tag{4.3.27}$$

The equilibrated inertia $\bar{i}_{\alpha k\ell}$, defined by (3.2.3), is frame-indifferent, since it transforms according to (4.1.7), *i.e.*

$$\bar{i}_{\alpha k\ell}^{*} = \frac{1}{\bar{\rho}_{\alpha}^{*}}\frac{1}{U}\sum_{\delta}\int_{U(\alpha\delta)} \rho^{*(\alpha\delta)}\xi_{k}^{*(\alpha\delta)}\xi_{\ell}^{*(\alpha\delta)}\,dU$$
$$= \frac{1}{\bar{\rho}_{\alpha}}\frac{1}{U}\sum_{\delta}\int_{U(\alpha\delta)} \rho^{(\alpha\delta)}Q_{km}\xi_{m}^{(\alpha\delta)}Q_{\ell n}\xi_{n}^{(\alpha\delta)}\,dU = Q_{km}\bar{i}_{\alpha mn}Q_{\ell n} \tag{4.3.28}$$

The equilibrated source inertia tensor, defined by (3.2.4), is also frame-indifferent, whereas the hyperinertia tensor, defined by (3.2.5), is not, since:

$$\hat{i}_{\alpha k\ell}^{*} = Q_{km}\hat{i}_{\alpha mn}Q_{\ell n} \tag{4.3.29}$$

$$\bar{\rho}_{\alpha}^{*}\hat{I}_{\alpha mk\ell}^{*} = Q_{ma}Q_{kb}Q_{\ell c}\bar{\rho}_{\alpha}\hat{I}_{\alpha abc}$$
$$+\dot{Q}_{ma}Q_{kb}Q_{\ell c}\frac{1}{U}\sum_{\delta}\int_{U(\alpha\delta)} \rho^{(\alpha\delta)}\xi_{b}^{(\alpha\delta)}\xi_{c}^{(\alpha\delta)}\xi_{a}^{(\alpha\delta)}\,dU \tag{4.3.30}$$

As discussed above, the forces are taken to be frame-indifferent. This requirement should then hold locally in the spatial configuration space in Figure 2.1 and be preserved in the mapped configuration space when produced by a suitable averaging procedure. The latter requirement is satisfied exactly by *volume averaging*, as may be easily proved by noting that $\mathbf{Q}$ depends only on time and is unaffected by a spatial integration at constant $t$. Thus, the body force and stress tensor of phase $\alpha$ transform as frame-indifferent variables, i.e.

$$\bar{b}_{\alpha i}^{*} = Q_{ij}\bar{b}_{\alpha j} \tag{4.3.31}$$

$$\bar{T}_{\alpha ij}^{*} = Q_{im}\bar{T}_{\alpha mn}Q_{jn} \tag{4.3.32}$$

In view of the result (4.3.21) and assumed frame-indifference of scalar variables, forces, and mass as discussed above, it readily follows from (3.2.11),

(3.2.12), (3.3.2), (3.3.7), (3.3.8), and (3.3.9) that *the following variables are frame-indifferent*

$$\bar{\lambda}^*_{\alpha kjm} = Q_{ka}Q_{jb}Q_{mc}\bar{\lambda}_{\alpha abc}, \quad \bar{S}^*_{\alpha jk} = Q_{ja}\bar{S}_{\alpha ab}Q_{kb}$$

$$\bar{t}^*_{\alpha i} = Q_{ia}\bar{t}_{\alpha a}, \quad \tilde{\ell}\epsilon^*_{\alpha m} = Q_{ma}\tilde{\ell}\epsilon_{\alpha a}, \quad \hat{\hat{\epsilon}}^*_{\alpha} = \hat{\hat{\epsilon}}_{\alpha}, \quad \bar{q}^*_{s\alpha} = \bar{q}_{s\alpha} \tag{4.3.33}$$

Note that $\tilde{\boldsymbol{\ell}}_\alpha$ is excluded from being frame-indifferent. The reason for this is that $\tilde{\boldsymbol{\ell}}_\alpha$ may include body moments that can be responsible for the particle rotations. A magnetic particle placed in a magnetic field, for example, will rotate due to the field, and $\tilde{\boldsymbol{\ell}}_\alpha$ needs to be suitably generalized to allow for this phenomenon in the presence of the electromagnetic field.

The momentum source $\hat{\mathbf{p}}_\alpha$ is expressed by (3.3.3) and is not, in general, frame-indifferent, unless the structural characteristics of the mixture (as considered above) are very special, or are ignored ($\boldsymbol{\nu}^{(\alpha)}=0, \bar{\mathbf{i}}_\alpha=0$). To determine the change of frame transformation property of the energy supply defined by (2.4.35) we may proceed as follows.

Using (4.3.16) it follows that

$$\begin{aligned} tr(\bar{\mathbf{T}}^T_\alpha \boldsymbol{\nabla}\tilde{\mathbf{v}}_\alpha) &= \bar{T}_{\alpha ij}\tilde{v}_{\alpha i,j} = \bar{T}_{\alpha ij}D_{\alpha ij} + \bar{T}_{\alpha ij}W_{\alpha ij} \\ &= \bar{T}_{\alpha ij}D_{\alpha ij} - \frac{1}{2}\hat{M}_{\alpha ji}W_{\alpha ij} = tr(\bar{\mathbf{T}}^T_\alpha \mathbf{D}_\alpha) - \frac{1}{2}tr(\hat{\mathbf{M}}_\alpha \mathbf{W}_\alpha) \end{aligned} \tag{4.3.34}$$

But, since $\bar{\mathbf{T}}_\alpha$ and $\mathbf{D}_\alpha$ are frame-indifferent, *i.e.*

$$tr(\bar{\mathbf{T}}^{*T}_\alpha \mathbf{D}^*_\alpha) = tr(\bar{\mathbf{T}}^T_\alpha \mathbf{D}_\alpha) \tag{4.3.35}$$

these equations can be used to derive the following result:

$$tr(\bar{\mathbf{T}}^{*T}_\alpha \boldsymbol{\nabla}^*\tilde{\mathbf{v}}^*_\alpha) = tr(\bar{\mathbf{T}}^T_\alpha \boldsymbol{\nabla}\tilde{\mathbf{v}}_\alpha) - \frac{1}{2}tr(\hat{\mathbf{M}}_\alpha \mathbf{Q}^T\dot{\mathbf{Q}}) \tag{4.3.36}$$

The heat flux vector is assumed to be frame-indifferent, transforming under the change of frame as

$$\bar{\mathbf{q}}^*_\alpha = \mathbf{Q}\bar{\mathbf{q}}_\alpha, \quad \boldsymbol{\nabla}^*\cdot\bar{\mathbf{q}}^*_\alpha = \boldsymbol{\nabla}\cdot\bar{\mathbf{q}}_\alpha \tag{4.3.37}$$

Using this result, equations (4.3.36) and $(4.3.10)_1$, and frame-indifference of scalar variables (4.3.1), it follows from the energy balance equation (2.4.35) and entropy inequality (2.4.45) that

$$\hat{\epsilon}^*_\alpha = \hat{\epsilon}_\alpha + \frac{1}{2}tr(\hat{\mathbf{M}}_\alpha \mathbf{Q}^T\dot{\mathbf{Q}}), \quad \hat{s}^*_\alpha = \hat{s}_\alpha \tag{4.3.38}$$

The energy supply $\hat{\epsilon}_\alpha$ is, therefore, *not* frame-indifferent, unless the angular momentum supply $\hat{\mathbf{M}}_\alpha$ is equal to zero. Note that in the Galilean frame of reference $d\mathbf{Q}/dt = 0$ and $\hat{\epsilon}_\alpha$ is then frame-indifferent. Moreover, by redefining the energy supply according to the relation (DOBRAN,1985)

$$\hat{\epsilon}_\alpha = \hat{\epsilon}_{\alpha R} + \frac{1}{2} tr(\hat{\mathbf{M}}_\alpha \mathbf{W}_\alpha) \tag{4.3.39}$$

yields a frame-indifferent variable $\hat{\epsilon}_{\alpha R}$, with the second term on the right of the equation representing the rate of work due to the couple $\hat{\mathbf{M}}_\alpha$.

The material derivative of the inertia tensor is not frame-indifferent and may be shown from (4.3.28) to transform according to

$$\dot{\bar{i}}^*_{\alpha k\ell} = Q_{km} \dot{\bar{i}}_{\alpha mn} Q_{\ell n} + (\dot{Q}_{km} Q_{\ell n} + Q_{km} \dot{Q}_{\ell n}) \bar{i}_{\alpha mn} \tag{4.3.40}$$

Furthermore, from (4.3.25) and (4.3.28) it follows that

$$\nu^{*(\alpha)}_{km} \bar{i}^*_{\alpha m\ell} = Q_{kj} \nu^{(\alpha)}_{jm} \bar{i}_{\alpha mn} Q_{\ell n} + \dot{Q}_{km} Q_{\ell n} \bar{i}_{\alpha mn} \tag{4.3.41}$$

$$\nu^{*(\alpha)}_{\ell m} \bar{i}^*_{\alpha km} = Q_{\ell j} \nu^{(\alpha)}_{jn} \bar{i}_{\alpha mn} Q_{km} + \dot{Q}_{\ell n} Q_{km} \bar{i}_{\alpha mn} \tag{4.3.42}$$

which when subtracted from (4.3.40) yields a frame-indifferent tensor

$$\dot{\bar{i}}^*_{\alpha k\ell} - \nu^{*(\alpha)}_{km} \bar{i}^*_{\alpha m\ell} - \nu^{*(\alpha)}_{\ell m} \bar{i}^*_{\alpha km} = Q_{km} (\dot{\bar{i}}_{\alpha mn} - \nu^{*(\alpha)}_{mi} \bar{i}_{\alpha in} - \nu^{*(\alpha)}_{ni} \bar{i}_{\alpha mi}) Q_{\ell n} \tag{4.3.43}$$

The above result can now be used in (3.2.2) to conclude that $(\bar{\rho}_\alpha \hat{I}_{\alpha mk\ell})_{,m}$ should be a frame-indifferent tensor

$$(\bar{\rho}^*_\alpha \hat{I}^*_{\alpha mk\ell})_{,m} = Q_{kb} (\bar{\rho}_\alpha \hat{I}_{\alpha abc})_{,a} Q_{\ell c} \tag{4.3.44}$$

since, as proved above, $\bar{\rho}_\alpha$, $\hat{c}_\alpha$, $\bar{i}_{\alpha k\ell}$, and $\hat{i}_{\alpha k\ell}$ are frame-indifferent. A third order tensor defined by

$$U_{abca} = \frac{1}{U} \sum_\delta \int_{U(\alpha\delta)} \rho^{(\alpha\delta)} \xi^{(\alpha\delta)}_b \xi^{(\alpha\delta)}_c \xi^{(\alpha\delta)}_a \, dU \tag{4.3.45}$$

can be shown to be frame-indifferent by using (4.3.21), *i.e.*

$$U^*_{\alpha ijk} = Q_{ia} Q_{jb} Q_{kc} U_{\alpha abc} \tag{4.3.46}$$

The hyperinertia tensor (3.2.5) can be expressed as

$$\bar{\rho}_\alpha \hat{I}_{\alpha mk\ell} = \nu^{(\alpha)}_{mn} U_{\alpha k\ell n} \tag{4.3.47}$$

which upon differentiation gives the result

$$(\bar{\rho}_\alpha \hat{I}_{\alpha mk\ell})_{,m} = \nu^{(\alpha)}_{mn,m} U_{\alpha k\ell n} + \nu^{(\alpha)}_{mn} U_{\alpha k\ell n,m} \tag{4.3.48}$$

By (4.3.44) this last result is required to be frame-indifferent, and since $\nu^{(\alpha)}_{mn,m}$ and $U_{\alpha k\ell n}$ are frame-indifferent (*cit.* (4.3.27) and (4.3.46)), it is sufficient and necessary that

$$\nu^{(\alpha)}_{mn} U_{\alpha k\ell n,m} = 0, \quad or \quad frame-indifferent \tag{4.3.49}$$

This condition is, of course, satisfied if the gyration tensor represents, physically, a pure dilatational effect of the material motion relative to the center of mass, describing a very special constitutive assumption whose full implication in the theory of structured multiphase mixtures will be discussed further in the following chapter. More generally, however, the condition (4.3.49) is satisfied if $\nu^{(\alpha)}_{mn}$ is a skew-symmetric tensor *and* $U_{\alpha k\ell n,m}$ a symmetric tensor in the indices $n$ and $m$.

The results on the frame-indifference of various multiphase variables presented above can be used to prove that the fields

$$k_{\alpha ijk}, \quad K_{\alpha ij}, \quad \hat{g}_{\alpha ij}, \quad \bar{h}_{\alpha ijk}, \quad \check{g}_{\alpha jk} \tag{4.3.50}$$

defined by (3.4.5), (3.4.6), (3.4.10), (3.4.11), and (3.4.13), respectively, *are not frame-indifferent.* The general results of the change of frame transformation properties of multiphase variables discussed in this chapter will be very useful in the following chapters dealing with special cases of the theory.

## 4.4 Remarks on the Closure of Conservation and Balance Equations by the Constitutive Relations

The conservation and balance equations of structured multiphase mixtures discussed in chapters 2 and 3 are assumed to model a wide variety of mixtures. Depending on the application, a multiphase mixture may be comprised of solid, liquid or gas phases, and the *diversity* of such mixtures of materials is represented in the theory by the *constitutive equations.* A constitutive assumption is, therefore, a restriction placed upon the material constituting the multiphase mixture and is expressed as a relation between the thermodynamic variables of the mixture at some instant of time and possibly at all instants of the mixture's past history, as discussed in detail in chapter 6. In effect, the applied mechanical and thermal fields (such as mechanical forces and heat flux) cause the material of each phase to undergo a motion and exchange energy with the surrounding phases. The thermodynamic state of a phase may, in general, depend on the past history which is simply a statement or principle of determinism or causality where the cause preceds the effect. A process defined by a time-dependent set of configurations, forces and moments of forces, temperature, internal energy, entropy, heat generation, heat flux, inertia, gyration, *etc.*, and compatible with the conservation and balance equations will be called a *thermodynamic process.* When a thermodynamic process is compatible with the constitutive assumptions under the consideration, then such a process will be called admissible. The basic notions of constitutive equations sketched above were laid down by NOLL (1958) and COLEMAN & NOLL (1963) for single phase materials, and by DOBRAN (1985) for multiphase mixtures without the phase change.

As an introduction to the discussion in chapter 6 involved with the theory of constitutive principles of multiphase mixtures, the thermodynamic process is defined by $\mathbf{x} = \boldsymbol{\chi}_{\alpha\boldsymbol{\kappa}}(\mathbf{X}_\alpha, t)$ and by the following fields which depend on $\mathbf{x}$ and $t$:

$$\begin{gathered}\bar{\rho}_\alpha,\ \hat{c}_\alpha,\ \bar{\mathbf{T}}_\alpha,\ \tilde{\mathbf{b}}_\alpha,\ \hat{\mathbf{M}}_\alpha\ (or\ \bar{\mathbf{T}}_\alpha^T),\ \tilde{\epsilon}_\alpha,\ \bar{\mathbf{q}}_\alpha,\ \tilde{r}_\alpha,\ \tilde{s}_\alpha,\ \bar{\bar{\theta}}_\alpha,\ \hat{s}_\alpha,\\ \tilde{\bar{\mathbf{i}}}_\alpha,\ \boldsymbol{\nu}^{(\alpha)},\ \hat{\mathbf{i}}_\alpha,\ \mathbf{U}_\alpha,\ \tilde{\boldsymbol{\ell}}_\alpha,\ \bar{\boldsymbol{\lambda}}_\alpha,\ \bar{\mathbf{S}}_\alpha,\ \bar{\mathbf{t}}_\alpha,\ \tilde{\boldsymbol{\ell}}\epsilon_\alpha,\ \hat{\hat{\epsilon}}_\alpha,\ \bar{q}_{s\alpha};\ \alpha = 1, \ldots, \gamma\end{gathered} \tag{4.4.1}$$

as may be seen by examining the governing field equations (2.4.14), (2.4.19), (2.4.31), (3.4.19), (3.4.8), (3.4.14), (3.3.3), (3.4.1), (3.4.5)-(3.4.7), (3.4.10)-(3.4.13), (4.3.47), (3.4.15), (3.4.16) and (2.4.45). The balance of mass equation (2.4.14), linear momentum (2.4.19), energy (3.4.19), equilibrated inertia (3.4.8), and equilibrated moments (3.4.14) provide $19\gamma$ scalar equations whose number is far less than $144\gamma$ unknown variables listed in (4.4.1). To solve the governing field equations requires, therefore, a specification of large number of constitutive relations in terms of independent variables in the thermodynamic process represented by the kinematic, thermal, and phase change conditions of the mixture. This "closure" problem will be studied further in much more detail in chapters 6-8 dealing with the constitutive equations.

## 4.5 Concluding Remarks

At this stage of the analysis it is possible to develop different models of multiphase flow based on different constitutive assumptions, or to postpone with this development of the theory and discuss special forms of field equations. This latter approach is desirable for two reasons: (1) for the purpose of developing simpler modeling approaches, and (2) for comparing the results of the theory with the *available* field equations of structured multiphase mixtures. As will be seen in the subsequent chapter, a special case of the theory that has been developed in this and preceding chapters will allow us to place into a direct correspondence the averaging and postulatory theories of multiphase mixtures and, therefore, fulfill the third objective set forth in section 1.2. The development of constitutive equations of structured multiphase mixtures is resumed in chapter 6.

CHAPTER 5

# SIMPLIFICATION OF THE THEORY OF STRUCTURED MULTIPHASE MIXTURES

The theory of multiphase mixtures with structure developed in chapters 2 and 3 involves the basic set of balance equations of mass, momentum, energy, and entropy inequality, and an additional set of transport equations for the equilibrated inertia and equilibrated moments for modeling the structural characteristics of each phase. The additional transport equations, expressed by (3.4.4) and (3.4.9), are second order tensor equations which, in general, consist of fourteen partial differential equations for the evolution of inertia, $\bar{i}_{\alpha ij}$, and gyration, $\nu_{ij}^{(\alpha)}$. This large number of equations together with the large number of phases that may be present in a mixture may be extremely difficult to solve. For this reason, a simplification of the theory of structured multiphase mixtures becomes of considerable practical necessity.

In this chapter, a simplification of the theory is discussed by invoking further restrictions on the material motion relative to the center of mass. This involves a decomposition of motion into rotational and dilatational effects. If, moreover, the rotational effect is ignored, then the resulting balance equations become similar to the equations advocated in the postulatory theories of mixtures discussed in chapter 1. The purpose of this chapter is, therefore, to present special cases or simplified forms of the field equations developed in previous chapters, and to compare the results with existing models of structured multiphase mixtures.

## 5.1 Further Restrictions on the Material Deformation Relative to the Center of Mass

The basic material deformation postulate (3.1.1) or (3.1.2) contains an assumption that the material of each phase can undergo only a homogeneous or affine deformation relative to the center of mass. The affine deformation has the property of preserving mappings of hypersurfaces so that an ellipsoid in the reference configuration space of material is mapped into an ellipsoid in the deformed or spatial configuration space, or vice versa.

A simple example of an affine deformation consists of rotation and dilatation that may be expressed as follows:

$$\nu_{ij}^{(\alpha)} = \hat{\nu}_{ij}^{(\alpha)} + \grave{\phi}_\alpha \delta_{ij} \tag{5.1.1}$$

where $\grave{\phi}_\alpha$ denotes the material derivative of a function $\phi_\alpha$ of phase $\alpha$ expressing the dilatational rate, whereas $\hat{\nu}_{ij}^{(\alpha)}$ is a skew-symmetric tensor

$$\hat{\nu}_{ij}^{(\alpha)} = -\hat{\nu}_{ji}^{(\alpha)}, \quad \hat{\nu}_{ii}^{(\alpha)} = 0 \tag{5.1.2}$$

expressing the rotational effect. These material deformation characteristics may be seen more clearly by noting that in a three-dimensional space every skew-symmetric second order tensor can be represented by an axial vector $\mu_k^{(\alpha)}$, *i.e.*

$$\mu_k^{(\alpha)} = \frac{1}{2} \epsilon_{kij} \hat{\nu}_{ji}^{(\alpha)} \tag{5.1.3}$$

where $\epsilon_{kij}$ is the alternating tensor, or equivalently as

$$\hat{\nu}_{ki}^{(\alpha)} = -\epsilon_{kij} \mu_j^{(\alpha)} \tag{5.1.4}$$

Using (5.1.2)-(5.1.4) in (3.1.11) results in an expression for the velocity of a particle of phase $\alpha$ in the spatial configuration space

$$\mathbf{v}^{(\alpha\delta)} = \mathbf{v}^{(\alpha)} + \boldsymbol{\mu}^{(\alpha)} \times \boldsymbol{\xi}^{(\alpha\delta)} + \grave{\phi}_\alpha \boldsymbol{\xi}^{(\alpha\delta)} \tag{5.1.5}$$

This result clearly shows the rotational and dilatational effects of material relative to the center of mass. The rotational rate or angular velocity is expressed by $\boldsymbol{\mu}^{(\alpha)}$, whereas the dilatational velocity is expressed by $\grave{\phi}_\alpha \boldsymbol{\xi}^{(\alpha\delta)}$. The scalar function $\phi_\alpha$ represents the dilatation or stretching effect and may be set equal to the *volumetric fraction* of phase $\alpha$ in the averaging volume $U$. The association of $\phi_\alpha$ with the volumetric fraction of phase $\alpha$ is, clearly, a *further* constitutive assumption, and may not be a valid choice as further discussed in section 6.12 dealing with the internal constraints.

To prove that $\phi_\alpha$ represents dilatation can be accomplished easily for the situation of no rotation. Setting $\hat{\nu}_{ij}^{(\alpha)} = 0$ in (5.1.1) and using (3.1.10) gives

$$\grave{\xi}_k^{(\alpha\delta)} = \grave{\phi}_\alpha \xi_k^{(\alpha\delta)} \tag{5.1.6}$$

which upon integration yields

$$\xi_k^{(\alpha\delta)} = exp(\phi_\alpha - \phi_{\alpha 0}) \, \xi_k^{(\alpha\delta)}(\phi_{\alpha 0}) \tag{5.1.7}$$

It then follows that

$$\frac{|\xi_k^{(\alpha\delta)}|}{|\xi_k^{(\alpha\delta)}(\phi_{\alpha 0})|} = exp(\phi_\alpha - \phi_{\alpha 0}) = \lambda_\alpha \tag{5.1.8}$$

This result shows that the stretching or extension of phase $\alpha$ occurs for $\lambda_\alpha > 1$ ($\phi_\alpha > \phi_{\alpha 0}$), whereas a compression occurs for $\lambda_\alpha < 1$ ($\phi_\alpha < \phi_{\alpha 0}$).

In simplifying the theory of structured multiphase mixtures developed above, an additional assumption is made that the material of each phase has an *isotropic inertia tensor*, *i.e.*

$$\bar{\imath}_{\alpha ij} = \bar{\imath}_\alpha \delta_{ij} \tag{5.1.9}$$

The assumptions (5.1.1) and (5.1.9) are, clearly, *the constitutive assumptions*, since they specify or restrict the motion of material in the mixture. They greatly reduce the complexity of the theory by reducing the number of transport equations for the equilibrated inertia and equilibrated moments from fourteen to eight, as further discussed in the next section.

## 5.2 Balance Equations of Multiphase Mixtures with Rotation and Dilatation

In this section use will be made of (5.1.1) and (5.1.9) to simplify the theory of structured multiphase mixtures discussed in previous chapters. Here, as in chapter 3, various forms of the field equations will be presented and their properties discussed. Of particular importance in this discussion will be the principle of material frame-indifference, since this principle restricts the change of frame transformation properties of the kinematic, dynamic, and energetic variables appearing in the conservation and balance equations of multiphase mixtures.

### 5.2.1 Balance Equation for the Equilibrated Inertia

Making use of (5.1.1) and (5.1.9) in (3.4.4) results in the following simplified form of the balance equation for the equilibrated inertia:

$$\bar{\rho}_\alpha \delta_{jn}(\grave{\bar{\imath}}_\alpha - 2\bar{\imath}_\alpha \grave{\phi}_\alpha) = -\hat{c}_\alpha \bar{\imath}_\alpha \delta_{jn} + \hat{k}_{\alpha jn} \tag{5.2.1}$$

from where it follows that

$$\bar{\rho}_\alpha(\grave{\bar{\imath}}_\alpha - 2\bar{\imath}_\alpha \grave{\phi}_\alpha) = -\hat{c}_\alpha \bar{\imath}_\alpha + \hat{k}_{\alpha jj}; \quad j = n \tag{5.2.2}$$

$$\hat{k}_{\alpha jn} = 0; \quad j \neq n \tag{5.2.3}$$

An equivalent form of (3.4.8), or of the above equations, is

$$\bar{\rho}_\alpha \grave{\bar{i}}_\alpha = -\hat{c}_\alpha \bar{i}_\alpha + K_{\alpha jj} + k_{\alpha jji,i} + \check{k}_{\alpha jj} \tag{5.2.4}$$

and from (3.4.3) and (3.4.5)-(3.4.7) it follows that

$$\hat{k}_{\alpha jj} = \hat{c}_\alpha \hat{i}_{\alpha jj} - (\bar{\rho}_\alpha \hat{I}_{\alpha mjj})_{,m} = \hat{c}_\alpha \hat{i}_{\alpha jj} - (\hat{\nu}^{(\alpha)}_{mn} U_{\alpha jjn})_{,m} - (\grave{\phi}_\alpha U_{\alpha jjm})_{,m} \tag{5.2.5}$$

$$k_{\alpha jji} = -\bar{\rho}_\alpha \hat{I}_{\alpha ijj} = -\hat{\nu}^{(\alpha)}_{in} U_{\alpha jjn} - \grave{\phi}_\alpha U_{\alpha jji} \tag{5.2.6}$$

$$K_{\alpha jj} = 2\bar{\rho}_\alpha \grave{\phi}_\alpha \bar{i}_\alpha \tag{5.2.7}$$

$$\check{k}_{\alpha jj} = \hat{c}_\alpha \hat{i}_{\alpha jj} \tag{5.2.8}$$

$$K_{\alpha jn} = 0, \quad \check{k}_{\alpha jn} + k_{\alpha jni,i} = 0 \quad for \quad j \neq n \tag{5.2.9}$$

where in obtaining (5.2.5) and (5.2.6) use was made of (4.3.47) and (5.1.1).

It is shown in chapter 4 that the basic material deformation principle (3.1.1) yields frame-indifference of $\hat{k}_{\alpha jn}$ and $\check{k}_{\alpha jn}$, but not of $k_{\alpha ijk}$ and $K_{\alpha ij}$. The additional material deformation assumptions (5.1.1) and (5.1.9) give also the frame-indifference of $K_{\alpha ij}$ (*cit.* (5.2.7)), since $\bar{\rho}_\alpha$, $\bar{i}_\alpha$ and $\grave{\phi}_\alpha$ are assumed to be frame-indifferent scalars. Notice, however, that $k_{\alpha jji}$ is not frame-indifferent unless the rotational effect of the material is ignored ($\hat{\boldsymbol{\nu}}^{(\alpha)} = \mathbf{0}$). The isotropy assumption (5.1.9) of the inertia tensor, when used in (4.3.40), gives

$$\grave{\bar{i}}^*_\alpha = \grave{\bar{i}}_\alpha \tag{5.2.10}$$

which upon using in (5.2.4) and noting the frame-indifference of $\bar{\rho}_\alpha$, $\hat{c}_\alpha$, $\bar{i}_\alpha$, $K_{\alpha jj}$, and $\check{k}_{\alpha jj}$ requires the frame-indifference of $k_{\alpha jji,i}$. This result is, of course, a special case of (4.3.49) that requires

$$\hat{\nu}^{(\alpha)}_{in} U_{\alpha jjn,i} = 0 \quad or \quad frame-indifferent \tag{5.2.11}$$

Substituting equation (5.1.1) into (4.3.25) and (4.3.27), and noting the frame-indifference of $\grave{\phi}_\alpha$ and $\nabla\grave{\phi}_\alpha$, produces the change of frame transformation properties of the rotation tensor $\hat{\boldsymbol{\nu}}^{(\alpha)}$, *i.e.*

$$\hat{\nu}^{*(\alpha)}_{i\ell} = Q_{ij}\hat{\nu}^{(\alpha)}_{jm} Q_{\ell m} + \dot{Q}_{im} Q_{\ell m}, \quad \hat{\nu}^{*(\alpha)}_{i\ell,k} = Q_{ij} Q_{\ell m} Q_{kn} \hat{\nu}^{(\alpha)}_{jm,n} \tag{5.2.12}$$

From (5.1.2), $\hat{\nu}^{(\alpha)}_{ii} = 0$, and equation (5.2.11) is satisfied identically, but if $i \neq n$ it is sufficient and necessary that $U_{\alpha jjn,i}$ is symmetric.

### 5.2.2 Balance Equations for Equilibrated Moments

Using (5.1.1)-(5.1.3) and (5.1.9) the symmetric and skew-symmetric parts of (3.2.13) are given, respectively, by

$$\begin{aligned}\bar{\rho}_\alpha \bar{i}_\alpha [((\grave{\phi}_\alpha)^{\grave{}} + \grave{\phi}_\alpha \grave{\phi}_\alpha - \mu_i^{(\alpha)} \mu_i^{(\alpha)}) \delta_{jk} + \mu_j^{(\alpha)} \mu_k^{(\alpha)}] \\ = \bar{S}_{\alpha(jk)} + \bar{\rho}_\alpha \tilde{\ell}_{\alpha(jk)} + \bar{\lambda}_{\alpha(jk)m,m} + \hat{\bar{T}}_{\alpha(jk)} \end{aligned} \quad (5.2.13)$$

$$\bar{\rho}_\alpha \bar{i}_\alpha (\grave{\mu}_\ell^{(\alpha)} + 2\grave{\phi}_\alpha \mu_\ell^{(\alpha)}) = \frac{1}{2} \epsilon_{\ell jk} (\bar{S}_{\alpha[jk]} + \bar{\rho}_\alpha \tilde{\ell}_{\alpha[jk]} + \bar{\lambda}_{\alpha[jk]m,m} + \hat{\bar{T}}_{\alpha[jk]}) \quad (5.2.14)$$

where parentheses and brackets enclosing pairs of indices denote, respectively, symmetric and skew-symmetric parts. For example, $\bar{\lambda}_{\alpha(jk)m} = \frac{1}{2}(\bar{\lambda}_{\alpha jkm} + \bar{\lambda}_{\alpha kjm})$ and $\bar{\lambda}_{\alpha[jk]m} = \frac{1}{2}(\bar{\lambda}_{\alpha jkm} - \bar{\lambda}_{\alpha kjm})$. $\hat{\bar{T}}_{\alpha jk}$ is defined as

$$\begin{aligned}\hat{\bar{T}}_{\alpha jk} &= -\bar{\rho}_\alpha \nu_{kn,m}^{(\alpha)} \hat{I}_{\alpha mjn} - \bar{\rho}_\alpha \nu_{mq}^{(\alpha)} \tilde{v}_{\alpha k,m} \bar{i}_{\alpha jq} \\ &= \hat{g}_{\alpha jk} - \hat{\nu}_{kn}^{(\alpha)} \hat{k}_{\alpha jn} - \grave{\phi}_\alpha \hat{k}_{\alpha jk} - \bar{\rho}_\alpha \hat{\nu}_{mj}^{(\alpha)} \tilde{v}_{\alpha k,m} \bar{i}_\alpha - \bar{\rho}_\alpha \grave{\phi}_\alpha \tilde{v}_{\alpha k,j} \bar{i}_\alpha \end{aligned} \quad (5.2.15)$$

$$\hat{\bar{T}}_{\alpha jj} = -\grave{\phi}_\alpha (\nu_{jn,\ell}^{(\alpha)} U_{\alpha jn\ell} + \bar{\rho}_\alpha \bar{i}_\alpha \tilde{v}_{\alpha j,j}) + \mu_i^{(\alpha)} \epsilon_{m\ell i} (\nu_{jn,m}^{(\alpha)} U_{\alpha jn\ell} + \bar{\rho}_\alpha \bar{i}_\alpha \tilde{v}_{\alpha \ell,m})$$

where use is made of (4.3.47), (5.1.1), and (5.1.4). Substituting $(5.2.15)_2$ into (5.2.13) and noting the frame-indifference of $(\grave{\phi}_\alpha)^{\grave{}}$, $\grave{\phi}_\alpha$, $\bar{\mathbf{S}}_\alpha$, $\bar{\boldsymbol{\lambda}}_\alpha$, $\nabla \hat{\boldsymbol{\nu}}^{(\alpha)}$, $\mathbf{U}_\alpha$, and $tr\mathbf{D}_\alpha$, establishes the symmetric part $\tilde{\ell}_{\alpha(jk)}$, of the body force moment $\tilde{\ell}_{\alpha jk}$, as

$$\begin{aligned}\bar{\rho}_\alpha \tilde{\ell}_{\alpha(jk)} &= \bar{\rho}_\alpha \bar{i}_\alpha (\mu_j^{(\alpha)} \mu_k^{(\alpha)} - \mu_i^{(\alpha)} \mu_i^{(\alpha)} \delta_{jk}) \\ &-\frac{1}{2} \mu_i^{(\alpha)} [\epsilon_{m\ell i} (\nu_{kn,m}^{(\alpha)} U_{\alpha jn\ell} + \nu_{jn,m}^{(\alpha)} U_{\alpha kn\ell}) + \bar{\rho}_\alpha \bar{i}_\alpha (\tilde{v}_{\alpha k,m} \epsilon_{mji} + \tilde{v}_{\alpha j,m} \epsilon_{mki})] \end{aligned} \quad (5.2.16)$$

with (5.2.13) reducing to

$$\begin{aligned}\bar{\rho}_\alpha \bar{i}_\alpha ((\grave{\phi}_\alpha)^{\grave{}} + \grave{\phi}_\alpha \grave{\phi}_\alpha) \delta_{jk} &= \bar{S}_{\alpha(jk)} + \bar{\lambda}_{\alpha(jk)m,m} - \bar{\rho}_\alpha \bar{i}_\alpha \grave{\phi}_\alpha D_{\alpha jk} \\ &-\frac{1}{2} \grave{\phi}_\alpha (\nu_{k\ell,i}^{(\alpha)} U_{\alpha j\ell i} + \nu_{j\ell,i}^{(\alpha)} U_{\alpha k\ell i}) \end{aligned} \quad (5.2.17)$$

Multiplying (5.2.1) by $\grave{\phi}_\alpha$ and adding the result to (5.2.17) results in the following equation:

$$\begin{aligned}\bar{\rho}_\alpha (\overline{\bar{i}_\alpha \grave{\phi}_\alpha}^{\grave{}} - \bar{i}_\alpha \grave{\phi}_\alpha \grave{\phi}_\alpha) \delta_{jk} &= \bar{S}_{\alpha(jk)} + \bar{\lambda}_{\alpha(jk)m,m} - \hat{c}_\alpha \bar{i}_\alpha \grave{\phi}_\alpha \delta_{jk} + \hat{k}_{\alpha ii} \grave{\phi}_\alpha \delta_{jk} \\ &-\bar{\rho}_\alpha \bar{i}_\alpha \grave{\phi}_\alpha D_{\alpha jk} - \frac{1}{2} \grave{\phi}_\alpha (\nu_{k\ell,i}^{(\alpha)} U_{\alpha j\ell i} + \nu_{j\ell,i}^{(\alpha)} U_{\alpha k\ell i}) \end{aligned} \quad (5.2.18)$$

Using $(5.2.15)_1$ in (5.2.14) gives an alternate form of this equation, *i.e.*

$$\begin{aligned}\bar{\rho}_\alpha \bar{i}_\alpha(\grave{\mu}_\ell^{(\alpha)} + 2\mu_\ell^{(\alpha)}\grave{\phi}_\alpha) &= \frac{1}{2}\epsilon_{\ell jk}[\bar{S}_{\alpha[jk]} + \bar{\rho}_\alpha \tilde{\ell}_{\alpha[jk]} + \bar{\lambda}_{\alpha[jk]m,m} \\ &+ \bar{\rho}_\alpha \bar{i}_\alpha \grave{\phi}_\alpha W_{\alpha jk} + \frac{1}{2}\mu_i^{(\alpha)}(\epsilon_{m\ell i}(\nu_{kn,m}^{(\alpha)} U_{\alpha jn\ell} - \nu_{jn,m}^{(\alpha)} U_{\alpha kn\ell}) \\ &+ \bar{\rho}_\alpha \bar{i}_\alpha(\tilde{v}_{\alpha k,m}\epsilon_{mji} - \tilde{v}_{\alpha j,m}\epsilon_{mki})) - \frac{1}{2}\grave{\phi}_\alpha(\nu_{k\ell,i}^{(\alpha)} U_{\alpha j\ell i} - \nu_{j\ell,i}^{(\alpha)} U_{\alpha k\ell i})]\end{aligned} \quad (5.2.19)$$

### 5.2.3 The Energy Balance Equation

The special material deformation assumptions (5.1.1), (5.1.2), and (5.1.9) reduce the energy supplies (3.4.16) and (3.4.17), and the energy balance equations (3.4.18) and (3.4.19), to

$$\check{e}_\alpha = \hat{e}_\alpha + \hat{c}_\alpha \frac{1}{2}(\tilde{v}_{\alpha i}\tilde{v}_{\alpha i} + \hat{\nu}_{im}^{(\alpha)}\hat{\nu}_{im}^{(\alpha)}\bar{i}_\alpha + \grave{\phi}_\alpha\grave{\phi}_\alpha\bar{i}_\alpha) + (\hat{\nu}_{im}^{(\alpha)} + \grave{\phi}_\alpha\delta_{im})\bar{S}_{\alpha mi} \quad (5.2.20)$$

$$\begin{aligned}\hat{\epsilon}_\alpha &= -\hat{p}_{\alpha i}\tilde{v}_{\alpha i} - \hat{c}_\alpha\tilde{\epsilon}_\alpha - \epsilon_{jmn}[\mu_n^{(\alpha)}(\bar{S}_{\alpha mj} + \bar{\rho}_\alpha\tilde{\ell}_{\alpha mj}) + (\mu_n^{(\alpha)}\bar{\lambda}_{\alpha mji} \\ &+ \epsilon_{nik}\mu_m^{(\alpha)}\mu_k^{(\alpha)}\bar{i}_\alpha\tilde{v}_{\alpha j}\bar{\rho}_\alpha)_{,i}] + [\grave{\phi}_\alpha(\bar{S}_{\alpha ii} + \bar{\rho}_\alpha\tilde{\ell}_{\alpha ii}) + (\grave{\phi}_\alpha\bar{\lambda}_{\alpha mmi} - \grave{\phi}_\alpha\grave{\phi}_\alpha\bar{\rho}_\alpha\tilde{v}_{\alpha i}\bar{i}_\alpha)_{,i}] \\ &- \bar{\rho}_\alpha(\overline{\bar{i}_\alpha\mu_m^{(\alpha)}\mu_m^{(\alpha)}})^{\grave{}} - \frac{1}{2}\bar{\rho}_\alpha(\overline{\bar{i}_\alpha\grave{\phi}_\alpha\grave{\phi}_\alpha})^{\grave{}} + \hat{e}_\alpha\end{aligned} \quad (5.2.21)$$

$$\begin{aligned}\bar{\rho}_\alpha(\grave{\tilde{\epsilon}}_\alpha + \frac{1}{2}\overline{\tilde{v}_{\alpha i}\tilde{v}_{\alpha i}}^{\grave{}} + \overline{\bar{i}_\alpha\mu_n^{(\alpha)}\mu_n^{(\alpha)}}^{\grave{}} + \frac{1}{2}\overline{\bar{i}_\alpha\grave{\phi}_\alpha\grave{\phi}_\alpha}^{\grave{}}) &= (\bar{T}_{\alpha ij}\tilde{v}_{\alpha i})_{,j} - \bar{q}_{\alpha i,i} + \bar{\rho}_\alpha\tilde{b}_{\alpha i}\tilde{v}_{\alpha i} \\ &- \hat{c}_\alpha\tilde{\epsilon}_\alpha - \epsilon_{jmn}[\mu_n^{(\alpha)}(\bar{S}_{\alpha mj} + \bar{\rho}_\alpha\tilde{\ell}_{\alpha mj}) + (\mu_n^{(\alpha)}\bar{\lambda}_{\alpha mji} + \epsilon_{nik}\mu_m^{(\alpha)}\mu_k^{(\alpha)}\bar{i}_\alpha\tilde{v}_{\alpha j}\bar{\rho}_\alpha)_{,i}] \\ &+ \bar{\rho}_\alpha\tilde{r}_\alpha + \grave{\phi}_\alpha(\bar{S}_{\alpha ii} + \bar{\rho}_\alpha\tilde{\ell}_{\alpha ii}) + (\grave{\phi}_\alpha\bar{\lambda}_{\alpha mmi} - \grave{\phi}_\alpha\grave{\phi}_\alpha\tilde{v}_{\alpha i}\bar{\rho}_\alpha\bar{i}_\alpha)_{,i} + \hat{e}_\alpha\end{aligned} \quad (5.2.22)$$

$$\begin{aligned}\bar{\rho}_\alpha\grave{\tilde{\epsilon}}_\alpha &= \bar{T}_{\alpha ij}\tilde{v}_{\alpha i,j} - \bar{q}_{\alpha i,i} + \bar{\rho}_\alpha\tilde{r}_\alpha + (\hat{\nu}_{im,q}^{(\alpha)} + \grave{\phi}_{\alpha,q}\delta_{im})\bar{h}_{\alpha miq} \\ &- \check{m}_{\alpha i}\tilde{v}_{\alpha i} - (\hat{\nu}_{im}^{(\alpha)} + \grave{\phi}_\alpha\delta_{im})(\check{g}_{\alpha mi} + \bar{\rho}_\alpha\tilde{f}_{\alpha mi}) - \hat{c}_\alpha[\tilde{\epsilon}_\alpha - \frac{1}{2}\tilde{v}_{\alpha i}\tilde{v}_{\alpha i} \\ &- \frac{1}{2}\bar{i}_\alpha(\hat{\nu}_{im}^{(\alpha)}\hat{\nu}_{im}^{(\alpha)} + \grave{\phi}_\alpha\grave{\phi}_\alpha)] + \frac{1}{2}\bar{\rho}_\alpha(\hat{\nu}_{im}^{(\alpha)}\hat{\nu}_{im}^{(\alpha)} + \grave{\phi}_\alpha\grave{\phi}_\alpha)\grave{\bar{i}}_\alpha + \check{e}_\alpha\end{aligned} \quad (5.2.23)$$

The energy balance equation (5.2.22) illustrates that the total energy of phase $\alpha$ consists of the internal energy, the kinetic energy of the center of mass, and the rotational and dilatational kinetic energies relative to the center of mass. The total energy is balanced by the work produced by surface and body forces, heat transfer and heat generation, phase change, and energy transfer resulting from the structural characteristics of the mixture.

The special material properties expressed by (5.1.1) and (5.1.9) have considerably simplified the equilibrated inertia, equilibrated moments, and energy equations to (5.2.4), (5.2.18), (5.2.19), and (5.2.23). These equations express the evolution of inertia $\bar{i}_\alpha$, dilatation $\phi_\alpha$, angular velocity $\mu^{(\alpha)}$, and energy $\tilde{\epsilon}_\alpha$, and must be supplemented by the balance of mass equation (2.4.14), balance of momentum equation (2.4.19), and by the appropriate constitutive relations. The entropy inequality (2.4.45) may be used to restrict the form of the assumed constitutive equations as will be further discussed in later chapters. An inspection of the special forms of balance equations in this chapter indicates that a considerable further simplification may be achieved by ignoring the rotational effect of material relative to the center of mass. The results obtained with this very special assumption of the material deformation are worth deriving and discussing, since, as will be seen below, such a special case of the theory produces simple, yet very powerful, results.

## 5.3 Balance Equations of Multiphase Mixtures with Dilatation

The additional assumption that the material deformation relative to the center of mass involves only dilatation, reduces the balance equations of multiphase mixtures discussed in the last section into simpler forms. This simplification is achieved by setting $\hat{\boldsymbol{\nu}}^{(\alpha)}=\mathbf{0}$ and $\boldsymbol{\mu}^{(\alpha)}=\mathbf{0}$ in (5.2.5), (5.2.6), and (5.2.13) to (5.2.23). The resulting model is, therefore, described by the following balance equations:

$$\grave{\bar{\rho}}_\alpha + \bar{\rho}_\alpha \tilde{v}_{\alpha i,i} = \hat{c}_\alpha \qquad (2.4.14)$$

$$\bar{\rho}_\alpha \grave{\tilde{v}}_{\alpha i} = \bar{T}_{\alpha ij,j} + \bar{\rho}_\alpha \bar{b}_{\alpha i} + \check{m}_{\alpha i} - \hat{c}_\alpha \tilde{v}_{\alpha i} \qquad (3.4.2)$$

$$\hat{M}_{\alpha ij} = \bar{T}_{\alpha ij} - \bar{T}_{\alpha ji} \qquad (2.4.31)$$

$$\bar{\rho}_\alpha \grave{\bar{i}}_\alpha = -\hat{c}_\alpha \bar{i}_\alpha + k_{\alpha jji,i} + K_{\alpha jj} + \check{k}_{\alpha jj} \qquad (5.2.4)$$

$$\bar{\rho}_\alpha \grave{\overline{(\bar{i}_\alpha \grave{\phi}_\alpha)}} \delta_{jk} = -\hat{c}_\alpha \bar{i}_\alpha \grave{\phi}_\alpha \delta_{jk} + \bar{\rho}_\alpha \bar{f}_{\alpha(jk)} + \bar{h}_{\alpha(jk)m,m} + \check{g}_{\alpha(jk)}$$
$$(\mathit{from}\ (5.2.18)) \qquad (5.3.1)$$

$$\bar{S}_{\alpha[jk]} + \bar{\rho}_\alpha \tilde{\ell}_{\alpha[jk]} + \bar{\lambda}_{\alpha[jk]m,m} + \bar{\rho}_\alpha \bar{\tilde{\imath}}_\alpha \grave{\phi}_\alpha W_{\alpha jk} - \grave{\phi}_\alpha (\grave{\phi}_\alpha)_{,i} U_{\alpha[jk]i} = 0$$
$$(from\ (5.2.19)) \qquad (5.3.2)$$

$$\bar{\rho}_\alpha \grave{\tilde{\varepsilon}}_\alpha = \bar{T}_{\alpha ij} \tilde{v}_{\alpha i,j} - \bar{q}_{\alpha k,k} + \bar{\rho}_\alpha \tilde{r}_\alpha - \check{m}_{\alpha i} \tilde{v}_{\alpha i} - \check{g}_{\alpha kk} \grave{\phi}_\alpha + \bar{h}_{\alpha kkm} \grave{\phi}_{\alpha,m}$$
$$+ \frac{1}{2} \bar{\rho}_\alpha \grave{\bar{\tilde{\imath}}}_\alpha \grave{\phi}_\alpha \grave{\phi}_\alpha - \bar{\rho}_\alpha \grave{\phi}_\alpha \tilde{f}_{\alpha kk} - \hat{c}_\alpha (\tilde{\varepsilon}_\alpha - \frac{1}{2} \bar{\tilde{\imath}}_\alpha \grave{\phi}_\alpha \grave{\phi}_\alpha - \frac{1}{2} \tilde{v}_{\alpha i} \tilde{v}_{\alpha i}) + \check{e}_\alpha \qquad (5.3.3)$$

$$\bar{\rho}_\alpha \grave{\tilde{s}}_\alpha + \nabla \cdot \left(\frac{\bar{\mathbf{q}}_\alpha}{\bar{\bar{\theta}}_\alpha}\right) - \frac{\bar{\rho}_\alpha \tilde{r}_\alpha}{\bar{\bar{\theta}}_\alpha} + \hat{c}_\alpha \tilde{s}_\alpha + \hat{s}_\alpha \geq 0 \qquad (2.4.45)$$

where

$$\check{m}_{\alpha i} = \hat{p}_{\alpha i} + \hat{c}_\alpha \tilde{v}_{\alpha i} \qquad (3.4.1)$$

$$\hat{p}_{\alpha i} = \bar{t}_{\alpha i} - \grave{\phi}_{\alpha,i} \bar{\rho}_\alpha \grave{\phi}_\alpha \bar{\tilde{\imath}}_\alpha \quad (from\ (3.3.3)) \qquad (5.3.4)$$

$$k_{\alpha jji} = -\grave{\phi}_\alpha U_{\alpha jji} \quad (from\ (5.2.6)) \qquad (5.3.5)$$

$$K_{\alpha jj} = 2 \bar{\rho}_\alpha \grave{\phi}_\alpha \bar{\tilde{\imath}}_\alpha \qquad (5.2.7)$$

$$\check{k}_{\alpha jj} = \hat{c}_\alpha \hat{\imath}_{\alpha jj} \qquad (5.2.8)$$

$$\bar{\rho}_\alpha \tilde{f}_{\alpha(jk)} = \bar{S}_{\alpha(jk)} \quad (from\ (3.4.12)) \qquad (5.3.6)$$

$$\bar{h}_{\alpha(jk)m} = \bar{\lambda}_{\alpha(jk)m} - \frac{1}{2} \bar{\rho}_\alpha \bar{\tilde{\imath}}_\alpha \grave{\phi}_\alpha (\tilde{v}_{\alpha k} \delta_{mj} + \tilde{v}_{\alpha j} \delta_{mk}) \quad (from\ (3.4.11)) \quad (5.3.7)$$

$$\check{g}_{\alpha(jk)} = \hat{g}_{\alpha(jk)} + \bar{\rho}_\alpha \bar{\tilde{\imath}}_\alpha \grave{\phi}_\alpha \grave{\phi}_\alpha \delta_{jk} + \frac{1}{2} [\tilde{v}_{\alpha k} (\bar{\rho}_\alpha \bar{\tilde{\imath}}_\alpha \grave{\phi}_\alpha)_{,j} + \tilde{v}_{\alpha j} (\bar{\rho}_\alpha \bar{\tilde{\imath}}_\alpha \grave{\phi}_\alpha)_{,k}]$$
$$(from\ (3.4.13)) \qquad (5.3.8)$$

$$\hat{g}_{\alpha(jk)} = \frac{1}{2} \hat{c}_\alpha \grave{\phi}_\alpha (\hat{\imath}_{\alpha jk} + \hat{\imath}_{\alpha kj}) - \frac{1}{2} [(\grave{\phi}_\alpha \grave{\phi}_\alpha U_{\alpha jkm})_{,m} + (\grave{\phi}_\alpha \grave{\phi}_\alpha U_{\alpha kjm})_{,m}]$$
$$(from\ (3.4.10)) \qquad (5.3.9)$$

$$\check{e}_\alpha = \hat{e}_\alpha + \frac{1}{2} \hat{c}_\alpha (\bar{\tilde{\imath}}_\alpha \grave{\phi}_\alpha \grave{\phi}_\alpha + \tilde{v}_{\alpha i} \tilde{v}_{\alpha i}) + \bar{S}_{\alpha ii} \grave{\phi}_\alpha \quad (from\ (5.2.20)) \quad (5.3.10)$$

Equation (5.3.2) may be used to produce restrictions on the body moment $\tilde{\ell}_{\alpha[jk]}$ as a necessary, but not sufficient, condition

$$\bar{\rho}_\alpha \tilde{\ell}_{\alpha[jk]} = -\bar{\rho}_\alpha \tilde{i}_\alpha \grave{\phi}_\alpha W_{\alpha jk} \tag{5.3.11}$$

and skew-symmetry of $\bar{\mathbf{S}}_\alpha$ and $\bar{\boldsymbol{\lambda}}_\alpha$,

$$(\bar{S}_{\alpha jk} - \bar{S}_{\alpha kj}) + (\bar{\lambda}_{\alpha jkm,m} - \bar{\lambda}_{\alpha kjm,m}) - \grave{\phi}_\alpha(\grave{\phi}_\alpha)_{,i}(U_{\alpha jki} - U_{\alpha kji}) = 0 \tag{5.3.12}$$

From the above equations, it may be noted that the variables

$$\hat{p}_{\alpha i},\quad k_{\alpha jji},\quad K_{\alpha jj},\quad \check{k}_{\alpha jj},\quad \tilde{f}_{\alpha(jk)},\quad \hat{g}_{\alpha(jk)},\quad \tilde{\ell}_{\alpha(jk)} = 0 \tag{5.3.13}$$

*are frame-indifferent*, whereas the following variables

$$\check{m}_{\alpha i},\quad \bar{h}_{\alpha(jk)m},\quad \check{g}_{\alpha(jk)},\quad \check{e}_\alpha,\quad \tilde{\ell}_{\alpha[jk]} \tag{5.3.14}$$

*are not.*

## 5.4 Comparison of Results with Other Theories of Structured Multiphase Mixtures

The special results of the theory of multiphase mixtures discussed in the previous section are useful for comparison with the postulatory theories of GOODMAN & COWIN (1972), PASSMAN (1977), PASSMAN *et al.* (1984), DRUMHELLER & BEDFORD (1980), and AHMADI (1985). As discussed in chapter 1, the GOODMAN and COWIN's theory of granular media includes additional balance equations for the equilibrated inertia and equilibrated force, whereas the theory of PASSMAN and co-workers is a generalization of the same theory that includes the additional effects of the inertia body forces. The field equations of multiphase mixtures of DRUMHELLER, BEDFORD, and AHMADI also involve additional transport equations and these models too will be compared with the results of the previous section.

The balance equations of multiphase mixtures advocated by PASSMAN and co-workers (1977,1984) are an extension of the work of GOODMAN & COWIN (1972) and are of the following form:

$$\grave{\rho}_\alpha + \rho_\alpha v_{\alpha i,i} = \check{c}_\alpha \tag{5.4.1}$$

$$\rho_\alpha \grave{v}_{\alpha i} = T_{\alpha ij,j} + \rho_\alpha b_{\alpha i} + \check{m}_{\alpha i} - \check{c}_\alpha v_{\alpha i} \tag{5.4.2}$$

$$\check{M}_{\alpha ij} = T_{\alpha ij} - T_{\alpha ji} \tag{5.4.3}$$

$$\rho_\alpha \grave{k}_\alpha = -\check{c}_\alpha k_\alpha + k_{\alpha i,i} + K_\alpha + \check{k}_\alpha \tag{5.4.4}$$

$$\rho_\alpha \overline{(k_\alpha \grave{\phi}_\alpha)}^{\grave{}} = -\check{c}_\alpha k_\alpha \grave{\phi}_\alpha + \rho_\alpha(\ell_\alpha + f_\alpha) + h_{\alpha i,i} + \check{g}_\alpha \tag{5.4.5}$$

$$\begin{aligned}\rho_\alpha \grave{\epsilon}_\alpha &= T_{\alpha ij} v_{\alpha i,j} - q_{\alpha i,i} + \rho_\alpha r_\alpha - \check{m}_{\alpha i} v_{\alpha i} - \check{g}_\alpha \grave{\phi}_\alpha + h_{\alpha i} \grave{\phi}_{\alpha,i} \\ &+ \frac{1}{2}\rho_\alpha \grave{k}_\alpha \grave{\phi}_\alpha \grave{\phi}_\alpha - \rho_\alpha f_\alpha \grave{\phi}_\alpha - \check{c}_\alpha(\epsilon_\alpha - \frac{1}{2} v_{\alpha i} v_{\alpha i} - \frac{1}{2} k_\alpha \grave{\phi}_\alpha \grave{\phi}_\alpha) + \check{e}_\alpha\end{aligned} \tag{5.4.6}$$

$$\rho_\alpha \grave{s}_\alpha + \nabla \cdot \frac{\mathbf{q}_\alpha}{\theta_\alpha} - \frac{\rho_\alpha r_\alpha}{\theta_\alpha} + \check{c}_\alpha s_\alpha = \check{\eta}_\alpha \tag{5.4.7}$$

where: $\rho_\alpha$ is the density, $\mathbf{v}_\alpha$ is the velocity, $\check{c}_\alpha$ is the growth of mass, $\mathbf{T}_\alpha$ is the stress tensor, $\mathbf{b}_\alpha$ is the body force, $\check{\mathbf{m}}_\alpha$ is the growth of linear momentum, $\check{\mathbf{M}}_\alpha$ is the growth of angular momentum, $k_\alpha$ is the uquilibrated inertia, $\mathbf{k}_\alpha$ is the equilibrated inertia force, $K_\alpha$ is the inertia body force, $\check{k}_\alpha$ is the growth of equilibrated inertia, $\phi_\alpha$ is the volumetric fraction, $\ell_\alpha$ is the equilibrated body force, $f_\alpha$ is the equilibrated force supply or intrinsic body force, $\mathbf{h}_\alpha$ is the equilibrated stress, $\check{g}_\alpha$ is the growth of equilibrated force, $\epsilon_\alpha$ is the internal energy, $\mathbf{q}_\alpha$ is the heat flux, $r_\alpha$ is the body heating or heat generation rate, $\check{e}_\alpha$ is the growth of energy, $s_\alpha$ is the entropy, $\theta_\alpha$ is the absolute temperature, and $\check{\eta}_\alpha$ is the growth of entropy of phase $\alpha$. Moreover, the growths $\check{c}_\alpha$, $\check{\mathbf{m}}_\alpha$, $\check{\mathbf{M}}_\alpha$, $\check{k}_\alpha$, $\check{g}_\alpha$ and $\check{e}_\alpha$ are required to satisfy the following conditions:

$$\sum_\alpha \check{c}_\alpha = 0 \tag{5.4.8}$$

$$\sum_\alpha \check{m}_{\alpha i} = 0 \tag{5.4.9}$$

$$\sum_\alpha \check{M}_{\alpha ij} = 0 \tag{5.4.10}$$

$$\sum_\alpha \check{k}_\alpha = 0 \tag{5.4.11}$$

$$\sum_\alpha \check{g}_\alpha = 0 \tag{5.4.12}$$

$$\sum_\alpha \check{e}_\alpha = 0 \tag{5.4.13}$$

$$\sum_{\alpha} \breve{\eta}_\alpha \geq 0 \tag{5.4.14}$$

Equations (5.4.1)-(5.4.7) are, clearly, identical to the special results of the theory presented in the last section and expressed by equations (2.4.14), (3.4.2), (2.4.31), (5.2.4), (5.3.1), and (5.3.3) *if* the following identification of phase variables is made:

$$\begin{gathered}
\rho_\alpha = \bar{\rho}_\alpha, \quad \mathbf{v}_\alpha = \tilde{\mathbf{v}}_\alpha, \quad \check{c}_\alpha = \hat{c}_\alpha, \quad \mathbf{T}_\alpha = \bar{\mathbf{T}}_\alpha, \quad \mathbf{b}_\alpha = \tilde{\mathbf{b}}_\alpha, \\
\check{\mathbf{M}}_\alpha = \hat{\mathbf{M}}_\alpha, \quad k_\alpha = \tilde{i}_\alpha, \quad k_{\alpha i} = k_{\alpha jji}, \quad K_\alpha = K_{\alpha jj}, \quad \check{k}_\alpha = \check{k}_{\alpha jj}, \\
\ell_\alpha = \tilde{\ell}_{\alpha kk}, \quad f_\alpha = \tilde{f}_{\alpha kk}, \quad h_{\alpha i} = \bar{h}_{\alpha kki}, \quad \check{g}_\alpha = \check{g}_{\alpha kk}, \\
\epsilon_\alpha = \tilde{\epsilon}_\alpha, \quad \mathbf{q}_\alpha = \bar{\mathbf{q}}_\alpha, \quad r_\alpha = \tilde{r}_\alpha, \quad s_\alpha = \tilde{s}_\alpha, \quad \theta_\alpha = \bar{\bar{\theta}}_\alpha
\end{gathered} \tag{5.4.15}$$

The balance equations of multiphase mixtures from the postulatory theory can be, therefore, derived from the volume averaging approach and special assumptions of the material deformation relative to the center of mass. This result is significant, since it *provides a direct correspondence between the averaging and postulatory theories of mixtures* and fulfills the third objective set forth in section 1.2. This correspondence between the averaging and postulatory theories of multiphase mixtures is, however, not total, since there are some very important differences between the two of them. These are:

1. Equation (5.3.2) does not appear in the works of PASSMAN and coworkers.

2. The phasic balance variables $k_{\alpha jji}$, $K_{\alpha jj}$, $\check{k}_{\alpha jj}$, $\tilde{\ell}_{\alpha kk}$, $\tilde{f}_{\alpha kk}$, $\bar{h}_{\alpha kki}$, and $\check{g}_{\alpha kk}$ *are of higher tensorial order* than the corresponding variables in the postulatory theory, and their physical meaning is, therefore, broader.

3. The entropy inequality of phase $\alpha$ (2.4.45) contains the entropy source $\hat{s}_\alpha$ and its left side is assumed to be positive semidefinite (validity of the local axiom of dissipation within each subvolume of phase $\alpha$), whereas (5.4.7) contains no such restriction.

4. The conditions on the growths expressed by (5.4.9)-(5.4.13) *are not necessarily satisfied in the volume-averaged field equations*, as discussed in sections 2.5 and 3.4.2.

The condition (5.4.8) is identical to the condition (2.4.17). Using (3.4.1) and (5.3.4) in (2.4.26) it may be seen that this does not yield $\hat{\mathbf{p}} = \mathbf{0}$ as stated by (5.4.9), unless the surface tension and structural characteristics of the mixture are ignored (*cit.* (2.4.25)). The condition (5.4.11) may be satisfied if there is no phase change (this is shown using (5.2.8) in (3.4.34)). The conditions (5.4.12) and (5.4.13) are also not satisfied as may be seen from (5.3.8), (5.3.9), (5.3.10), (3.4.33), and (3.4.35). The global entropy production condition (5.4.14) is, of course, satisfied as can be observed from (2.4.50). The claim that the growth variables $\check{c}_\alpha$, $\check{\mathbf{m}}_\alpha$, $\check{\mathbf{M}}_\alpha$, $\check{k}_\alpha$, $\check{g}_\alpha$ and $\check{e}_\alpha$, when summed over $\alpha$, reduce to zero is based on a principle of TRUESDELL (1984, page 221) that the motion of the mixture is governed by the same equations as is a single body. The motivation behind this principle is discussed in section 2.5. It basically requires that the growth of source terms for the mixture as a whole produce no effect due to the mutual cancellations. The reason that the volume-averaged theory of multiphase mixtures together with the special deformation assumptions involving dilatation does not yield such restrictions on these growth or interaction variables should be clear. In view of the demonstration that the postulatory theory of multiphase mixtures described by equations (5.4.1)-(5.4.6) has an existence within the very *special* case in the theory of mixtures based on the volume averaging approach, it is unlikely that the conditions (5.4.9)-(5.4.13) are satisfied in general. An additional important difference between the special theory of multiphase mixtures and that of PASSMAN *et al.* (1984) is the assumption in the latter works that the phase variables $\mathbf{h}_\alpha$, $\check{\mathbf{m}}_\alpha$, $\check{g}_\alpha$, and $\check{e}_\alpha$ are frame-indifferent. According to (5.3.14) these variables are not frame-indifferent separately, but $\nabla \cdot \bar{\mathbf{h}}_\alpha + \check{\mathbf{g}}_\alpha$ *is*, as may be seen from (5.3.7)-(5.3.9).

DRUMHELLER & BEDFORD (1980) used a variational formulation to develop a theory of immiscible mixtures. Their theory is based on the "HAMILTON's extended variational principle" of the following form:

$$\int_{t_1}^{t_2} (\delta t - \delta V + \delta W)\, dt = 0 \tag{5.4.16}$$

where $T$ and $V$ are the mixture's kinetic and potential energies, $\delta W$ is the virtual work, and $t_1$ and $t_2$ are arbitrary times. The total kinetic energy is assumed to consist of the translational energies of mass and virtual mass, and of the dilatational energy which accounts for expansion and contraction of each phase. The dilatational kinetic energy is modeled in terms of the inertia $I_\alpha$, densities $\bar{\rho}_\alpha$ and $\gamma_\alpha = \bar{\rho}_\alpha / \phi_\alpha$, and volumetric fraction $\phi_\alpha$, *i.e.*

$$\frac{1}{2} \bar{\rho}_\alpha I_\alpha \grave{\gamma}_\alpha \grave{\gamma}_\alpha \tag{5.4.17}$$

with $I_\alpha$ (and virtual mass) assumed to be given by the following constitutive equation:

$$I_\alpha = I_\alpha(\gamma_1, \ldots, \gamma_\gamma, \phi_1, \ldots, \phi_\gamma) \tag{5.4.18}$$

The virtual work and potential energy terms involve works due to the stress tensor, body force, momentum supply or interaction of each phase, and due to the variations of density (with the generalized force $P_\alpha$) and phase change (with the generalized force $S_\alpha$). By ignoring the virtual mass effect in the theory (BEDFORD & DRUMHELLER,1983), the resulting variational formulation produces the following equations (in the notation of this monograph):

$$\grave{\bar{\rho}}_\alpha - \bar{\rho}_\alpha \tilde{v}_{\alpha i,i} = \hat{c}_\alpha \tag{5.4.19}$$

$$\sum_\alpha \hat{c}_\alpha = 0 \tag{5.4.20}$$

$$\begin{aligned} \bar{\rho}_\alpha \grave{\tilde{v}}_{\alpha i} = & -\hat{c}_\alpha \tilde{v}_{\alpha i} + \bar{T}_{\alpha ij,j} + \bar{\rho}_\alpha \tilde{b}_{\alpha i} + \hat{p}_{\alpha i} + \lambda \phi_{\alpha,i} - \mu_{\alpha,i} + \hat{c}_\alpha \tau_{,i} \\ & + \frac{1}{2} \sum_\eta \Big( \bar{\rho}_\alpha \frac{\partial I_\alpha}{\partial \gamma_\eta} \grave{\gamma}_\alpha \grave{\gamma}_\alpha \gamma_{\eta,i} - \bar{\rho}_\eta \frac{\partial I_\eta}{\partial \gamma_\alpha} \grave{\gamma}_\eta \grave{\gamma}_\eta \gamma_{\alpha,i} + \bar{\rho}_\alpha \frac{\partial I_\alpha}{\partial \phi_\eta} \grave{\gamma}_\alpha \grave{\gamma}_\alpha \phi_{\eta,i} \\ & - \bar{\rho}_\eta \frac{\partial I_\eta}{\partial \phi_\alpha} \grave{\gamma}_\eta \grave{\gamma}_\eta \phi_{\alpha,i} \Big) \end{aligned} \tag{5.4.21}$$

$$\bar{\rho}_\alpha (\grave{\overline{I_\alpha \grave{\gamma}_\alpha}}) = -\hat{c}_\alpha I_\alpha \grave{\gamma}_\alpha + \frac{1}{2} \sum_\eta \bar{\rho}_\eta \frac{\partial I_\eta}{\partial \bar{\rho}_\alpha} \grave{\gamma}_\eta \grave{\gamma}_\eta - \frac{P_\alpha}{\gamma_\alpha} + \frac{\mu_\alpha}{\gamma_\alpha} \tag{5.4.22}$$

$$\frac{1}{2} \sum_\eta \bar{\rho}_\eta \frac{\partial I_\eta}{\partial \phi_\alpha} \grave{\gamma}_\eta \grave{\gamma}_\eta + \frac{\mu_\alpha}{\phi_\alpha} - \lambda = 0 \tag{5.4.23}$$

$$\frac{1}{2} \tilde{v}_{\alpha i} \tilde{v}_{\alpha i} + \frac{1}{2} I_\alpha \grave{\gamma}_\alpha \grave{\gamma}_\alpha + S_\alpha - \frac{\mu_\alpha}{\bar{\rho}_\alpha} = \left( \frac{d\tau}{dt} \right)_{X_\alpha} \tag{5.4.24}$$

where $\lambda$, $\tau$, $\mu_\alpha$, $P_\alpha$, and $S_\alpha$ are the Lagrangian multipliers that include the constraint equations (5.4.19), (5.4.20), and

$$\sum_\alpha \phi_\alpha = 0 \tag{5.4.25}$$

The energy balance equation is assumed to be of the following form:

$$\bar{\rho}_\alpha \grave{\tilde{\epsilon}}_\alpha = -\hat{c}_\alpha \tilde{\epsilon}_\alpha + \bar{T}^T_{\alpha ij} \tilde{v}_{\alpha j,i} + P_\alpha \frac{\grave{\gamma}_\alpha}{\gamma_\alpha} - S_\alpha \hat{c}_\alpha - \bar{q}_{\alpha,i} + \bar{\rho}_\alpha \tilde{r}_\alpha + \grave{e}_\alpha \tag{5.4.26}$$

where $\grave{e}_\alpha$ is the energy interaction. In addition to the above equations, the following conditions and entropy inequality were also assumed:

$$\sum_\alpha \hat{p}_{\alpha i} = 0 \tag{5.4.27}$$

$$\sum_{\alpha}(\hat{p}_{\alpha i}\tilde{v}_{\alpha i} + \grave{e}_{\alpha}) = 0 \tag{5.4.28}$$

$$\sum_{\alpha}(\bar{\rho}_{\alpha}\grave{\bar{s}}_{\alpha} + \nabla\cdot\frac{\mathbf{q}_{\alpha}}{\bar{\bar{\theta}}_{\alpha}} + \hat{c}_{\alpha}\bar{s}_{\alpha} - \frac{\bar{\rho}_{\alpha}\tilde{r}_{\alpha}}{\bar{\bar{\theta}}_{\alpha}}) \geq 0 \tag{5.4.29}$$

An interesting result of the above theory is equation (5.4.22) which describes the evolution of structural characteristics of the mixture: the inertia and volumetric fraction of each phase. When this equation is compared with the result of the special theory of mixtures expressed by (5.3.1), it is seen that its form depends on the constitutive assumption (5.4.18) which does not clearly express the material deformation assumption, as contrasted with (5.3.1) which is a result only of a restriction imposed on the material deformation *relative to the center of mass.* Although the theory of DRUMHELLER and BEDFORD and the special theory of multiphase mixtures discussed in section 5.3 are both assumed to model the dilatation or the expansion and contraction of each phase in the mixture, the results of the variational and volume averaging approaches are considerably diffferent. These differences may be summarized as follows:

1. The variational theory yields only one additional equation to model the structural characteristics of mixture, whereas the averaging theory produces many more (*cit.* (5.2.4), (5.3.1), and (5.3.2)).
2. The variational theory is based on an additional restriction expressed by (5.4.18).
3. The volume averaging theory employs the local axiom of dissipation which is assumed valid within each subvolume of each phase (see chapter 2), whereas the variational theory is based on an entropy inequality for the mixture as a whole.
4. The conditions (5.4.27) and (5.4.28) are not necessarily satisfied in the volume averaging theory.
5. The momentum equation (5.4.21) appears to be more general than (3.4.2). This is, of course, misleading, since the momentum interactions $\hat{\mathbf{p}}_{\alpha}$ in (3.4.2) and (5.4.21) are not necessarily equivalent.
6. The volume averaging theory allows for the specification of intrinsic and extrinsic forces and couples in the additional field equations which must be described by the constitutive relations, whereas the variational theory is constructed on the basis of the constitutive assumption (5.4.18) and, as such, it does not allow for the specification

of arbitrary sources. In effect, there is no clear separation between the balance and constitutive equations in this theory. This point was also criticized by PASSMAN *et al.* (1984).

AHMADI's (1985) study of multiphase suspension flows without the phase change is also made with the balance equations for the equilibrated force and equilibrated inertia. His balance equation for the equilibrated force is similar to (5.4.5) (with $\check{c}_\alpha = 0$ and $\rho_\alpha f_\alpha$ included in $h_{\alpha i}$), whereas his balance equation for the equilibrated inertia has the form of (5.2.2) with $\check{c}_\alpha = 0$ and $\hat{k}_{\alpha jj} = 0$. In addition to the conservation of mass and balance of linear momentum, energy, equilibrated force, and equilibrated inertia, AHMADI also includes within the theory of multiphase mixtures the conservation equations for the microinertia of each phase and balance equations for the angular momentum. The latter equations express the time rates of change of the microgyration in terms of microinertia, couple stress tensor, body and internal body couples, such that these equations have some similarity with (5.2.19) with $\bar{\imath}_\alpha$ set equal to the microinertia and a constant. AHMADI's justification for the use of such a varied set of field equations comes partially from the granular media equations of GOODMAN & COWIN (1972) and partially from the micromorphic theory of ERINGEN (1964). His energy equation of each phase includes the work done by the rotation of particles, but this expression differs considerably from (5.2.23) and precludes, therefore, a detailed comparison with the present work.

## 5.5 Concluding Remarks

In this chapter the theory of multiphase mixtures developed in previous chapters was simplified by invoking further restrictions of the material deformation relative to the center of mass. By decomposing the material motion relative to the center of mass into rotation and dilatation, and by assuming that the inertia tensor is isotropic, produces considerable simplification of the additional balance equations. By ignoring the rotational effect has then allowed a comparison of results with the existing structured theories of mixtures.

*The existing theories of structured multiphase mixtures are very special*, since they have an existence property within a very special material deformation assumption relative to the center of mass in the volume averaging approach. It also appears that these theories assume incorrectly the frame-indifference of many phase variables and, consequently, the constitutive equations constructed on such assumptions are incorrect. The volume averaging approach together with a basic assumption of the mate-

**rial deformation relative to the center of mass of each phase have shown to be very powerful to derive additional field equations for modeling the mixture's structural properties.**

CHAPTER 6

# CONCEPTS AND PRINCIPLES OF CONSTITUTIVE THEORY

## 6.1 Purpose of the Chapter

Useful models of multiphase mixtures are constructed by determining the constitutive relations for different types of mixtures. A multiphase mixture may contain fluid- and solid-like phases, each of which responds differently to mechanical loading and heat transfer. Consequently, a constitutive assumption should reflect as closely as possible the true material behavior over the anticipated range of thermomechanical processes. For this reason, the construction of constitutive equations should be based upon a rational set of constitutive methods or principles possessing a sufficient mathematical structure to be useful in the development of practical models. In contrast to the single phase multicomponent mixture theory, the constitutive concepts and principles for multiphase mixtures are not sufficiently developed. Our objective in this chapter will be to develop these concepts for multiphase mixtures by assuming that there exists a rigorous set of field equations as presented in chapters 2 and 3. The concepts and principles of the constitutive theory presented in this chapter will form a foundation for the study of constitutive relations of special mixtures in subsequent chapters. As a basis for this development, many concepts will be borrowed from single phase (single component and multicomponent) continuum theories that may be found in the works of NOLL (1958), TRUESDELL & NOLL (1965), and BOWEN (1976).

## 6.2 Principle of Determinism

The thermodynamic variables in the conservation and balance equations of multiphase mixtures define the *thermodynamic process* for the mixture. For the field equations discussed in chapters 2 and 3, these variables are summarized by (4.4.1), *i.e.*

$$P_\alpha = (\mathbf{x} = \chi_{\alpha\boldsymbol{\kappa}}(\mathbf{X}_\alpha, t),\ \bar{\rho}_\alpha,\ \hat{c}_\alpha,\ \bar{\mathbf{T}}_\alpha,\ \tilde{\mathbf{b}}_\alpha,\ \hat{\mathbf{M}}_\alpha,\ \tilde{\epsilon}_\alpha,\ \bar{\mathbf{q}}_\alpha,\ \tilde{r}_\alpha,\ \tilde{s}_\alpha,\ \bar{\bar{\theta}}_\alpha,\ \hat{s}_\alpha,$$
$$\bar{\bar{\mathbf{i}}}_\alpha,\ \boldsymbol{\nu}^{(\alpha)},\ \hat{\mathbf{i}}_\alpha,\ \mathbf{U}_\alpha,\ \tilde{\boldsymbol{\ell}}_\alpha,\ \bar{\boldsymbol{\lambda}}_\alpha,\ \bar{\mathbf{S}}_\alpha,\ \bar{\mathbf{t}}_\alpha,\ \tilde{\boldsymbol{\ell}}\epsilon_\alpha,\ \hat{\hat{\epsilon}}_\alpha,\ \bar{q}_{s\alpha});\ \alpha = 1, \ldots, \gamma \quad (6.2.1)$$

The *constitutive equations or relations* are a collection from this set of thermodynamic variables which satisfy the principle of the material frame-indifference and which are expressed in terms of independent thermodynamic variables. The number of independent variables depends on the constraining conditions of the mixture, such as no phase change or constant density of some phases. For a mixture in motion and exchanging energy between the phases, the independent thermodynamic variables reflect this mechanical and thermal conditions of the mixture.

In the absence of structural effects and thermal and phase change processes, the dynamic conditions of the mixture (such as the stress tensor) can be expressed in terms of the kinematic variables. The field equations (2.4.14), (2.4.19), and (2.4.31) may in this instance provide a solution for the density and velocity fields for prescribed values of the body force $\tilde{\mathbf{b}}_\alpha$ and constitutive equations $\bar{\mathbf{T}}_\alpha$ and $\hat{\mathbf{p}}_\alpha$, with the latter given in terms of the kinematic variables. With the allowance for heat transfer between the phases of the mixture we need to adjoin to the kinematical and mechanical set of field equations the energy equation (3.4.19) which introduces the energetic variables the internal energy $\tilde{\epsilon}_\alpha$, heat flux $\bar{\mathbf{q}}_\alpha$, heat generation rate $\tilde{r}_\alpha$, and the energy interaction field $\check{e}_\alpha$. If, however, the prescription of the mechanical and thermal fields were only in terms of kinematic variables, then the additional energy equation may not, in general, be satisfied for arbitrary values of the heat generation rate $\tilde{r}_\alpha$. We would in this instance have an overdetermined system of equations that would not reflect the additional degree of freedom brought about by the allowance for heat transfer in the mixture. To bring this additional degree of freedom within the theory use should be made of a physical base or experience which tells us that the heat transfer occurs due to the temperature gradient within the mixture. For this reason, we need to adjoin to the independent kinematic variables the thermal variables in the form of phase temperatures for the determination of joint dynamic and thermal response of the mixture under the applied mechanical forces and thermal gradients.

When an allowance is made for the phase change or chemical reaction processes occurring within a mixture, additional degrees of freedom or independent variables must be introduced into the theory. To see this, use can be made of the result (TRUESDELL & TOUPIN,1960, sec. 76.8)

$$\nabla \cdot \tilde{\mathbf{v}}_\alpha = \frac{\overline{|det\mathbf{F}_\alpha|}^{\grave{}}}{|det\mathbf{F}_\alpha|} \tag{6.2.2}$$

in order to express the balance of mass equation (2.4.14) in the following form:

$$\overline{\grave{\bar{\rho}_\alpha |det\mathbf{F}_\alpha|}} = |det\mathbf{F}_\alpha|\hat{c}_\alpha = \overline{\grave{\bar{\rho}_{\alpha\boldsymbol{\kappa}}(\mathbf{X}_\alpha,t)|det(GRAD\chi_{\alpha\boldsymbol{\kappa}}(\mathbf{X}_\alpha,t))|}}$$
$$= \grave{\mathfrak{R}}_{\alpha\boldsymbol{\kappa}}(\mathbf{X}_\alpha,t) \qquad (6.2.3)$$

where $\mathfrak{R}_{\alpha\boldsymbol{\kappa}}(\mathbf{X}_\alpha,t) = \bar{\rho}_\alpha|det\mathbf{F}_\alpha|$. This result shows that the density and motion (or deformation gradient) of phase $\alpha$ are independent when $\hat{c}_\alpha$ is different from zero, and that the constitutive relations should reflect this independence through the independent variables representing the motion $\mathbf{x}=\chi_{\alpha\boldsymbol{\kappa}}(\mathbf{X}_\alpha,t)$ and one of the fields $\bar{\rho}_{\alpha\boldsymbol{\kappa}}(\mathbf{X}_\alpha,t)$ or $\mathfrak{R}_{\alpha\boldsymbol{\kappa}}(\mathbf{X}_\alpha,t)$. Moreover, to integrate the differential equations (6.2.3) it is only necessary to specify the initial conditions $\mathfrak{R}_{\alpha\boldsymbol{\kappa}}(\mathbf{X}_\alpha,t_0)$[1], and motions $\mathbf{x} = \chi_{\alpha\boldsymbol{\kappa}}(\mathbf{X}_\alpha,t)$ and temperatures for $t \geq 0$ and $\alpha = 1,\ldots,\gamma$, since $\hat{c}_\alpha$ is a constitutive relation which, in the absence of structural characteristics of mixture, depends on $\chi_{\alpha\boldsymbol{\kappa}}(\mathbf{X}_\alpha,t)$, $\bar{\bar{\theta}}_\alpha(\mathbf{x},t)$ and $\mathfrak{R}_{\alpha\boldsymbol{\kappa}}(\mathbf{X}_\alpha,t)$ for $t \geq 0$ and $\alpha = 1,\ldots,\gamma$.

The introduction of structural characteristics into the theory of multiphase mixtures as discussed in chapter 3 requires the consideration of additional independent variables in the constitutive equations. The equilibrated inertia equation (3.4.8) and equilibrated moments equation (3.4.14) introduce into the theory the inertia tensor $\bar{\mathbf{i}}_\alpha$ and gyration tensor $\boldsymbol{\nu}^{(\alpha)}$, respectively, as variables for the description of material motion relative to the center of mass of each phase. As such, these variables too should be considered as independent in the constitutive equations, since, otherwise, equations (3.4.8) and (3.4.14) may not be satisfied in general with $\bar{\mathbf{i}}_\alpha$ and $\boldsymbol{\nu}^{(\alpha)}$ included in the list of constitutive relations. Notice, however, in (3.4.8) and (2.4.14) that if: (1) there is no phase change ($\hat{c}_\alpha = 0$), (2) the third order moment effect is ignored ($\hat{I}_{\alpha ijk} = 0$), (3) assuming the mixture to be isotropic[2] ($\bar{i}_{\alpha ij} = \bar{i}_\alpha \delta_{ij}$), and (4) $\nu_{ij}^{(\alpha)} = -\nu_{ji}^{(\alpha)}$, then it implies that $\bar{i}_\alpha$ and $\mathfrak{R}_{\alpha\boldsymbol{\kappa}}$ are constant on a material line and would not appear as independent variables in the constitutive relations. The assumption that $\nu_{ij}^{(\alpha)} = -\nu_{ji}^{(\alpha)}$ implies that the mixture is micropolar where the material motion relative to the center of mass of each phase can only exhibit rotational effects.

From the list of thermodynamic variables in (6.2.1) and the relation expressed by (6.2.3), it will be assumed that the thermokinetic variables

$$\mathbf{x} = \chi_{\beta\boldsymbol{\kappa}}(\mathbf{X}_\beta,t),\ \mathfrak{R}_{\beta\boldsymbol{\kappa}}(\mathbf{X}_\beta,t),\ \bar{\bar{\theta}}_\beta(\mathbf{x},t),\ \bar{\mathbf{i}}_\beta(\mathbf{x},t),\ \boldsymbol{\nu}^{(\beta)}(\mathbf{x},t);\ \beta = 1,\ldots,\gamma \qquad (6.2.4)$$

[1] The choice of $\mathfrak{R}_{\alpha\boldsymbol{\kappa}}(\mathbf{X}_\alpha,t)$ instead of $\bar{\rho}_{\alpha\boldsymbol{\kappa}}(\mathbf{X}_\alpha,t)$ as an independent variable is a matter only of computational efficiency and not of any other restriction. By (6.2.3) it is irrelevant which of these variables are chosen as independent.

[2] Isotropic mixtures are discussed in section 6.8.

specify the *thermokinetic process* for multiphase mixtures, with $\chi_{\beta K}$ and $\Re_{\beta K}$ defined for all $X_\beta$ in the body $\mathcal{B}_\beta$, and $\bar{\bar{\theta}}_\beta$, $\bar{\mathbf{i}}_\beta$, and $\nu^{(\beta)}$ defined for all $\mathbf{x}$ delivered by $\chi_{\beta K}$, and all defined for some given interval of time $t$. For a given region of space $\mathcal{E}^3$ in which the thermokinetic process is occurring, the history of this process will be assumed to define a *calorodynamic process or constitutive relations* for multiphase mixtures. To determine the thermodynamic process (6.2.1), it is, therefore, sufficient to assign the constitutive equations for $105\gamma$ scalar variables

$$\Upsilon_\alpha = (\hat{c}_\alpha,\ \bar{\mathbf{T}}_\alpha,\ \hat{\mathbf{M}}_\alpha,\ \tilde{\epsilon}_\alpha,\ \bar{\mathbf{q}}_\alpha,\ \bar{s}_\alpha,\ \hat{s}_\alpha,\ \hat{\mathbf{i}}_\alpha,\ \mathbf{U}_\alpha,\ \bar{\boldsymbol{\lambda}}_\alpha,\ \bar{\mathbf{S}}_\alpha,\ \hat{\mathbf{t}}_\alpha,\ \tilde{\boldsymbol{\ell}}\epsilon_\alpha,\ \hat{\hat{\epsilon}}_\alpha,\ \bar{q}_{s\alpha});\quad \alpha = 1,\dots,\gamma \tag{6.2.5}$$

and determine the remaining $23\gamma$ fields $\bar{\rho}_\alpha$, $\tilde{\mathbf{b}}_\alpha$, $\tilde{r}_\alpha$, $\bar{\mathbf{i}}_\alpha$, and $\nu^{(\alpha)}$ from the $19\gamma$ balance equations (2.4.14), (2.4.19), (3.4.19), (3.4.8), and (3.4.14), and $4\gamma$ independent variables of inertia and gyration tensor. The body force and heat generation rate ($\tilde{\mathbf{b}}_\alpha$ and $\tilde{r}_\alpha$) are the *assigned* quantities in usual experiments conducted in a laboratory, and the above statement that these fields are determined by the field equations may seem unnatural. To get around this apparent conflict between theory and experimental method it should be noted that we may choose the fields $\chi_{\beta K}(\mathbf{X}_\beta, t)$ and $\bar{\bar{\theta}}_\beta(\mathbf{x}, t)$ in such a way that the values of $\tilde{\mathbf{b}}_\alpha$ and $\tilde{r}_\alpha$, when determined from the conservation and balance equations, are consistent with the values assigned in the laboratory experiment. The body force moment $\tilde{\boldsymbol{\ell}}_\alpha$ is assumed to be given or determined through the additional equations (such as the electromagnetic field equations when magnetic particles are present in a magnetic field).

The assumption that the calorodynamic process is determined by the history of the thermokinetic process is a statement of *determinism or principle of causality*, where the cause (the history of the thermokinetic process) determines the effect (the calorodynamic process). The principle of determinism applied to the constitutive variables in the set $\Upsilon_\alpha$ as given by (6.2.5) may be stated as follows:

$$\Upsilon_\alpha(\mathbf{x}, t) = G_\alpha[\chi^t_{\beta K}(\mathbf{Z}_\beta, s),\ \Re^t_{\beta K}(\mathbf{Z}_\beta, s),\ \bar{\bar{\theta}}^t_\beta(\mathbf{z}, s),\ \bar{\mathbf{i}}^t_\beta(\mathbf{z}, s),\ \nu^{(\beta)t}(\mathbf{z}, s),\ X_\beta,\ t];\quad \alpha = 1,\dots,\gamma \tag{6.2.6}$$

where $G_\alpha$ is the *functional* over the fields of real functions defined as

$$\begin{aligned}
&\chi^t_{\beta K}(\mathbf{Z}_\beta, s) = \chi_{\beta K}(\mathbf{Z}_\beta, t-s),\quad \Re^t_{\beta K}(\mathbf{Z}_\beta, s) = \Re_{\beta K}(\mathbf{Z}_\beta, t-s),\\
&\bar{\bar{\theta}}^t_\beta(\mathbf{z}, s) = \bar{\bar{\theta}}_\beta(\mathbf{z}, t-s),\quad \bar{\mathbf{i}}^t_\beta(\mathbf{z}, s) = \bar{\mathbf{i}}_\beta(\mathbf{z}, t-s),\\
&\nu^{(\beta)t}(\mathbf{z}, s) = \nu^{(\beta)}(\mathbf{z}, t-s);\quad \beta = 1,\dots,\gamma
\end{aligned} \tag{6.2.7}$$

for all $s \geq 0$ such that the set of all these variables represents possible motion, phase change, temperature, inertia, and gyration *histories* of each phase in a multiphase mixture. $\chi^t_{\beta K}(\mathbf{Z}_\beta, s)$ in the list (6.2.7) is the value of $\chi_{\beta K}$ at a time $s$ units before the present time and is undefined for $s<0$. A similar conclusion is also valid for the remaining variables in this list. The functional notation in (6.2.6) implies that for each $\alpha$, $\alpha=1,\ldots,\gamma$, $\beta$ ranges over all values from 1 to $\gamma$. That is, the value of the functional $G_\alpha$ is defined by the histories of the thermokinetic processes of the material particles of *all* phases in the mixture and also explicitly by the present time $t$.

The principle of determinism effectively excludes the material behavior on any point outside of the mixture and any future event. The reverse statement, that the present thermokinetic processes determine the past calorodynamic process, is not generally implied by (6.2.6) due to the necessity of accounting within the theory for the irreversibility of macroscopic processes. This irreversibility is inherent in the entropy inequality (or the second law of thermodynamics) expressed by (2.4.45). In the continuum theory of multiphase mixtures the entropy inequality is required to be satisfied for all physical thermodynamic processes, and a constitutive assumption or calorodynamic process that is consistent with this inequality forms an *admissible thermodynamic process.* The entropy inequality places, therefore, restrictions on the constitutive response functionals and will be used in later chapters after studying the properties of the functional $G_\alpha$ in (6.2.6)

## 6.3 Principle of Local Action

In the development of the theory of multiphase mixtures in chapters 2 and 3 it has been alluded several times to the nonlocal character of the field equations. The region of this nonlocality is at least as large as the volume averaging region $U$, and a theory of multiphase mixtures based on volume averaging approach should account for the size of $U$ in the constitutive equations. In this way the size of the averaging volume $U$ can re-enter into the theory and produce, possibly, a minimum loss of information of the mixture brought about by the averaging process. In the local constitutive theory of mixtures which employs the *principle of local action*, the thermokinetic response of material particles of each phase is determined by the conditions in arbitrary small neighborhoods of these particles and not by the conditions on the outside of them. The validity of this assumption depends, of course, on the degree of dispersiveness of the mixture. A mixture containing micron-size particles is expected to be adequately modeled by a local theory, whereas a mixture with particle sizes with the same order of magnitude as the characteristic flow geometry may require a nonlocal theory of mixtures for modeling the flow processes. Notwithstanding the importance of a nonlocal theoretical structure to model many real multiphase mixtures, in this book only a local theory will be developed, as this is the first step towards studying the additional complications brought about by the nonlocal material response.

For this purpose, define by $N_\beta(X_\beta)$ to be a neighborhood of the particle $X_\beta$ of phase $\beta$ and let $Z_\beta$ denote another particle of phase $\beta$ contained in $N_\beta(X_\beta)$. With the assumed differentiablity of $\chi_{\beta\boldsymbol{\kappa}}$, $\Re_{\beta\boldsymbol{\kappa}}$, $\bar{\bar{\theta}}_\beta$, $\bar{\mathbf{i}}_\beta$, and $\boldsymbol{\nu}^{(\beta)}$, the Taylor series expansion up to and including the second order terms in space variables can be employed to show that

$$\chi^t_{\beta\boldsymbol{\kappa}i}(\mathbf{Z}_\beta, s) \approx \chi^t_{\beta\boldsymbol{\kappa}i}(\mathbf{X}_\beta, s) + F^t_{\beta iJ}(\mathbf{X}_\beta, s)\, dX_{\beta J} + F^t_{\beta iJ,K}(\mathbf{X}_\beta, s)\, dX_{\beta J}\, dX_{\beta K} \tag{6.3.1}$$

$$\Re^t_{\beta\boldsymbol{\kappa}}(\mathbf{Z}_\beta, s) \approx \Re^t_{\beta\boldsymbol{\kappa}}(\mathbf{X}_\beta, s) + \Re^t_{\beta\boldsymbol{\kappa},I}(\mathbf{X}_\beta, s)\, dX_{\beta I} + \Re^t_{\beta\boldsymbol{\kappa},IJ}(\mathbf{X}_\beta, s)\, dX_{\beta I}\, dX_{\beta J} \tag{6.3.2}$$

$$\bar{\bar{\theta}}^t_\beta(\mathbf{z}, s) \approx \bar{\bar{\theta}}^t_\beta(\mathbf{x}, s) + \bar{\bar{\theta}}^t_{\beta,i}(\mathbf{x}, s)\, dx_i + \bar{\bar{\theta}}^t_{\beta,ij}(\mathbf{x}, s)\, dx_i\, dx_j \tag{6.3.3}$$

$$\bar{i}^t_{\beta ij}(\mathbf{z}, s) \approx \bar{i}^t_{\beta ij}(\mathbf{x}, s) + \bar{i}^t_{\beta ij,k}(\mathbf{x}, s)\, dx_k + \bar{i}^t_{\beta ij,k\ell}(\mathbf{x}, s)\, dx_k\, dx_\ell \tag{6.3.4}$$

$$\nu^{(\beta)t}_{ij}(\mathbf{z}, s) \approx \nu^{(\beta)t}_{ij}(\mathbf{x}, s) + \nu^{(\beta)t}_{ij,k}(\mathbf{x}, s)\, dx_k + \nu^{(\beta)t}_{ij,k\ell}(\mathbf{x}, s)\, dx_k\, dx_\ell \tag{6.3.5}$$

where use was made of (2.4.5) and where $d\mathbf{X}_\beta$ and $d\mathbf{x}$ are defined by

$$d\mathbf{X}_\beta = \mathbf{Z}_\beta - \mathbf{X}_\beta \tag{6.3.6}$$

$$d\mathbf{x} = \chi_{\beta\boldsymbol{\kappa}}(\mathbf{Z}_\beta, t) - \chi_{\beta\boldsymbol{\kappa}}(\mathbf{X}_\beta, t) = \mathbf{z} - \mathbf{x} \tag{6.3.7}$$

To simplify the development of the theory it will be assumed that the second order terms in (6.3.2)-(6.3.5) can be ignored in comparison to the first two terms in the series expansion. With this simplification and substitution of (6.3.1)-(6.3.5) into (6.2.6), yields

$$\begin{aligned}\Upsilon_\alpha(\mathbf{x}, t) = G_\alpha[&\chi^t_{\beta\boldsymbol{\kappa}}(\mathbf{X}_\beta, s),\ \mathbf{F}^t_\beta(\mathbf{X}_\beta, s),\ GRAD\mathbf{F}^t_\beta(\mathbf{X}_\beta, s),\ \Re^t_{\beta\boldsymbol{\kappa}}(\mathbf{X}_\beta, s),\\ &GRAD\Re^t_{\beta\boldsymbol{\kappa}}(\mathbf{X}_\beta, s),\ \bar{\bar{\theta}}^t_\beta(\mathbf{x}, s),\ \nabla\bar{\bar{\theta}}^t_\beta(\mathbf{x}, s),\ \bar{\mathbf{i}}^t_\beta(\mathbf{x}, s),\\ &\nabla\bar{\mathbf{i}}^t_\beta(\mathbf{x}, s),\ \boldsymbol{\nu}^{(\beta)t}(\mathbf{x}, s),\ \nabla\boldsymbol{\nu}^{(\beta)t}(\mathbf{x}, s),\ d\mathbf{X}_\beta,\ t]\end{aligned} \tag{6.3.8}$$

The presence of $d\mathbf{X}_\beta$ in the above equation comes from (6.3.1)-(6.3.5) and signifies a possible directional dependence of the material properties or material anisotropy at $\mathbf{X}_\beta$.

The material dependence on the first $N$ gradients of the independent variables in the constitutive relations is referred to as the *material of grade* $N$ (TRUESDELL & NOLL,1965). In the constitutive assumption (6.3.8) $N=2$ for $\chi^t_{\beta\mathcal{K}}$, and $N=1$ for $\mathfrak{R}^t_{\beta\mathcal{K}}$, $\bar{\bar{\theta}}^t_\beta$, $\tilde{\mathbf{i}}^t_\beta$, and $\boldsymbol{\nu}^{(\beta)t}$. The reason for retaining within the constitutive assumption the gradient of the deformation function, $GRAD\mathbf{F}^t_\beta$, will become clear in the next chapter where it will be seen that this allows for the dependence of constitutive variables on the density gradients in fluid mixtures. Without the retention of this term, the resulting mixture models may be too special to model real multiphase flows. The retention of first gradients of independent variables in the series expansion does not necessarily imply that the resulting constitutive equations will be linear, since other principles of constitutive theory will require that only certain combination of independent variables form valid expressions of these equations. This valid combination of independent variables produces, in general, nonlinear constitutive relations and, consequently, a nonlinear theory of multiphase mixtures. The constitutive assumption may also contain restrictive conditions on the local material behavior called the internal constraints. This particular aspect of the theory will be discussed further in section 6.12.

## 6.4 Principle of Smooth and Local Memory

The *principle of local memory* asserts that the thermokinetic processes at distant past do not appreciably affect the present value of the constitutive response functional $G_\alpha$. The memory of past motions, phase change processes, temperature, inertia, and gyration is thus assumed to decay rapidly. In a sense, the principle of local memory is the counterpart in the domain of time to the principle of local action in the domain of space.

*Smooth memory* is an additional assumption to the local memory as it allows an approximation of the thermokinetic variables (6.2.7) by the Taylor series expansion and an assumption that the constitutive response functional $G_\alpha$ depends smoothly on the functions occurring in this expansion. For this purpose, it may be assumed that the histories (6.2.7) can be approximated in the interval $|t-s| \leq \delta$ by a suitable number of terms in the Taylor series expansion which will be taken to be only up to the first time derivative, *i.e.*

$$\chi^t_{\beta\mathcal{K}}(\mathbf{Z}_\beta, s) \approx \chi_{\beta\mathcal{K}}(\mathbf{Z}_\beta, t) + (t-s)\dot{\chi}_{\beta\mathcal{K}}(\mathbf{Z}_\beta, t) \tag{6.4.1}$$

$$\mathfrak{R}^t_{\beta\boldsymbol{\kappa}}(\mathbf{Z}_\beta, s) \approx \mathfrak{R}_{\beta\boldsymbol{\kappa}}(\mathbf{Z}_\beta, t) + (t-s)\dot{\mathfrak{R}}_{\beta\boldsymbol{\kappa}}(\mathbf{Z}_\beta, t) \tag{6.4.2}$$

$$\bar{\bar{\theta}}^t_\beta(\mathbf{z}, s) \approx \bar{\bar{\theta}}_\beta(\mathbf{z}, t) + (t-s)\dot{\bar{\bar{\theta}}}_\beta(\mathbf{z}, t) \tag{6.4.3}$$

$$\bar{\mathbf{i}}^t_\beta(\mathbf{z}, s) \approx \bar{\mathbf{i}}_\beta(\mathbf{z}, t) + (t-s)\dot{\bar{\mathbf{i}}}_\beta(\mathbf{z}, t) \tag{6.4.4}$$

$$\boldsymbol{\nu}^{(\beta)t}(\mathbf{z}, s) \approx \boldsymbol{\nu}^{(\beta)}(\mathbf{z}, t) + (t-s)\dot{\boldsymbol{\nu}}^{(\beta)}(\mathbf{z}, t) \tag{6.4.5}$$

Substituting (6.4.1)-(6.4.5) into (6.2.6) results in an expression for the constitutive response functional obeying the smooth and local memory assumptions. Hence,

$$\begin{aligned}\Upsilon_\alpha(\mathbf{x}, t) = G_\alpha[&\chi_{\beta\boldsymbol{\kappa}}(\mathbf{Z}_\beta, t),\ \dot{\chi}_{\beta\boldsymbol{\kappa}}(\mathbf{Z}_\beta, t),\ \mathfrak{R}_{\beta\boldsymbol{\kappa}}(\mathbf{Z}_\beta, t),\ \dot{\mathfrak{R}}_{\beta\boldsymbol{\kappa}}(\mathbf{Z}_\beta, t),\\ &\bar{\bar{\theta}}_\beta(\mathbf{z}, t),\ \dot{\bar{\bar{\theta}}}_\beta(\mathbf{z}, t),\ \bar{\mathbf{i}}_\beta(\mathbf{z}, t),\ \dot{\bar{\mathbf{i}}}_\beta(\mathbf{z}, t),\ \boldsymbol{\nu}^{(\beta)}(\mathbf{z}, t),\ \dot{\boldsymbol{\nu}}^{(\beta)}(\mathbf{z}, t),\ X_\beta,\ t]\end{aligned} \tag{6.4.6}$$

## 6.5 Principles of Local Action and Smooth and Local Memory Combined

The combination of results obtained for the constitutive response functionals satisfying the principle of local action (equation (6.3.8)) and principle of smooth and local memory (equation (6.4.6)) yields the following result:

$$\begin{aligned}\Upsilon_\alpha(\mathbf{x}, t) = G_\alpha[&\chi_{\beta\boldsymbol{\kappa}}(\mathbf{X}_\beta, t),\ \dot{\chi}_{\beta\boldsymbol{\kappa}}(\mathbf{X}_\beta, t),\ \mathbf{F}_\beta(\mathbf{X}_\beta, t),\ \dot{\mathbf{F}}_\beta(\mathbf{X}_\beta, t),\\ &GRAD\mathbf{F}_\beta(\mathbf{X}_\beta, t),\ GRAD\dot{\mathbf{F}}_\beta(\mathbf{X}_\beta, t),\ \mathfrak{R}_{\beta\boldsymbol{\kappa}}(\mathbf{X}_\beta, t),\ \dot{\mathfrak{R}}_{\beta\boldsymbol{\kappa}}(\mathbf{X}_\beta, t),\\ &GRAD\mathfrak{R}_{\beta\boldsymbol{\kappa}}(\mathbf{X}_\beta, t),\ GRAD\dot{\mathfrak{R}}_{\beta\boldsymbol{\kappa}}(\mathbf{X}_\beta, t),\ \bar{\bar{\theta}}_\beta(\mathbf{x}, t),\ \dot{\bar{\bar{\theta}}}_\beta(\mathbf{x}, t),\ \boldsymbol{\nabla}\bar{\bar{\theta}}_\beta(\mathbf{x}, t),\\ &\boldsymbol{\nabla}\dot{\bar{\bar{\theta}}}_\beta(\mathbf{x}, t),\ \bar{\mathbf{i}}_\beta(\mathbf{x}, t),\ \dot{\bar{\mathbf{i}}}_\beta(\mathbf{x}, t),\ \boldsymbol{\nabla}\bar{\mathbf{i}}_\beta(\mathbf{x}, t),\ \boldsymbol{\nabla}\dot{\bar{\mathbf{i}}}_\beta(\mathbf{x}, t),\\ &\boldsymbol{\nu}^{(\beta)}(\mathbf{x}, t),\ \dot{\boldsymbol{\nu}}^{(\beta)}(\mathbf{x}, t),\ \boldsymbol{\nabla}\boldsymbol{\nu}^{(\beta)}(\mathbf{x}, t),\ \boldsymbol{\nabla}\dot{\boldsymbol{\nu}}^{(\beta)}(\mathbf{x}, t),\ d\mathbf{X}_\beta,\ t]\end{aligned} \tag{6.5.1}$$

Notice that $G_\alpha$ above is not anymore a functional, but a tensor-valued function of the corresponding arguments. The result (6.5.1) is clearly very special and sufficiently simple to study further constitutive properties of $G_\alpha$. In what follows, the *a priori* validity of the smooth and local memory in approximating $G_\alpha$ will not be assumed, however, and use will be made of the less restrictive form satisfying only the principle of local action to obtain as many results as possible before simplifying the result in section 6.10.

## 6.6 Principle of Material Frame-Indifference

The principle of material frame-indifference is discussed at length in chapter 4 where it is stated that a change of frame of reference from $\mathcal{F}$ to $\mathcal{F}^*$ of an observer observing the same event $(\mathbf{x}, \tau)$ in $\mathcal{F}$ and $(\mathbf{x}^*, \tau^*)$ in $\mathcal{F}^*$ may be represented by

$$\tau^* = \tau + a \tag{6.6.1}$$

$$\mathbf{x}^* = \boldsymbol{\chi}^*_{\beta\boldsymbol{\kappa}}(\mathbf{Z}_\beta, \tau^*) = \mathbf{Q}(\tau)\boldsymbol{\chi}_{\beta\boldsymbol{\kappa}}(\mathbf{Z}_\beta, \tau) + \mathbf{c}(\tau) \tag{6.6.2}$$

Here $\tau$ denotes the time, $a$ is a reference time, $\mathbf{c}(\tau)$ is the time-dependent translation vector, $\mathbf{Q}(\tau)$ in an orthogonal tensor satisfying (as in chapter 4) $det\mathbf{Q}(t) = +1$, and both $\mathbf{c}$ and $\mathbf{Q}$ are smooth functions of time. Such a change of frame transformation makes the two observers in reference frames $\mathcal{F}$ and $\mathcal{F}^*$ agree on the distance between a pair of points in space, time elapsed between a pair of events, the order in which two distant events occur, and the orientation in space. Frame-indifferent scalars, vectors, and second and third order tensors are required to transform according to (4.1.6)-(4.1.8), respectively. As demonstrated in chapter 4, the constitutive relations given by (6.2.5) all satisfy the principle of material frame-indifference.

In the reference frames $\mathcal{F}$ and $\mathcal{F}^*$, equation (6.2.6) may be written as follows:

$$\begin{aligned}\Upsilon_\alpha(\mathbf{x}, t) = G_\alpha[&\boldsymbol{\chi}^t_{\beta\boldsymbol{\kappa}}(\mathbf{Z}_\beta, \tau),\ \mathfrak{R}^t_{\beta\boldsymbol{\kappa}}(\mathbf{Z}_\beta, \tau),\ \bar{\bar{\theta}}^t_\beta(\mathbf{z}, \tau),\\ &\bar{\mathbf{i}}^t_\beta(\mathbf{z}, \tau),\ \boldsymbol{\nu}^{(\beta)t}(\mathbf{z}, \tau),\ X_\beta,\ t]\end{aligned} \tag{6.6.3}$$

$$\begin{aligned}\Upsilon^*_\alpha(\mathbf{x}^*, t^*) = G_\alpha[&\boldsymbol{\chi}^{*t^*}_{\beta\boldsymbol{\kappa}}(\mathbf{Z}_\beta, \tau^*),\ \mathfrak{R}^{*t^*}_{\beta\boldsymbol{\kappa}}(\mathbf{Z}_\beta, \tau^*),\ \bar{\bar{\theta}}^{*t^*}_\beta(\mathbf{z}^*, \tau^*),\\ &\bar{\mathbf{i}}^{*t^*}_\beta(\mathbf{z}^*, \tau^*),\ \boldsymbol{\nu}^{*(\beta)t^*}(\mathbf{z}^*, \tau^*),\ X_\beta,\ t^*]\end{aligned} \tag{6.6.4}$$

for $\tau \geq 0$ and $\tau^* \geq 0$. In order to determine the restrictions on the form of functional $G_\alpha$ implied by the principle of material frame-indifference, an approach used by TRUESDELL & NOLL (1965), section 26, will be followed, whereby the position, time, and orientation of a moving observer are successfully transformed away.

### 6.6.1 Rigid Translation of the Spatial Frame

By selecting $\mathbf{Q}(\tau) = \mathbf{I}$, $a = 0$, and $\mathbf{c}(\tau) = -\boldsymbol{\chi}_{\beta\boldsymbol{\kappa}}(\mathbf{X}_\beta, \tau)$ in (6.6.1) and (6.6.2) produces a rigid translation of the spatial frame of reference in which a material particle remains at the origin, *i.e.*

$$\tau^* = \tau, \qquad t^* = t \tag{6.6.5}$$

$$\chi^*_{\beta\boldsymbol{\kappa}}(\mathbf{Z}_\beta, \tau^*) = \chi_{\beta\boldsymbol{\kappa}}(\mathbf{Z}_\beta, \tau) - \chi_{\beta\boldsymbol{\kappa}}(\mathbf{X}_\beta, \tau) \tag{6.6.6}$$

From the frame-indifference of scalar variables $\mathfrak{R}^t_{\beta\boldsymbol{\kappa}}$ and $\overline{\overline{\theta}}{}^t_\beta$, tensor $\overline{\mathbf{i}}{}^t_\beta$ (*cit.* (4.3.29)), and the transformation property of the tensor $\boldsymbol{\nu}^{(\beta)t}$ (*cit.* (4.3.25)), it follows from (6.6.3) and (6.6.4) that

$$\begin{aligned}\Upsilon^*_\alpha(\mathbf{x}^*, t^*) &= G_\alpha[\chi^t_{\beta\boldsymbol{\kappa}}(\mathbf{Z}_\beta, \tau) - \chi^t_{\beta\boldsymbol{\kappa}}(\mathbf{X}_\beta, \tau),\ \mathfrak{R}^t_{\beta\boldsymbol{\kappa}}(\mathbf{Z}_\beta, \tau),\ \overline{\overline{\theta}}{}^t_\beta(\mathbf{z}, \tau),\\ &\quad \overline{\mathbf{i}}{}^t_\beta(\mathbf{z}, \tau),\ \boldsymbol{\nu}^{(\beta)t}(\mathbf{z}, \tau),\ X_\beta,\ t] = \mathbf{Q}(t)\Upsilon_\alpha(\mathbf{x}, t) = \Upsilon_\alpha(\mathbf{x}, t)\end{aligned} \tag{6.6.7}$$

### 6.6.2 Shift of Time

The shift of time is obtained by selecting $\mathbf{Q}(\tau) = \mathbf{I}$, $\mathbf{c}(\tau) = \mathbf{0}$, and $a = -t$. This makes the present time $t$ the reference time after the change of frame. Thus, from (6.6.1)-(6.6.4), (4.3.29), and (4.3.25) it is obtained

$$\tau^* = \tau - t, \qquad t^* = t - t = 0 \tag{6.6.8}$$

$$\mathbf{x}^* = \chi^*_{\beta\boldsymbol{\kappa}}(\mathbf{Z}_\beta, \tau^*) = \chi_{\beta\boldsymbol{\kappa}}(\mathbf{Z}_\beta, \tau^* + t) = \mathbf{x} \tag{6.6.9}$$

$$\Upsilon^*_\alpha(\mathbf{x}^*, 0) = \Upsilon_\alpha(\mathbf{x}, t) \tag{6.6.10}$$

But from the definition of the history of motion (6.2.7) and $(6.6.8)_1$ it follows that

$$\chi^{*t^*}_{\beta\boldsymbol{\kappa}}(\mathbf{Z}_\beta, s) = \chi^*_{\beta\boldsymbol{\kappa}}(\mathbf{Z}_\beta, t^* - s) = \chi^*_{\beta\boldsymbol{\kappa}}(\mathbf{Z}_\beta, -s) \tag{6.6.11}$$

and upon taking $s = -\tau^*$ and using $(6.6.8)_2$ and (6.6.9) this becomes

$$\begin{aligned}\chi^{*t^*}_{\beta\boldsymbol{\kappa}}(\mathbf{Z}_\beta, s) &= \chi^*_{\beta\boldsymbol{\kappa}}(\mathbf{Z}_\beta, \tau^*) = \chi_{\beta\boldsymbol{\kappa}}(\mathbf{Z}_\beta, t + \tau^*)\\ &= \chi_{\beta\boldsymbol{\kappa}}(\mathbf{Z}_\beta, t - s) = \chi^t_{\beta\boldsymbol{\kappa}}(\mathbf{Z}_\beta, s)\end{aligned} \tag{6.6.12}$$

Also,

$$\begin{aligned}\mathfrak{R}^{*t^*}_{\beta\boldsymbol{\kappa}}(\mathbf{Z}_\beta, s) &= \mathfrak{R}^*_{\beta\boldsymbol{\kappa}}(\mathbf{Z}_\beta, -s) = \mathfrak{R}^*_{\beta\boldsymbol{\kappa}}(\mathbf{Z}_\beta, \tau^*)\\ &= \mathfrak{R}_{\beta\boldsymbol{\kappa}}(\mathbf{Z}_\beta, \tau) = \mathfrak{R}^t_{\beta\boldsymbol{\kappa}}(\mathbf{Z}_\beta, s)\end{aligned} \tag{6.6.13}$$

with similar results holding for $\overline{\overline{\theta}}_\beta$, $\overline{\mathbf{i}}_\beta$, and $\boldsymbol{\nu}^{(\beta)}$. These results may now be used in (6.6.10) and (6.6.4), *i.e.*

$$\begin{aligned}\Upsilon_\alpha(\mathbf{x}, t) = \Upsilon^*_\alpha(\mathbf{x}^*, 0) &= G_\alpha[\chi^t_{\beta\boldsymbol{\kappa}}(\mathbf{Z}_\beta, s),\ \mathfrak{R}^t_{\beta\boldsymbol{\kappa}}(\mathbf{Z}_\beta, s),\\ &\quad \overline{\overline{\theta}}{}^t_\beta(\mathbf{z}, s),\ \overline{\mathbf{i}}{}^t_\beta(\mathbf{z}, s),\ \boldsymbol{\nu}^{(\beta)t}(\mathbf{z}, s),\ X_\beta,\ 0]\end{aligned} \tag{6.6.14}$$

from where it follows that $\Upsilon_\alpha(\mathbf{x},t)$ does not depend explicitly on time $t$.

Combining the results (6.3.8), (6.6.7), and (6.6.14) gives

$$\begin{aligned}\Upsilon_\alpha(\mathbf{x},t) = G_\alpha[&\mathbf{F}^t_\beta(\mathbf{X}_\beta,s),\ GRAD\mathbf{F}^t_\beta(\mathbf{X}_\beta,s),\ \mathfrak{R}^t_{\beta\boldsymbol{\kappa}}(\mathbf{X}_\beta,s),\\ &GRAD\mathfrak{R}^t_{\beta\boldsymbol{\kappa}}(\mathbf{X}_\beta,s),\ \bar{\bar{\theta}}^t_\beta(\mathbf{x},s),\ \nabla\bar{\bar{\theta}}^t_\beta(\mathbf{x},s),\ \bar{\mathbf{i}}^t_\beta(\mathbf{x},s),\ \nabla\bar{\mathbf{i}}^t_\beta(\mathbf{x},s),\\ &\boldsymbol{\nu}^{(\beta)t}(\mathbf{x},s),\ \nabla\boldsymbol{\nu}^{(\beta)t}(\mathbf{x},s),\ d\mathbf{X}_\beta]\end{aligned} \tag{6.6.15}$$

from where it is evident that the dependence of $G_\alpha$ on $\chi^t_{\beta\boldsymbol{\kappa}}(\mathbf{Z}_\beta,s)$ and $t$ has been eliminated.

### 6.6.3 Rigid Rotation of the Spatial Frame

The rigid rotation of the spatial frame of reference is accomplished by taking $a=0$, $\mathbf{c}(\tau)=\mathbf{0}$, and $\mathbf{Q}(t)$ arbitrary in (6.6.1) and (6.6.2). Thus,

$$\tau^* = \tau, \qquad t^* = t \tag{6.6.16}$$

$$\chi^*_{\beta\boldsymbol{\kappa}}(\mathbf{Z}_\beta,\tau^*) = \mathbf{Q}(\tau)\chi_{\beta\boldsymbol{\kappa}}(\mathbf{Z}_\beta,\tau) \tag{6.6.17}$$

and if $\Upsilon_\alpha$ is a vector, for example, (4.1.6) gives that

$$\Upsilon^*_\alpha(\mathbf{x}^*,t^*) = \mathbf{Q}(t)\Upsilon_\alpha(\mathbf{x},t) \tag{6.6.18}$$

Using these results, (6.6.14), (6.6.4), and the change of frame transformation properties of $\bar{\bar{\theta}}_\beta$, $\bar{\mathbf{i}}_\beta$, and $\boldsymbol{\nu}^{(\beta)}$ gives

$$\begin{aligned}\Upsilon^*_\alpha(\mathbf{x}^*,t^*) &= \mathbf{Q}(t)\,G_\alpha[\chi^t_{\beta\boldsymbol{\kappa}}(\mathbf{Z}_\beta,s),\ \mathfrak{R}^t_{\beta\boldsymbol{\kappa}}(\mathbf{Z}_\beta,s),\\ &\qquad \bar{\bar{\theta}}^t_\beta(\mathbf{z},s),\ \bar{\mathbf{i}}^t_\beta(\mathbf{z},s),\ \boldsymbol{\nu}^{(\beta)t}(\mathbf{z},s),\ X_\beta]\\ &= G_\alpha[\mathbf{Q}^t(s)\,\chi^t_{\beta\boldsymbol{\kappa}}(\mathbf{Z}_\beta,s),\ \mathfrak{R}^t_{\beta\boldsymbol{\kappa}}(\mathbf{Z}_\beta,s),\ \bar{\bar{\theta}}^t_\beta(\mathbf{z},s),\ \mathbf{Q}^t(s)\,\bar{\mathbf{i}}^t_\beta(\mathbf{z},s)\,\mathbf{Q}^t(s)^T,\\ &\qquad \mathbf{Q}^t(s)\,\boldsymbol{\nu}^{(\beta)t}(\mathbf{z},s)\,\mathbf{Q}^t(s)^T + \dot{\mathbf{Q}}^t(s)\,\mathbf{Q}^t(s)^T,\ X_\beta]\end{aligned} \tag{6.6.19}$$

where $\mathbf{Q}^t(s)$ is the history of $\mathbf{Q}$, and $\mathbf{Q}(t) = \mathbf{Q}^t(0)$. Choosing in this equation a necessary condition that $\mathbf{Q}^t(s) = \mathbf{I}$ and $d\mathbf{Q}^t(s)/dt = -\nabla\tilde{\mathbf{v}}^t_\beta(\mathbf{z},s)$, $\beta=1,\ldots,\gamma$, gives

$$\begin{aligned}\Upsilon^*_\alpha(\mathbf{x}^*,t^*) &= \mathbf{Q}(t)\Upsilon_\alpha(\mathbf{x},t) = \mathbf{Q}(t)G_\alpha[\chi^t_{\beta\boldsymbol{\kappa}}(\mathbf{Z}_\beta,s),\\ &\qquad \mathfrak{R}^t_{\beta\boldsymbol{\kappa}}(\mathbf{Z}_\beta,s),\ \bar{\bar{\theta}}^t_\beta(\mathbf{z},s),\ \bar{\mathbf{i}}^t_\beta(\mathbf{z},s),\ \boldsymbol{\nu}^{(\beta)t}(\mathbf{z},s),\ X_\beta]\\ &= G_\alpha[\mathbf{Q}^t(s)\chi^t_{\beta\boldsymbol{\kappa}}(\mathbf{Z}_\beta,s),\ \mathfrak{R}^t_{\beta\boldsymbol{\kappa}}(\mathbf{Z}_\beta,s),\ \bar{\bar{\theta}}^t_\beta(\mathbf{z},s),\ \mathbf{Q}^t(s)\bar{\mathbf{i}}^t_\beta(\mathbf{z},s)\mathbf{Q}^t(s)^T,\\ &\qquad \mathbf{Q}^t(s)(\boldsymbol{\nu}^{(\beta)t}(\mathbf{z},s) - \nabla\tilde{\mathbf{v}}^t_\beta(\mathbf{z},s))\mathbf{Q}^t(s)^T,\ X_\beta]\end{aligned} \tag{6.6.20}$$

Equation (6.6.20) can now be combined with the result (6.6.15) to produce the following result:

$$
\begin{gathered}
G_{\alpha\kappa}[\mathbf{Q}^t(s)\mathbf{F}^t_\beta(\mathbf{X}_\beta,s),\ \mathbf{Q}^t(s)GRAD\mathbf{F}^t_\beta(\mathbf{X}_\beta,s),\ \Re^t_{\beta\kappa}(\mathbf{X}_\beta,s),\\
GRAD\Re^t_{\beta\kappa}(\mathbf{X}_\beta,s),\ \bar{\bar{\theta}}^t_\beta(\mathbf{x},s),\ \mathbf{Q}^t(s)\nabla\bar{\bar{\theta}}^t_\beta(\mathbf{x},s),\ \mathbf{Q}^t(s)\bar{\mathbf{i}}^t_\beta(\mathbf{x},s)\mathbf{Q}^t(s)^T,\\
\mathbf{Q}^t(s)\nabla\bar{\mathbf{i}}^t_\beta(\mathbf{x},s)\mathbf{Q}^t(s)^T\mathbf{Q}^t(s)^T,\ \mathbf{Q}^t(s)\mathbf{b}^{(\beta)t}(\mathbf{x},s)\mathbf{Q}^t(s)^T,\\
\mathbf{Q}^t(s)\nabla\boldsymbol{\nu}^{(\beta)t}(\mathbf{x},s)\mathbf{Q}^t(s)^T\mathbf{Q}^t(s)^T]\\
=\mathbf{Q}(t)G_{\alpha\kappa}[\mathbf{F}^t_\beta(\mathbf{X}_\beta,s),\ GRAD\mathbf{F}^t_\beta(\mathbf{X}_\beta,s),\ \Re^t_{\beta\kappa}(\mathbf{X}_\beta,s),\ \nabla\boldsymbol{\nu}^{(\beta)t}(\mathbf{x},s),\\
GRAD\Re^t_{\beta\kappa}(\mathbf{X}_\beta,s),\ \bar{\bar{\theta}}^t_\beta(\mathbf{x},s),\ \nabla\bar{\bar{\theta}}^t_\beta(\mathbf{x},s),\ \bar{\mathbf{i}}^t_\beta(\mathbf{x},s),\ \nabla\bar{\mathbf{i}}^t_\beta(\mathbf{x},s),\ \mathbf{b}^{(\beta)t}(\mathbf{x},s)]
\end{gathered}
\tag{6.6.21}
$$

where use is made of the definition of $\mathbf{b}^{(\beta)}$ given by (4.3.26) and the fact that $\mathbf{F}^*_\beta = \mathbf{QF}_\beta$ and $F^*_{\beta kK} = Q_{ki}F_{\beta iK}$ ($GRAD\mathbf{F}^*_\beta = \mathbf{Q}GRAD\mathbf{F}_\beta$). The dependence of $G_\alpha$ on $d\mathbf{X}_\beta$ appearing in (6.6.15) is represented implicitly in the above equation by $G_{\alpha\kappa}$, where the subscript $\alpha\kappa$ denotes the reference configuration spanned by $d\mathbf{X}_\beta$. Notice in (6.6.21) that $\nabla\bar{\mathbf{i}}_\beta$, $\mathbf{b}^{(\beta)}$, and $\nabla\boldsymbol{\nu}^{(\beta)}$ are frame-indifferent quantities as proved in section 4.3.

It is useful to derive an equivalent form of (6.6.21) when the principles of local action and smooth and local memory are combined. To derive this result we can start from (6.5.1) by expressing this equation in the frame $\mathcal{F}^*$ and, to simplify analysis, ignore the terms $GRAD\grave{\mathbf{F}}_\beta$, $GRAD\grave{\Re}_{\beta\kappa}$, $\nabla\grave{\bar{\bar{\theta}}}_\beta$, $\grave{\boldsymbol{\nu}}^{(\beta)}$, $\nabla\grave{\bar{\mathbf{i}}}_\beta$, and $\nabla\grave{\boldsymbol{\nu}}^{(\beta)}$, *i.e.*

$$
\begin{gathered}
\Upsilon^*_\alpha(\mathbf{x}^*,t^*) = G_\alpha[\chi^*_{\beta\kappa}(\mathbf{X}_\beta,t^*),\ \grave{\chi}^*_{\beta\kappa}(\mathbf{X}_\beta,t^*),\ \mathbf{F}^*_\beta(\mathbf{X}_\beta,t^*),\ \grave{\mathbf{F}}^*_\beta(\mathbf{X}_\beta,t^*),\\
GRAD\mathbf{F}^*_\beta(\mathbf{X}_\beta,t^*),\ \Re^*_{\beta\kappa}(\mathbf{X}_\beta,t^*),\ \grave{\Re}^*_{\beta\kappa}(\mathbf{X}_\beta,t^*),\ GRAD\Re^*_{\beta\kappa}(\mathbf{X}_\beta,t^*),\\
\bar{\bar{\theta}}^*_\beta(\mathbf{x}^*,t^*),\ \grave{\bar{\bar{\theta}}}^*_\beta(\mathbf{x}^*,t^*),\ \nabla^*\bar{\bar{\theta}}^*_\beta(\mathbf{x}^*,t^*),\ \bar{\mathbf{i}}^*_\beta(\mathbf{x}^*,t^*),\ \grave{\bar{\mathbf{i}}}^*_\beta(\mathbf{x}^*,t^*),\\
\nabla^*\bar{\mathbf{i}}^*_\beta(\mathbf{x}^*,t^*),\ \boldsymbol{\nu}^{*(\beta)}(\mathbf{x}^*,t^*),\ \nabla^*\boldsymbol{\nu}^{*(\beta)}(\mathbf{x}^*,t^*),\ d\mathbf{X}_\beta,\ t^*]
\end{gathered}
\tag{6.6.22}
$$

Choosing the necessary conditions $\mathbf{Q}(t)=\mathbf{I}$, $\dot{\mathbf{Q}}=\mathbf{0}$, $\ddot{\mathbf{Q}}=\mathbf{0}$, $a=0$, and $\mathbf{c}(t)=-\chi_{\gamma\kappa}(\mathbf{X}_\gamma,t)$ in (6.6.1) and (6.6.2), and using the following change of frame transformation properties of multiphase variables as developed in section 4.3

$$
\mathbf{F}^*_\beta = \mathbf{QF}_\beta
$$

$$
\grave{\mathbf{F}}^*_\beta = \mathbf{L}^*_\beta\mathbf{F}^*_\beta = (\mathbf{QD}_\beta\mathbf{Q}^T + \mathbf{QW}_\beta\mathbf{Q}^T + \dot{\mathbf{Q}}\mathbf{Q}^T)\mathbf{QF}_\beta = \mathbf{QL}_\beta\mathbf{F}_\beta + \dot{\mathbf{Q}}\mathbf{F}_\beta
$$

$$\mathfrak{R}^*_{\beta\boldsymbol{\kappa}} = \mathfrak{R}_{\beta\boldsymbol{\kappa}}, \quad \grave{\mathfrak{R}}^*_{\beta\boldsymbol{\kappa}} = \grave{\mathfrak{R}}_{\beta\boldsymbol{\kappa}}, \quad \nabla^*\bar{\bar{\theta}}^*_\beta = \mathbf{Q}\nabla\bar{\bar{\theta}}_\beta, \quad \bar{\mathbf{i}}^*_\beta = \mathbf{Q}\bar{\mathbf{i}}_\beta\mathbf{Q}^T \tag{6.6.23}$$

$$\grave{\bar{\mathbf{i}}}^*_\beta = \mathbf{Q}\grave{\bar{\mathbf{i}}}_\beta\mathbf{Q}^T + (\dot{\mathbf{Q}}\bar{\mathbf{i}}_\beta\mathbf{Q}^T + \mathbf{Q}\bar{\mathbf{i}}_\beta\dot{\mathbf{Q}}^T), \quad \boldsymbol{\nu}^{*(\beta)} = \mathbf{Q}\boldsymbol{\nu}^{(\beta)}\mathbf{Q}^T + \dot{\mathbf{Q}}\mathbf{Q}^T$$

$$\nabla^*\bar{\mathbf{i}}^*_\beta = \mathbf{Q}\nabla\bar{\mathbf{i}}_\beta\mathbf{Q}^T\mathbf{Q}^T, \quad \nabla^*\boldsymbol{\nu}^{*(\beta)} = \mathbf{Q}\boldsymbol{\nu}^{(\beta)}\mathbf{Q}^T\mathbf{Q}^T$$

it follows from (6.6.22) that

$$\begin{aligned} \Upsilon^*_\alpha(\mathbf{x}^*,t^*) = G_\alpha[&\tilde{\mathbf{v}}_\beta - \tilde{\mathbf{v}}_\gamma,\ \mathbf{F}_\beta,\ \grave{\mathbf{F}}_\beta,\ GRAD\mathbf{F}_\beta,\ \mathfrak{R}_{\beta\boldsymbol{\kappa}},\ \grave{\mathfrak{R}}_{\beta\boldsymbol{\kappa}},\\ &GRAD\mathfrak{R}_{\beta\boldsymbol{\kappa}},\ \bar{\bar{\theta}}_\beta,\ \grave{\bar{\bar{\theta}}}_\beta,\ \nabla\bar{\bar{\theta}}_\beta,\ \bar{\mathbf{i}}_\beta,\ \grave{\bar{\mathbf{i}}}_\beta,\ \nabla\bar{\mathbf{i}}_\beta,\ \boldsymbol{\nu}^{(\beta)},\ \nabla\boldsymbol{\nu}^{(\beta)},\ d\mathbf{X}_\beta,\ t]\\ &= \mathbf{Q}\Upsilon_\alpha(\mathbf{x},t) = \Upsilon_\alpha(\mathbf{x},t) \end{aligned} \tag{6.6.24}$$

Selecting now a shift of time $a=-t$, $\mathbf{Q}(t)=\mathbf{I}$, and $\mathbf{c}=\mathbf{0}$ it follows from the above equation that the explicit dependence of $G_\alpha$ on time is eliminated, *i.e.*

$$\begin{aligned} \Upsilon^*_\alpha(\mathbf{x}^*,0) = G_\alpha[&\tilde{\mathbf{v}}_\beta - \tilde{\mathbf{v}}_\gamma,\ \mathbf{F}_\beta,\ \grave{\mathbf{F}}_\beta,\ GRAD\mathbf{F}_\beta,\ \mathfrak{R}_{\beta\boldsymbol{\kappa}},\ \grave{\mathfrak{R}}_{\beta\boldsymbol{\kappa}},\\ &GRAD\mathfrak{R}_{\beta\boldsymbol{\kappa}},\ \bar{\bar{\theta}}_\beta,\ \grave{\bar{\bar{\theta}}}_\beta,\ \nabla\bar{\bar{\theta}}_\beta,\ \bar{\mathbf{i}}_\beta,\ \grave{\bar{\mathbf{i}}}_\beta,\ \nabla\bar{\mathbf{i}}_\beta,\ \boldsymbol{\nu}^{(\beta)},\ \nabla\boldsymbol{\nu}^{(\beta)},\ d\mathbf{X}_\beta]\\ &= \mathbf{Q}\Upsilon_\alpha(\mathbf{x},t) = \Upsilon_\alpha(\mathbf{x},t) \end{aligned} \tag{6.6.25}$$

Requiring again that (6.6.25) holds in the frame $\mathcal{F}^*$ and using the change of frame transformation properties (6.6.23) with another set of necessary conditions $\mathbf{Q}=\mathbf{I}$ and $d\mathbf{Q}/dt=-\mathbf{W}_\gamma$ it follows that

$$\begin{aligned} \Upsilon^*_\alpha(\mathbf{x}^*,t^*) = G_\alpha[&\tilde{\mathbf{v}}_\beta - \tilde{\mathbf{v}}_\gamma,\ \mathbf{F}_\beta,\ \mathbf{D}_\beta + (\mathbf{W}_\beta - \mathbf{W}_\gamma),\ GRAD\mathbf{F}_\beta,\\ &\mathfrak{R}_{\beta\boldsymbol{\kappa}},\ \grave{\mathfrak{R}}_{\beta\boldsymbol{\kappa}},\ GRAD\mathfrak{R}_{\beta\boldsymbol{\kappa}},\ \bar{\bar{\theta}}_\beta,\ \grave{\bar{\bar{\theta}}}_\beta,\ \nabla\bar{\bar{\theta}}_\beta,\ \bar{\mathbf{i}}_\beta,\ \grave{\bar{\mathbf{i}}}_\beta - \mathbf{W}_\gamma\bar{\mathbf{i}}_\beta - \bar{\mathbf{i}}_\beta\mathbf{W}^T_\gamma,\ \nabla\bar{\mathbf{i}}_\beta,\\ &\boldsymbol{\nu}^{(\beta)} - \mathbf{W}_\beta,\ \nabla\boldsymbol{\nu}^{(\beta)},\ d\mathbf{X}_\beta] = \mathbf{Q}\Upsilon_\alpha(\mathbf{x},t) = \Upsilon_\alpha(\mathbf{x},t) \end{aligned} \tag{6.6.26}$$

Notice in the above equation that

$$\boldsymbol{\nu}^{(\beta)} - \mathbf{W}_\gamma = \mathbf{b}^{(\beta)} + (\mathbf{W}_\beta - \mathbf{W}_\gamma) + \mathbf{D}_\beta$$

$$\grave{\bar{\mathbf{i}}}_\beta - \mathbf{W}_\gamma\bar{\mathbf{i}}_\beta - \bar{\mathbf{i}}_\beta\mathbf{W}^T_\gamma = \grave{\bar{\mathbf{i}}}_\beta - \mathbf{W}_\beta\bar{\mathbf{i}}_\beta - \bar{\mathbf{i}}_\beta\mathbf{W}^T_\beta + (\mathbf{W}_\beta - \mathbf{W}_\gamma)\bar{\mathbf{i}}_\beta + \bar{\mathbf{i}}_\beta(\mathbf{W}_\beta - \mathbf{W}_\gamma)^T$$

which implies that (6.6.26) can be written in the following *frame-indifferent form*:

$$\begin{aligned} \Upsilon_\alpha(\mathbf{x},t) = G_{\alpha\boldsymbol{\kappa}}[&\tilde{\mathbf{v}}_\beta - \tilde{\mathbf{v}}_\gamma,\ \mathbf{F}_\beta,\ \mathbf{D}_\beta,\ \mathbf{W}_\beta - \mathbf{W}_\gamma,\ GRAD\mathbf{F}_\beta,\ \mathfrak{R}_{\beta\boldsymbol{\kappa}},\ \grave{\mathfrak{R}}_{\beta\boldsymbol{\kappa}},\\ &GRAD\mathfrak{R}_{\beta\boldsymbol{\kappa}},\ \bar{\bar{\theta}}_\beta,\ \grave{\bar{\bar{\theta}}}_\beta,\ \nabla\bar{\bar{\theta}}_\beta,\ \bar{\mathbf{i}}_\beta,\ \grave{\bar{\mathbf{i}}}_\beta - \mathbf{W}_\beta\bar{\mathbf{i}}_\beta - \bar{\mathbf{i}}_\beta\mathbf{W}^T_\beta,\ \nabla\bar{\mathbf{i}}_\beta,\ \mathbf{b}^{(\beta)},\ \nabla\boldsymbol{\nu}^{(\beta)}] \end{aligned} \tag{6.6.27}$$

The necessary conditions considered above are also sufficient for (6.6.27) to hold for all $\mathbf{Q}\mathbf{Q}^T = \mathbf{I}$ and $\mathbf{Q}\dot{\mathbf{Q}}^T = -(\mathbf{Q}\dot{\mathbf{Q}}^T)^T$. To prove this, require (6.6.27) to hold in the frame $\mathcal{F}^*$, use the change of frame transformation properties of the independent variables, and by substituting $\mathbf{Q}(t)=\mathbf{I}$ show that (6.6.27) is recovered or remains frame-indifferent. Thus, for $G_{\alpha\boldsymbol{\kappa}}$ a vector, the equivalent form of (6.6.21) that includes the principle of local action and smooth and local memory is

$$\begin{aligned} &G_{\alpha\boldsymbol{\kappa}}[\mathbf{Q}(t)(\tilde{\mathbf{v}}_\beta(\mathbf{x},t)-\tilde{\mathbf{v}}_\gamma(\mathbf{x},t)),\ \mathbf{Q}(t)\mathbf{F}_\beta(\mathbf{X}_\beta,t),\ \mathbf{Q}(t)\mathbf{D}_\beta(\mathbf{x},t)\mathbf{Q}(t)^T,\\ &\mathbf{Q}(t)(\mathbf{W}_\beta(\mathbf{x},t)-\mathbf{W}_\gamma(\mathbf{x},t))\mathbf{Q}(t)^T,\ GRAD\mathbf{Q}(t)\mathbf{F}_\beta(\mathbf{X}_\beta,t),\ \mathfrak{R}_{\beta\boldsymbol{\kappa}}(\mathbf{X}_\beta,t),\\ &\grave{\mathfrak{R}}_{\beta\boldsymbol{\kappa}}(\mathbf{X}_\beta,t),\ GRAD\mathfrak{R}_{\beta\boldsymbol{\kappa}}(\mathbf{X}_\beta,t),\ \bar{\bar{\theta}}_\beta(\mathbf{x},t),\ \grave{\bar{\bar{\theta}}}_\beta(\mathbf{x},t),\ \mathbf{Q}(t)\nabla\bar{\bar{\theta}}_\beta(\mathbf{x},t),\\ &\mathbf{Q}(t)(\grave{\bar{\mathbf{i}}}_\beta(\mathbf{x},t)-\mathbf{W}_\beta(\mathbf{x},t)\bar{\mathbf{i}}_\beta(\mathbf{x},t)-\bar{\mathbf{i}}_\beta(\mathbf{x},t)\mathbf{W}_\beta^T(\mathbf{x},t))\mathbf{Q}(t)^T,\\ &\mathbf{Q}(t)\bar{\mathbf{i}}_\beta(\mathbf{x},t)\mathbf{Q}(t)^T,\ \mathbf{Q}(t)\nabla\bar{\mathbf{i}}_\beta(\mathbf{x},t)\mathbf{Q}(t)^T\mathbf{Q}(t)^T,\\ &\mathbf{Q}(t)\mathbf{b}^{(\beta)}(\mathbf{x},t)\mathbf{Q}(t)^T,\ \mathbf{Q}(t)\nabla\boldsymbol{\nu}^{(\beta)}(\mathbf{x},t)\mathbf{Q}(t)^T\mathbf{Q}(t)^T]\\ &=\mathbf{Q}(t)G_{\alpha\boldsymbol{\kappa}}[\tilde{\mathbf{v}}_\beta(\mathbf{x},t)-\tilde{\mathbf{v}}_\gamma(\mathbf{x},t),\ \mathbf{F}_\beta(\mathbf{X}_\beta,t),\ \mathbf{D}_\beta(\mathbf{x},t),\ \mathbf{W}_\beta(\mathbf{x},t)-\mathbf{W}_\gamma(\mathbf{x},t),\\ &GRAD\mathbf{F}_\beta(\mathbf{X}_\beta,t),\ \mathfrak{R}_{\beta\boldsymbol{\kappa}}(\mathbf{X}_\beta,t),\ \grave{\mathfrak{R}}_{\beta\boldsymbol{\kappa}}(\mathbf{X}_\beta,t),\ GRAD\mathfrak{R}_{\beta\boldsymbol{\kappa}}(\mathbf{X}_\beta,t),\\ &\bar{\bar{\theta}}_\beta(\mathbf{x},t),\ \grave{\bar{\bar{\theta}}}_\beta(\mathbf{x},t),\ \nabla\bar{\bar{\theta}}_\beta(\mathbf{x},t),\ \bar{\mathbf{i}}_\beta(\mathbf{x},t),\ \nabla\bar{\mathbf{i}}_\beta(\mathbf{x},t),\\ &\grave{\bar{\mathbf{i}}}_\beta(\mathbf{x},t)-\mathbf{W}_\beta(\mathbf{x},t)\bar{\mathbf{i}}_\beta(\mathbf{x},t)-\bar{\mathbf{i}}_\beta(\mathbf{x},t)\mathbf{W}_\beta(\mathbf{x},t)^T,\ \mathbf{b}^{(\beta)}(\mathbf{x},t),\ \nabla\boldsymbol{\nu}^{(\beta)}(\mathbf{x},t)] \end{aligned} \tag{6.6.28}$$

The important observations that need to be made regarding the above result are that the constitutive response functional $G_{\alpha\boldsymbol{\kappa}}$ depends on $\gamma-1$ velocity differences $(\tilde{\mathbf{v}}_\beta - \tilde{\mathbf{v}}_\gamma)$ and spin tensors $(\mathbf{W}_\beta - \mathbf{W}_\gamma)$, and that the structural properties of the mixture involve the inertia and gyration tensors and their gradients. Notice that if $\bar{\mathbf{i}}_\beta$ are isotropic (*cit.* (5.1.9)) then $G_{\alpha\boldsymbol{\kappa}}$ would depend on the frame-indifferent quantity $\grave{\bar{\mathbf{i}}}_\beta$.

## 6.7 Material Isomorphisms

The word "isomorphism" relates to structure as the root "morph" indicates, and "isomorphism" is equivalent to "having the same structure." A multiphase mixture may contain materials with different properties, such as "solids" and "fluids" which may or may not have the structural ordering of material points. The structural ordering of material pertains to the physical arrangement of atoms which give rise to certain crystallographic symmetries that should be accounted by the constitutive relations, since these equations must be form invariant with respect to the material symmetry transformations. In developing the concepts underlying the symmetry properties of multiphase material mixtures, use will be made of some of the basic concepts of single phase, single and multicomponent material theories as may be found in NOLL (1958), TRUESDELL & NOLL (1965), and CROSS (1973).

Let $X_{\delta 1}$ and $X_{\delta 2}$ be two different material points of phase $\delta$ with reference configurations $\boldsymbol{\kappa}_{\delta 1}$ and $\boldsymbol{\kappa}_{\delta 2}$, *i.e.* $\mathbf{X}_{\delta 1} = \boldsymbol{\kappa}_{\delta 1}(X_{\delta 1}, t)$ and $\mathbf{X}_{\delta 2} = \boldsymbol{\kappa}_{\delta 2}(X_{\delta 2}, t)$. If for any two neighborhoods $N(X_{\delta 1})$ and $N(X_{\delta 2})$ there exist reference configurations $\boldsymbol{\kappa}_{\delta 1}$ and $\boldsymbol{\kappa}_{\delta 2}$, respectively, in which the deformation, temperature and phase change histories give rise to exactly the same constitutive response functionals of phase $\alpha$ with respect to these phase $\delta$ particles, *i.e.*

$$G_{\alpha 1 \boldsymbol{\kappa}} = G_{\alpha 2 \boldsymbol{\kappa}} \tag{6.7.1}$$

then the two particles or material points of phase $\delta$ will be said to be *materially isomorphic.* Under these circumstances, no experimental measurement of $G_{\alpha \boldsymbol{\kappa}}$, when determined by the deformation, temperature and phase change, can detect whether a measurement started with $N(X_{\delta 1})$ or $N(X_{\delta 2})$. Expressed in different words, the above defined material isomorphism refers to the statement that the mechanical, thermal, and chemical properties of the two particles of phase $\delta$ are the same and that these particles belong to the same material.

When all material points of a phase are materially isomorphic to one another, then every neighborhood has the same properties as every other neighborhood and it may be said that the material of the phase is *uniform.* Uniform material phases may be, therefore, endowed with defects and dislocations, since uniformity of a phase places no restriction of choosing the same reference configuration for different particles in the same phase. When this choice is possible, however, it will be said that the phase is *homogeneous*, since then the isomorphism of all neighborhoods of a uniform phase may be demonstrated by the use of a single reference configuration. In a homogeneous phase $\delta$ every particle of the phase responds in just the same way as every other particle to mechanical, thermal, and phase change histories with respect to this configuration, and $G_{\alpha \boldsymbol{\kappa}}$ is independent of $\mathbf{X}_\delta$. Thus, a homogeneous phase is uniform whereas the reverse condition is not necessarily true.

Using (6.6.21) and (6.7.1) it follows that

$$G_{\alpha 1 \boldsymbol{\kappa}}[\mathbf{F}_1^t(\mathbf{X}_1, s), \ldots, \mathbf{F}_\delta^t(\mathbf{X}_{\delta 1}, s), \ldots, \mathbf{F}_\gamma^t(\mathbf{X}_\gamma, s), \; GRAD\mathbf{F}_1^t(\mathbf{X}_1, s), \ldots,$$
$$GRAD\mathbf{F}_\delta^t(\mathbf{X}_{\delta 1}, s), \ldots, GRAD\mathbf{F}_\gamma^t(\mathbf{X}_\gamma, s), \; \mathfrak{R}_{1\boldsymbol{\kappa}}^t(\mathbf{X}_1, s), \ldots,$$

$$
\begin{aligned}
&\mathfrak{R}^t_{\delta\boldsymbol{\kappa}}(\mathbf{X}_{\delta 1}, s), \ldots, \mathfrak{R}^t_{\gamma\boldsymbol{\kappa}}(\mathbf{X}_\gamma, s), GRAD\mathfrak{R}^t_{1\boldsymbol{\kappa}}(\mathbf{X}_1, s), \ldots, \\
&\quad GRAD\mathfrak{R}^t_{\delta\boldsymbol{\kappa}}(\mathbf{X}_{\delta 1}, s), \ldots, GRAD\mathfrak{R}^t_{\gamma\boldsymbol{\kappa}}(\mathbf{X}_\gamma, s), \\
&\quad \overline{\bar{\theta}}^t_\beta(\mathbf{x}, s), \nabla\overline{\bar{\theta}}^t_\beta(\mathbf{x}, s), \bar{\mathbf{i}}^t_\beta(\mathbf{x}, s), \nabla\bar{\mathbf{i}}^t_\beta(\mathbf{x}, s), \mathbf{b}^{(\beta)t}(\mathbf{x}, s), \nabla\boldsymbol{\nu}^{(\beta)t}(\mathbf{x}, s)] \\
= &\, G_{\alpha 2\boldsymbol{\kappa}}[\mathbf{F}^t_1(\mathbf{X}_1, s), \ldots, \mathbf{F}^t_\delta(\mathbf{X}_{\delta 2}, s), \ldots, \mathbf{F}^t_\gamma(\mathbf{X}_\gamma, s), GRAD\mathbf{F}^t_1(\mathbf{X}_1, s), \ldots, \\
&\quad GRAD\mathbf{F}^t_\delta(\mathbf{X}_{\delta 2}, s), \ldots, GRAD\mathbf{F}^t_\gamma(\mathbf{X}_\gamma, s), \mathfrak{R}^t_{1\boldsymbol{\kappa}}(\mathbf{X}_1, s), \ldots, \\
&\quad \mathfrak{R}^t_{\delta\boldsymbol{\kappa}}(\mathbf{X}_{\delta 2}, s), \ldots, \mathfrak{R}^t_{\gamma\boldsymbol{\kappa}}(\mathbf{X}_\gamma, s), GRAD\mathfrak{R}^t_{1\boldsymbol{\kappa}}(\mathbf{X}_1, s), \ldots, \\
&\quad GRAD\mathfrak{R}^t_{\delta\boldsymbol{\kappa}}(\mathbf{X}_{\delta 2}, s), \ldots, GRAD\mathfrak{R}^t_{\gamma\boldsymbol{\kappa}}(\mathbf{X}_\gamma, s), \\
&\quad \overline{\bar{\theta}}^t_\beta(\mathbf{x}, s), \nabla\overline{\bar{\theta}}^t_\beta(\mathbf{x}, s), \bar{\mathbf{i}}^t_\beta(\mathbf{x}, s), \nabla\bar{\mathbf{i}}^t_\beta(\mathbf{x}, s), \mathbf{b}^{(\beta)t}(\mathbf{x}, s), \nabla\boldsymbol{\nu}^{(\beta)t}(\mathbf{x}, s)]
\end{aligned}
\tag{6.7.2}
$$

where again the notation used implies that for each phase $\alpha$, $\alpha = 1, \ldots, \gamma$, the index $\beta$ ranges from 1 to $\gamma$. Introducing now the mapping $\lambda_\delta$ defined by

$$
\mathbf{X}_{\delta 2} = \lambda_\delta(\mathbf{X}_{\delta 1}), \quad \mathbf{P}_\delta = GRAD\lambda_\delta, \quad \mathbf{P}_\delta^{-1} = \mathbf{H}_\delta \tag{6.7.3}
$$

as a mapping from $\boldsymbol{\kappa}_{\delta 1}$ into $\boldsymbol{\kappa}_{\delta 2}$ and using (2.4.5) it follows that

$$
F_{\delta kJ}(\mathbf{X}_{\delta 1}, \tau) = \frac{\partial x_k}{\partial X_{\delta 1 J}} = \frac{\partial x_k}{\partial X_{\delta 2 M}} \frac{\partial X_{\delta 2 M}}{\partial X_{\delta 1 J}} = F_{\delta k M}(\mathbf{X}_{\delta 2}, \tau) P_{\delta M J}
$$

or

$$
\mathbf{F}^t_\delta(\mathbf{X}_{\delta 1}, s) = \mathbf{F}^t_\delta(\mathbf{X}_{\delta 2}, s)\, \mathbf{P}_\delta \tag{6.7.4}
$$

$$
\mathbf{F}^t_\delta(\mathbf{X}_{\delta 2}, s) = \mathbf{F}^t_\delta(\mathbf{X}_{\delta 1}, s)\, \mathbf{H}_\delta \tag{6.7.5}
$$

and

$$
F_{\delta kL,U}(\mathbf{X}_{\delta 2}, \tau) = F_{\delta kJ,M}(\mathbf{X}_{\delta 1}, \tau) H_{\delta JL} H_{\delta MU} + F_{\delta kJ}(\mathbf{X}_{\delta 1}, \tau) H_{\delta JL,U}
$$

or

$$
GRAD\mathbf{F}^t_\delta(\mathbf{X}_{\delta 2}, s) = (GRAD\mathbf{F}^t_\delta(\mathbf{X}_{\delta 1}, s))\mathbf{H}_\delta\mathbf{H}_\delta + (\mathbf{F}^t_\delta(\mathbf{X}_{\delta 1}, s))GRAD\mathbf{H}_\delta \tag{6.7.6}
$$

Moreover, from (6.2.3) and (6.7.5) it also follows that

$$
\begin{aligned}
\mathfrak{R}^t_{\delta\boldsymbol{\kappa}}(\mathbf{X}_{\delta 2}, s) &= \bar{\rho}^t_\delta(\mathbf{x}, s)\, |det\mathbf{F}^t_\delta(\mathbf{X}_{\delta 2}, s)| \\
&= \bar{\rho}^t_\delta(\mathbf{x}, s)\, |det\mathbf{F}^t_\delta(\mathbf{X}_{\delta 1}, s)||det\mathbf{H}_\delta| \\
&= \mathfrak{R}^t_{\delta\boldsymbol{\kappa}}(\mathbf{X}_{\delta 1}, s)\, |det\mathbf{H}_\delta|
\end{aligned}
\tag{6.7.7}
$$

and

$$GRAD\mathfrak{R}^t_{\delta\boldsymbol{\kappa}}(\mathbf{X}_{\delta 2}, s) = GRAD(\mathfrak{R}^t_{\delta\boldsymbol{\kappa}}(\mathbf{X}_{\delta 1}, s)|det\mathbf{H}_\delta) \tag{6.7.8}$$

We can now substitute (6.7.5)-(6.7.8) into (6.7.2) and dispense with the explicit notation that $\mathbf{F}_\beta$ and $\mathfrak{R}_\beta$ depend on the reference configuration $\mathbf{X}_\beta$, and the remaining independent constitutive variables on the position $\mathbf{x}$. Thus, by letting in (6.7.2) $\mathbf{F}^t_\beta = \mathbf{F}^t_\beta(\mathbf{X}_\beta, s)$, $\mathfrak{R}^t_{\beta\boldsymbol{\kappa}} = \mathfrak{R}^t_{\beta\boldsymbol{\kappa}}(\mathbf{X}_\beta, s)$, and $\phi^t_\beta = \phi^t_\beta(\mathbf{x}, s)$, where $\phi$ is any tensor-valued function such as $\bar{\bar{\theta}}_\beta$, we obtain

$$\begin{gathered}
G_{\alpha 1\boldsymbol{\kappa}}[\mathbf{F}^t_1, \ldots, \mathbf{F}^t_\delta, \ldots, \mathbf{F}^t_\gamma, GRAD\mathbf{F}^t_1, \ldots, GRAD\mathbf{F}^t_\delta, \ldots, GRAD\mathbf{F}^t_\gamma, \\
\mathfrak{R}^t_{1\boldsymbol{\kappa}}, \ldots, \mathfrak{R}^t_{\delta\boldsymbol{\kappa}}, \ldots, \mathfrak{R}^t_{\gamma\boldsymbol{\kappa}}, GRAD\mathfrak{R}^t_{1\boldsymbol{\kappa}}, \ldots, GRAD\mathfrak{R}^t_{\delta\boldsymbol{\kappa}}, \ldots, GRAD\mathfrak{R}^t_{\gamma\boldsymbol{\kappa}}, \\
\bar{\bar{\theta}}^t_\beta, \nabla\bar{\bar{\theta}}^t_\beta, \bar{\mathbf{i}}^t_\beta, \nabla\bar{\mathbf{i}}^t_\beta, \mathbf{b}^{(\beta)t}, \nabla\boldsymbol{\nu}^{(\beta)t}] \\
= G_{\alpha 2\boldsymbol{\kappa}}[\mathbf{F}^t_1, \ldots, \mathbf{F}^t_\delta\mathbf{H}_\delta, \ldots, \mathbf{F}^t_\gamma, GRAD\mathbf{F}^t_1, \ldots, \\
(GRAD\mathbf{F}^t_\delta)\mathbf{H}_\delta\mathbf{H}_\delta + \mathbf{F}^t_\delta GRAD\mathbf{H}_\delta, \ldots, \\
GRAD\mathbf{F}^t_\gamma, \mathfrak{R}^t_{1\boldsymbol{\kappa}}, \ldots, \mathfrak{R}^t_{\delta\boldsymbol{\kappa}}|det\mathbf{H}_\delta|, \ldots, \mathfrak{R}^t_{\gamma\boldsymbol{\kappa}}, GRAD\mathfrak{R}^t_{1\boldsymbol{\kappa}}, \ldots, \\
GRAD(\mathfrak{R}^t_{\delta\boldsymbol{\kappa}}|det\mathbf{H}_\delta|), \ldots, GRAD\mathfrak{R}^t_{\gamma\boldsymbol{\kappa}}, \bar{\bar{\theta}}^t_\beta, \nabla\bar{\bar{\theta}}^t_\beta, \bar{\mathbf{i}}^t_\beta, \nabla\bar{\mathbf{i}}^t_\beta, \mathbf{b}^{(\beta)t}, \nabla\boldsymbol{\nu}^{(\beta)t}]
\end{gathered} \tag{6.7.9}$$

In the special case when there is no phase change it can be seen from (6.2.3) that $\bar{\rho}_\alpha|det\mathbf{F}_\alpha| = \bar{\rho}_{\alpha\boldsymbol{\kappa}} = constant$, and by using (6.7.1), (6.7.5), and (6.7.3) the following result is obtained:

$$|det\mathbf{F}^t_\delta(\mathbf{X}_{\delta 1}, s)| = |det\mathbf{F}^t_\delta(\mathbf{X}_{\delta 2}, s)| = |det\mathbf{F}^t_\delta(\mathbf{X}_{\delta 1}, s)||det\mathbf{H}_\delta|$$

or

$$det\mathbf{H}_\delta = \pm 1 \tag{6.7.10}$$

which shows that $\mathbf{H}_\delta$ is a *unimodular tensor* and that, therefore, $\mathbf{H}_\delta$ represents a unimodular mapping which leaves the constitutive response functional $G_{\alpha\boldsymbol{\kappa}}$ invariant with respect to phase $\delta$. With the phase change, however, the density in the reference configuration need not be preserved and (6.7.10) need not hold.

We will demonstrate next that the set of pairs $(\mathbf{H}_\delta, \mathbf{J}_\delta)$, where

$$\mathbf{J}_\delta = GRAD\mathbf{H}_\delta$$

which satisfy (6.7.9) for all

$$(\mathbf{F}_\beta, GRAD\mathbf{F}_\beta, \mathfrak{R}_{\beta\boldsymbol{\kappa}}, GRAD\mathfrak{R}_{\beta\boldsymbol{\kappa}}, \bar{\bar{\theta}}_\beta, \nabla\bar{\bar{\theta}}_\beta, \bar{\mathbf{i}}_\beta, \nabla\bar{\mathbf{i}}_\beta, \mathbf{b}^{(\beta)}, \nabla\boldsymbol{\nu}^{(\beta)})$$

forms a group. If $(\mathbf{H}_\delta, \mathbf{J}_\delta)$ and $(\bar{\mathbf{H}}_\delta, \bar{\mathbf{J}}_\delta)$ are given pairs, their composition can be defined as follows (CROSS,1973):

$$(\mathbf{H}_\delta, \mathbf{J}_\delta) \circ (\bar{\mathbf{H}}_\delta, \bar{\mathbf{J}}_\delta) = (\hat{\mathbf{H}}_\delta, \hat{\mathbf{J}}_\delta) \tag{6.7.11}$$

where

$$\hat{\mathbf{H}}_\delta = \mathbf{H}_\delta \bar{\mathbf{H}}_\delta \tag{6.7.12}$$

$$\hat{\mathbf{J}}_\delta = \mathbf{J}_\delta \bar{\mathbf{H}}_\delta \bar{\mathbf{H}}_\delta + \mathbf{H}_\delta \bar{\mathbf{J}}_\delta \tag{6.7.13}$$

so that (6.7.9) is satisfied for the pair $(\hat{\mathbf{H}}_\delta, \hat{\mathbf{J}}_\delta)$ whenever it is satisfied by $(\mathbf{H}_\delta, \mathbf{J}_\delta)$ and $(\bar{\mathbf{H}}_\delta, \bar{\mathbf{J}}_\delta)$. This proof is left to the reader. The pairs $(\mathbf{H}_\delta, \mathbf{J}_\delta)$ have the identity element $(\mathbf{H}_\delta = \mathbf{I}, \mathbf{J}_\delta = \mathbf{0}) = (\mathbf{I}, \mathbf{0})$, since from (6.7.11)-(6.7.13)

$$(\mathbf{I}, \mathbf{0}) \circ (\mathbf{H}_\delta, \mathbf{J}_\delta) = (\mathbf{H}_\delta, \mathbf{J}_\delta) \tag{6.7.14}$$

$$(\mathbf{H}_\delta, \mathbf{J}_\delta) \circ (\mathbf{I}, \mathbf{0}) = (\mathbf{H}_\delta, \mathbf{J}_\delta) \tag{6.7.15}$$

The inverse pair $(\mathbf{H}_\delta^{-1}, \mathbf{J}_\delta^{-1})$ defined by

$$(\mathbf{H}_\delta, \mathbf{J}_\delta)^{-1} = (\mathbf{H}_\delta^{-1}, \mathbf{J}_\delta^{-1}) \tag{6.7.16}$$

also belongs to the pairs $(\mathbf{H}_\delta, \mathbf{J}_\delta)$, since

$$\begin{aligned}(\mathbf{H}_\delta, \mathbf{J}_\delta) \circ (\mathbf{H}_\delta, \mathbf{J}_\delta)^{-1} &= (\mathbf{H}_\delta, \mathbf{J}_\delta) \circ (\mathbf{H}_\delta^{-1}, \mathbf{J}_\delta^{-1}) \\ &= (\mathbf{H}_\delta \mathbf{H}_\delta^{-1},\ \mathbf{J}_\delta \mathbf{H}_\delta^{-1} \mathbf{H}_\delta^{-1} + \mathbf{H}_\delta \mathbf{J}_\delta^{-1}) = (\mathbf{I}, \mathbf{0})\end{aligned} \tag{6.7.17}$$

$$\begin{aligned}(\mathbf{H}_\delta, \mathbf{J}_\delta)^{-1} \circ (\mathbf{H}_\delta, \mathbf{J}_\delta) &= (\mathbf{H}_\delta^{-1}, \mathbf{J}_\delta^{-1}) \circ (\mathbf{H}_\delta, \mathbf{J}_\delta) \\ &= (\mathbf{H}_\delta^{-1} \mathbf{H}_\delta,\ \mathbf{J}_\delta^{-1} \mathbf{H}_\delta \mathbf{H}_\delta + \mathbf{H}_\delta^{-1} \mathbf{J}_\delta) = (\mathbf{I}, \mathbf{0})\end{aligned} \tag{6.7.18}$$

where $\mathbf{H}_\delta^{-1}$ is the inverse of $\mathbf{H}_\delta$, and $\mathbf{J}_\delta^{-1}$ is defined from the above equations as

$$\mathbf{J}_\delta^{-1} = -\mathbf{H}_\delta^{-1} \mathbf{J}_\delta \mathbf{H}_\delta^{-1} \mathbf{H}_\delta^{-1} \tag{6.7.19}$$

or

$$J_{\delta IJK}^{-1} = -H_{\delta IL}^{-1} J_{\delta LMN} H_{\delta MJ}^{-1} H_{\delta NK}^{-1} \tag{6.7.20}$$

The associativity property of the pairs $(\mathbf{H}_\delta, \mathbf{J}_\delta)$, $(\bar{\mathbf{H}}_\delta, \bar{\mathbf{J}}_\delta)$, and $(\bar{\bar{\mathbf{H}}}_\delta, \bar{\bar{\mathbf{J}}}_\delta)$, *i.e.*

$$(\mathbf{H}_\delta, \mathbf{J}_\delta) \circ ((\bar{\mathbf{H}}_\delta, \bar{\mathbf{J}}_\delta) \circ (\bar{\bar{\mathbf{H}}}_\delta, \bar{\bar{\mathbf{J}}})) = ((\mathbf{H}_\delta, \mathbf{J}_\delta) \circ (\bar{\mathbf{H}}_\delta, \bar{\mathbf{J}}_\delta)) \circ (\bar{\bar{\mathbf{H}}}_\delta, \bar{\bar{\mathbf{J}}}_\delta) \tag{6.7.21}$$

can be established from (6.7.11) by noting that

$$\mathbf{J}_\delta(\bar{\mathbf{H}}_\delta\bar{\bar{\mathbf{H}}}_\delta)(\bar{\mathbf{H}}_\delta\bar{\bar{\mathbf{H}}}_\delta) = (\mathbf{J}_\delta\bar{\mathbf{H}}_\delta\bar{\mathbf{H}}_\delta)(\bar{\bar{\mathbf{H}}}_\delta\bar{\bar{\mathbf{H}}}_\delta)$$

or

$$J_{\delta IJK}\bar{H}_{\delta JM}\bar{\bar{H}}_{\delta MQ}\bar{H}_{\delta KU}\bar{\bar{H}}_{\delta UL} = J_{\delta IJK}\bar{H}_{\delta JM}\bar{H}_{\delta KU}\bar{\bar{H}}_{\delta MQ}\bar{\bar{H}}_{\delta UL}$$

The identity properties (6.7.14) and (6.7.15), inverse properties (6.7.17) and (6.7.18), and the associativity property (6.7.21) prove that the pair $(\mathbf{H}_\delta, \mathbf{J}_\delta)$ does indeed form a group which will be called the *isotropy or symmetry group of the δ'th phase* and denoted by $g_{\boldsymbol{\kappa}\delta}$. Notice that this group depends on the constitutive relation for $G_{\alpha\boldsymbol{\kappa}}$, and thus on the reference configuration for the δ'th phase.

As discussed above, the tensor $\mathbf{H}_\delta$ need not be unimodular or orthogonal. If $\mathbf{Q}$, an orthogonal tensor, belongs to the isotropy group $g_{\boldsymbol{\kappa}\delta}$, then its inverse $\mathbf{Q}^{-1} = \mathbf{Q}^T$ also belongs to $g_{\boldsymbol{\kappa}\delta}$, since $g_{\boldsymbol{\kappa}\delta}$ is a group. By selecting

$$\mathbf{F}^t_\beta = \mathbf{Q}\mathbf{F}^t_\beta, \quad \nabla\bar{\bar{\theta}}^t_\beta = \mathbf{Q}\nabla\bar{\bar{\theta}}^t_\beta, \quad \bar{\mathbf{i}}^t_\beta = \mathbf{Q}\bar{\mathbf{i}}^t_\beta\mathbf{Q}^T,$$
$$\nabla\bar{\mathbf{i}}^t_\beta = \mathbf{Q}\nabla\bar{\mathbf{i}}^t_\beta\mathbf{Q}^T\mathbf{Q}^T, \quad \mathbf{b}^{(\beta)t} = \mathbf{Q}\mathbf{b}^{(\beta)t}\mathbf{Q}^T, \quad \nabla\boldsymbol{\nu}^{(\beta)t} = \mathbf{Q}\nabla\boldsymbol{\nu}^{(\beta)t}\mathbf{Q}^T\mathbf{Q}^T$$

and $\mathbf{H}_\delta = \mathbf{Q}^T$ in equation (6.7.9), yields:

$$\begin{aligned}
&G_{\alpha\boldsymbol{\kappa}}[\mathbf{Q}\mathbf{F}^t_1,\ldots, \mathbf{Q}\mathbf{F}^t_\delta\mathbf{Q}^T,\ldots, \mathbf{Q}\mathbf{F}^t_\gamma, \mathbf{Q}GRAD\mathbf{F}^t_1,\ldots,\\
&\mathbf{Q}(GRAD\mathbf{F}^t_\delta)\mathbf{Q}^T\mathbf{Q}^T,\ldots, \mathbf{Q}GRAD\mathbf{F}^t_\gamma, \mathfrak{R}^t_{1\boldsymbol{\kappa}},\ldots, \mathfrak{R}^t_{\delta\boldsymbol{\kappa}},\ldots, \mathfrak{R}^t_{\gamma\boldsymbol{\kappa}},\\
&GRAD\mathfrak{R}^t_{1\boldsymbol{\kappa}},\ldots, GRAD\mathfrak{R}^t_{\delta\boldsymbol{\kappa}},\ldots, GRAD\mathfrak{R}^t_{\gamma\boldsymbol{\kappa}}, \bar{\bar{\theta}}^t_\beta, \mathbf{Q}\nabla\bar{\bar{\theta}}^t_\beta, \mathbf{Q}\bar{\mathbf{i}}^t_\beta\mathbf{Q}^T,\\
&\mathbf{Q}\nabla\bar{\mathbf{i}}^t_\beta\mathbf{Q}^T\mathbf{Q}^T, \mathbf{Q}\mathbf{b}^{(\beta)t}\mathbf{Q}^T, \mathbf{Q}\nabla\boldsymbol{\nu}^{(\beta)t}\mathbf{Q}^T\mathbf{Q}^T]\\
&= G_{\alpha\boldsymbol{\kappa}}[\mathbf{Q}\mathbf{F}^t_\beta, \mathbf{Q}GRAD\mathbf{F}^t_\beta, \mathfrak{R}^t_{\beta\boldsymbol{\kappa}}, GRAD\mathfrak{R}^t_{\beta\boldsymbol{\kappa}}, \bar{\bar{\theta}}^t_\beta, \mathbf{Q}\nabla\bar{\bar{\theta}}^t_\beta,\\
&\mathbf{Q}\bar{\mathbf{i}}^t_\beta\mathbf{Q}^T, \mathbf{Q}\nabla\bar{\mathbf{i}}^t_\beta\mathbf{Q}^T\mathbf{Q}^T, \mathbf{Q}\mathbf{b}^{(\beta)t}\mathbf{Q}^T, \mathbf{Q}\nabla\boldsymbol{\nu}^{(\beta)t}\mathbf{Q}^T\mathbf{Q}^T]
\end{aligned} \tag{6.7.22}$$

Now, if $G_{\alpha\boldsymbol{\kappa}}$ is a vector, use can be made of its restriction imposed by the principle of material frame-indifference as expressed by (6.6.21). Choosing in this equation $\mathbf{Q}^t(s) = \mathbf{Q}(t)$, independent of s, gives

$$\begin{aligned}
&G_{\alpha\boldsymbol{\kappa}}[\mathbf{Q}(t)\mathbf{F}^t_\beta, \mathbf{Q}(t)GRAD\mathbf{F}^t_\beta, \mathfrak{R}^t_{\beta\boldsymbol{\kappa}}, GRAD\mathfrak{R}^t_{\beta\boldsymbol{\kappa}}, \bar{\bar{\theta}}^t_\beta, \mathbf{Q}(t)\nabla\bar{\bar{\theta}}^t_\beta,\\
&\mathbf{Q}(t)\bar{\mathbf{i}}^t_\beta\mathbf{Q}(t)^T, \mathbf{Q}(t)\nabla\bar{\mathbf{i}}^t_\beta\mathbf{Q}(t)^T\mathbf{Q}(t)^T,\\
&\mathbf{Q}(t)\mathbf{b}^{(\beta)t}\mathbf{Q}(t)^T, \mathbf{Q}(t)\nabla\boldsymbol{\nu}^{(\beta)t}\mathbf{Q}(t)^T\mathbf{Q}(t)^T]\\
&= \mathbf{Q}(t)G_{\alpha\boldsymbol{\kappa}}[\mathbf{F}^t_\beta, GRAD\mathbf{F}^t_\beta, \mathfrak{R}^t_{\beta\boldsymbol{\kappa}}, GRAD\mathfrak{R}^t_{\beta\boldsymbol{\kappa}},\\
&\bar{\bar{\theta}}^t_\beta, \nabla\bar{\bar{\theta}}^t_\beta, \bar{\mathbf{i}}^t_\beta, \nabla\bar{\mathbf{i}}^t_\beta, \mathbf{b}^{(\beta)t}, \nabla\boldsymbol{\nu}^{(\beta)t}]
\end{aligned} \tag{6.7.23}$$

a relation which holds for all orthogonal $\mathbf{Q}(t)$, whereas (6.7.22) holds only for those $\mathbf{Q}$ that belong to the isotropy group $g_{\kappa\delta}$. Combining these two relations yields the following result:

$$\begin{aligned}
&G_{\alpha\kappa}[\mathbf{Q}(t)\mathbf{F}_1^t,\ldots,\ \mathbf{Q}(t)\mathbf{F}_\delta^t\mathbf{Q}(t)^T,\ldots,\ \mathbf{Q}(t)\mathbf{F}_\gamma^t,\ \mathbf{Q}(t)GRAD\mathbf{F}_1^t,\ldots,\\
&\mathbf{Q}(t)(GRAD\mathbf{F}_\delta^t)\mathbf{Q}(t)^T\mathbf{Q}(t)^T,\ldots,\ \mathbf{Q}(t)GRAD\mathbf{F}_\gamma^t,\ \mathfrak{R}_{\beta\kappa}^t,\ GRAD\mathfrak{R}_{\beta\kappa}^t,\\
&\quad \overline{\overline{\theta}}_\beta^t,\ \mathbf{Q}(t)\nabla\overline{\overline{\theta}}_\beta^t,\ \mathbf{Q}(t)\overline{\overline{\mathbf{i}}}_\beta^t\mathbf{Q}(t)^T,\ \mathbf{Q}(t)\nabla\overline{\overline{\mathbf{i}}}_\beta^t\mathbf{Q}(t)^T\mathbf{Q}(t)^T,\\
&\quad \mathbf{Q}(t)\mathbf{b}^{(\beta)t}\mathbf{Q}(t)^T,\ \mathbf{Q}(t)\nabla\boldsymbol{\nu}^{(\beta)t}\mathbf{Q}(t)^T\mathbf{Q}(t)^T]\\
&= \mathbf{Q}(t)G_{\alpha\kappa}[\mathbf{F}_\beta^t,\ GRAD\mathbf{F}_\beta^t,\ \mathfrak{R}_{\beta\kappa}^t,\ GRAD\mathfrak{R}_{\beta\kappa}^t,\ \overline{\overline{\theta}}_\beta^t,\\
&\quad \nabla\overline{\overline{\theta}}_\beta^t,\ \overline{\overline{\mathbf{i}}}_\beta^t,\ \nabla\overline{\overline{\mathbf{i}}}_\beta^t,\ \mathbf{b}^{(\beta)t},\ \nabla\boldsymbol{\nu}^{(\beta)t}]
\end{aligned} \tag{6.7.24}$$

Notice that (6.7.24) is a sufficient and necessary condition for the orthogonal tensor $\mathbf{Q}(t)$ to belong to the isotropy group.

The orthogonal tensor or transformation may be *proper* (with the determinant=+1) or *improper* (with the determinant=-1). The set of all proper and improper orthogonal transformations belong to the orthogonal group $O_\delta$, and the set of all proper orthogonal transformations forms a subgroup, whereas the set of improper orthogonal transformations does not. In the case of no phase change, it is noted above that $\mathbf{H}_\delta$ is unimodular and that it need not be orthogonal. If a material possesses a local reference configuration $\kappa_\delta$ such that the isotropy group $g_{\kappa\delta}$ contains the full orthogonal group $O_\delta$ then such a material is referred as *isotropic* and $\kappa_\delta$ an undisturbed state of the material. An isotropic material may be arbitrarily rotated before deformation, heat transfer, and phase change, with the result that subsequent deformation, heat transfer, and phase change will not detect or be dependent on this initial rotation. In other words, an isotropic material has no preferred orientations. An anisotropic material possesses an isotropy group relative to an undisturbed state which is only a proper subgroup of the full orthogonal group.

## 6.8 Reduced Constitutive Equations and Isotropic Multiphase Mixtures

By definition, all material points or particles of a homogeneous phase must have the same material symmetry. A phase $\delta$ will be defined as *isotropic* if there is an undisturbed reference configuration $\kappa_\delta$ such that the isotropy group $g_{\kappa\delta}$ contains the orthogonal group $O_\delta$

$$O_\delta \subset g_{\kappa\delta} \tag{6.8.1}$$

If the phase $\delta$ is isotropic, (6.7.24) holds for all $\mathbf{Q}(t)$ and not just for some orthogonal $\mathbf{Q}(t)$, and our next goal is to seek the form of constitutive relation (6.6.21) for an isotropic phase $\delta$ in the mixture. Towards this objective we will first derive a reduced form of the constitutive relation (6.6.21) where the occurring functions are not subject to further restrictions, except for the material symmetry.

We will start by choosing in (6.6.21)

$$\mathbf{F}^t_\beta(\mathbf{X}_\beta, s) = \mathbf{R}^t(s)\mathbf{U}^t_\beta(\mathbf{X}_\beta, s) \tag{6.8.2}$$

where $\mathbf{R}^t(s)$ is orthogonal, *i.e.*

$$\mathbf{R}^t(s)\mathbf{R}^t(s)^T = \mathbf{I}, \quad det\mathbf{R}^t(s) = \pm 1$$

and called the *rotation tensor*, whereas $\mathbf{U}^t_\beta(\mathbf{X}_\beta, s)$ is symmetric and called the *right stretch tensor*. For a nonsingular linear transformation $\mathbf{F}_\beta$, the decomposition (6.8.2) is unique and usually referred to as the polar *decomposition theorem of* CAUCHY (TRUESDELL & TOUPIN,1960, section 43 of the Appendix). Moreover, another form of this theorem is

$$\mathbf{F}^t_\beta(\mathbf{X}_\beta, s) = \mathbf{V}^t_\beta(\mathbf{X}_\beta, s)\mathbf{R}^t(s) \tag{6.8.3}$$

where $\mathbf{V}^t_\beta(\mathbf{X}_\beta, s)$ is the *left stretch tensor* that is also symmetric. If

$$\mathbf{U}^t_\beta(\mathbf{X}_\beta, s)\mathbf{V}^t_\beta(\mathbf{X}_\beta, s) = \mathbf{I}$$

then the local deformation is only a rotation, *i.e.*

$$\mathbf{F}^t_\beta(\mathbf{X}_\beta, s) = \mathbf{R}^t(s)$$

From (6.8.2) and (6.8.3) it follows that

$$\mathbf{F}^t_\beta(\mathbf{X}_\beta, s)^T\mathbf{F}^t_\beta(\mathbf{X}_\beta, s) = \mathbf{U}^t_\beta(\mathbf{X}_\beta, s)^2 = \mathbf{C}^t_\beta(\mathbf{X}_\beta, s) \tag{6.8.4}$$

$$\mathbf{F}^t_\beta(\mathbf{X}_\beta, s)\mathbf{F}^t_\beta(\mathbf{X}_\beta, s)^T = \mathbf{V}^t_\beta(\mathbf{x}, s)^2 = \mathbf{B}^t_\beta(\mathbf{x}, s) \tag{6.8.5}$$

where $\mathbf{C}^t_\beta(\mathbf{X}_\beta, s)$ and $\mathbf{B}^t_\beta(\mathbf{x}, s)$ can be called the *right* and *left* CAUCHY-GREEN *tensors for phase $\beta$*, respectively. Since the calculation of $\mathbf{C}_\beta$ and $\mathbf{B}_\beta$ follows directly from $\mathbf{F}_\beta$, their use is more desirable than the use of $\mathbf{U}_\beta$ and $\mathbf{V}_\beta$ which, in general, can be irrational functions.

Returning to (6.6.21) with the choice for $\mathbf{F}^t_\beta(\mathbf{X}_\beta, s)$ from (6.8.2), $\mathbf{Q}^t(s) = \mathbf{R}^t(s)^T$ and $\mathbf{Q}(t) = \mathbf{R}(t)^T$, the following result is obtained:

$$\mathbf{R}(t)G_{\alpha\kappa}[\mathbf{U}^t_\beta(\mathbf{X}_\beta, s),\ GRAD\mathbf{U}^t_\beta(\mathbf{X}_\beta, s),\ \mathfrak{R}^t_{\beta\kappa}(\mathbf{X}_\beta, s),$$

$$GRAD\mathfrak{R}^t_{\beta\kappa}(\mathbf{X}_\beta, s),\ \overline{\overline{\theta}}^t_\beta(\mathbf{x}, s),\ \mathbf{R}^t(s)^T\nabla\overline{\overline{\theta}}^t_\beta(\mathbf{x}, s),\ \mathbf{R}^t(s)^T\overline{\overline{\mathbf{i}}}^t_\beta(\mathbf{x}, s)\mathbf{R}^t(s),$$

$$
\begin{aligned}
&\mathbf{R}^t(s)^T\nabla\bar{\mathbf{i}}^t_\beta(\mathbf{x},s)\mathbf{R}^t(s)\mathbf{R}^t(s),\ \mathbf{R}^t(s)^T\mathbf{b}^{(\beta)t}(\mathbf{x},s)\mathbf{R}^t(s)\mathbf{R}^t(s),\\
&\qquad\mathbf{R}^t(s)^T\nabla\boldsymbol{\nu}^{(\beta)t}(\mathbf{x},s)\mathbf{R}^t(s)\mathbf{R}^t(s)]\\
&= G_{\alpha\boldsymbol{\kappa}}[\mathbf{F}^t_\beta(\mathbf{X}_\beta,s),\ GRAD\mathbf{F}^t_\beta(\mathbf{X}_\beta,s),\ \mathfrak{R}^t_{\beta\boldsymbol{\kappa}}(\mathbf{X}_\beta,s),\ GRAD\mathfrak{R}^t_{\beta\boldsymbol{\kappa}}(\mathbf{X}_\beta,s),\\
&\qquad\bar{\bar{\theta}}^t_\beta(\mathbf{x},s),\ \nabla\bar{\bar{\theta}}^t_\beta(\mathbf{x},s),\ \bar{\mathbf{i}}^t_\beta(\mathbf{x},s),\ \nabla\bar{\mathbf{i}}^t_\beta(\mathbf{x},s),\ \mathbf{b}^{(\beta)t}(\mathbf{x},s),\ \nabla\boldsymbol{\nu}^{(\beta)t}(\mathbf{x},s)]
\end{aligned}
\tag{6.8.6}
$$

This equation shows that if the thermal effects and mixture's structural characteristics are absent ($\nabla\bar{\bar{\theta}}_\beta=\mathbf{0}$, $\bar{\mathbf{i}}_\beta=\mathbf{0}$, $\nabla\bar{\mathbf{i}}_\beta=\mathbf{0}$, $\mathbf{b}^{(\beta)}=\mathbf{0}$, $\nabla\boldsymbol{\nu}^{(\beta)}=\mathbf{0}$) then: (1) the constitutive functional $G_{\alpha\boldsymbol{\kappa}}$ at time $t$ is affected only by the present value of the rotation $\mathbf{R}(t)$ and not by the previous history of rotations, and (2) $G_{\alpha\boldsymbol{\kappa}}$ can only depend on one of the strain measures, such as $\mathbf{U}^t_\beta(\mathbf{X}_\beta,s)$ or $\mathbf{C}^t_\beta(\mathbf{X}_\beta,s)$. Put in different words, $G_{\alpha\boldsymbol{\kappa}}$ at time $t$ arises from the stretch history $\mathbf{U}^t_\beta(\mathbf{X}_\beta,s)$ relative to a reference configuration, followed by a rotation at time $t$. Such a result is important, of course, for the experimental determination of $G_{\alpha\boldsymbol{\kappa}}$ can be carried out only on the stretch, temperature, and phase change histories. The presence of temperature gradient and structural properties of the mixture in the constitutive relations dictate, however, the consideration of the rotation history in the determination of the constitutive response functional $G_{\alpha\boldsymbol{\kappa}}$. Since from (6.8.2) and (6.8.3) $\mathbf{U}_\beta=\mathbf{R}^T\mathbf{V}_\beta\mathbf{R}$, notice that this substitution into (6.8.6) would not eliminate the rotation history, even if temperature gradients and mixture's structural properties are absent.

Fluids have no preferred reference configurations and the configuration at the present time may be choosen as a reference configuration. For this reason, it is useful to explore the properties of (6.8.6) further to determine just to what extent does the reference placement play a role in determining the constitutive response functional $G_{\alpha\boldsymbol{\kappa}}$.

Consider the motion of a particle of phase $\beta$ at two different times, $t$ and $\tau$, *i.e.*

$$\mathbf{x}=\chi_\beta(\mathbf{X}_\beta,t) \tag{6.8.7}$$

$$\boldsymbol{\zeta}=\chi_\beta(\mathbf{X}_\beta,\tau) \tag{6.8.8}$$

where $\mathbf{x}$ is the place occupied by the particle at time $t$ and $\boldsymbol{\zeta}$ is the place occupied at time $\tau$. Thus

$$\boldsymbol{\zeta}=\chi_\beta(\mathbf{X}_\beta^{-1}(\mathbf{x},t),\tau)=\chi_{\beta t}(\mathbf{x},\tau) \tag{6.8.9}$$

where the function $\chi_{\beta t}$ is called the *relative placement* (TRUESDELL,1977) of particle $\beta$. Using (2.4.5) and the above equations, yields

$$F_{\beta iI}(\mathbf{X}_\beta,\tau)=\frac{\partial\zeta_i}{\partial X_{\beta I}}=\frac{\partial\chi_{\beta ti}}{\partial x_j}\frac{\partial x_j}{\partial X_{\beta I}}=F_{\beta tij}(\mathbf{x},\tau)F_{\beta jI}(\mathbf{X}_\beta,t)$$

or

$$\mathbf{F}_\beta(\mathbf{X}_\beta,\tau)=\mathbf{F}_{\beta t}(\mathbf{x},\tau)\,\mathbf{F}_\beta(\mathbf{X}_\beta,t) \tag{6.8.10}$$

where

$$\mathbf{F}_{\beta t}(\mathbf{x},\tau)=\nabla\chi_{\beta t} \tag{6.8.11}$$

is the *local relative deformation.*

From (6.8.10) and (6.8.2) we have

$$\mathbf{F}_\beta(\mathbf{X}_\beta,\tau)=\mathbf{R}_t(\tau)\mathbf{U}_{\beta t}(\mathbf{X}_\beta,\tau)\mathbf{R}(t)\mathbf{U}_\beta(\mathbf{X}_\beta,t) \tag{6.8.12}$$

or

$$\mathbf{F}_\beta(\mathbf{X}_\beta,\tau)=\mathbf{R}_t(\tau)\mathbf{R}(t)(\mathbf{R}(t)^T\mathbf{U}_{\beta t}(\mathbf{X}_\beta,\tau)\mathbf{R}(t))\mathbf{U}_\beta(\mathbf{X}_\beta,t) \tag{6.8.13}$$

Putting $\tau=t-s$ in (6.8.13) gives

$$\mathbf{F}^t_\beta(\mathbf{X}_\beta,s)=\mathbf{R}^t_t(s)\mathbf{R}(t)(\mathbf{R}(t)^T\mathbf{U}^t_{\beta t}(\mathbf{X}_\beta,s)\mathbf{R}(t))\mathbf{U}_\beta(\mathbf{X}_\beta,t) \tag{6.8.14}$$

and if we select

$$\mathbf{Q}^t(s)=\mathbf{R}(t)\mathbf{R}^t_t(s)=(\mathbf{R}^t_t(s)\mathbf{R}(t))^T \tag{6.8.15}$$

equation (6.8.14) is reduced to the form

$$\mathbf{Q}^t(s)\mathbf{F}^t_\beta(\mathbf{X}_\beta,s)=\mathbf{R}(t)^T\mathbf{U}^t_{\beta t}(\mathbf{X}_\beta,s)\mathbf{R}(t)\mathbf{U}_\beta(\mathbf{X}_\beta,t) \tag{6.8.16}$$

which will be used shortly.

The equivalent form of equation (6.6.21) satisfying the principle of the material frame-indifference is

$$\begin{aligned}
&G_{\alpha K}[\mathbf{F}^t_\beta(\mathbf{X}_\beta,s),\ GRAD\mathbf{F}^t_\beta(\mathbf{X}_\beta,s),\ \mathfrak{R}^t_{\beta K}(\mathbf{X}_\beta,s),\ GRAD\mathfrak{R}^t_{\beta K}(\mathbf{X}_\beta,s),\\
&\quad\bar{\bar{\theta}}^t_\beta(\mathbf{x},s),\ \nabla\bar{\bar{\theta}}^t_\beta(\mathbf{x},s),\ \bar{\mathbf{i}}^t_\beta(\mathbf{x},s),\ \nabla\bar{\mathbf{i}}^t_\beta(\mathbf{x},s),\ \mathbf{b}^{(\beta)t}(\mathbf{x},s),\ \nabla\boldsymbol{\nu}^{(\beta)t}(\mathbf{x},s)]\\
&=\mathbf{Q}(t)^TG_{\alpha K}[\mathbf{Q}^t(s)\mathbf{F}^t_\beta(\mathbf{X}_\beta,s),\ \mathbf{Q}^t(s)GRAD\mathbf{F}^t_\beta(\mathbf{X}_\beta,s),\ \mathfrak{R}^t_{\beta K}(\mathbf{X}_\beta,s),\\
&\quad GRAD\mathfrak{R}^t_{\beta K}(\mathbf{X}_\beta,s),\ \bar{\bar{\theta}}^t_\beta(\mathbf{x},s),\ \mathbf{Q}^t(s)\nabla\bar{\bar{\theta}}^t_\beta(\mathbf{x},s),\ \mathbf{Q}^t(s)\bar{\mathbf{i}}^t_\beta(\mathbf{x},s)\mathbf{Q}^t(s)^T,\\
&\quad\mathbf{Q}^t(s)\nabla\bar{\mathbf{i}}^t_\beta(\mathbf{x},s)\mathbf{Q}^t(s)^T\mathbf{Q}^t(s)^T,\ \mathbf{Q}^t(s)\mathbf{b}^{(\beta)t}(\mathbf{x},s)\mathbf{Q}^t(s)^T,\\
&\quad\mathbf{Q}^t(s)\nabla\boldsymbol{\nu}^{(\beta)t}(\mathbf{x},s)\mathbf{Q}^t(s)^T\mathbf{Q}^t(s)^T]
\end{aligned} \tag{6.8.17}$$

Using now in this equation (6.8.15) and (6.8.16) yields

$$\begin{aligned}
&G_{\alpha K}[\mathbf{F}^t_\beta(\mathbf{X}_\beta,s),\ GRAD\mathbf{F}^t_\beta(\mathbf{X}_\beta,s),\ \mathfrak{R}^t_{\beta K}(\mathbf{X}_\beta,s),\ GRAD\mathfrak{R}^t_{\beta K}(\mathbf{X}_\beta,s),\\
&\quad\bar{\bar{\theta}}^t_\beta(\mathbf{x},s),\ \nabla\bar{\bar{\theta}}^t_\beta(\mathbf{x},s),\ \bar{\mathbf{i}}^t_\beta(\mathbf{x},s),\ \nabla\bar{\mathbf{i}}^t_\beta(\mathbf{x},s),\ \mathbf{b}^{(\beta)t}(\mathbf{x},s),\ \nabla\boldsymbol{\nu}^{(\beta)t}(\mathbf{x},s)]
\end{aligned}$$

$$
\begin{gathered}
= \mathbf{R}(t) G_{\alpha\boldsymbol{\kappa}}[\mathbf{R}(t)^T \mathbf{U}^t_{\beta t}(\mathbf{X}_\beta, s)\mathbf{R}(t)\mathbf{U}_\beta(\mathbf{X}_\beta, t), \\
GRAD(\mathbf{R}(t)^T \mathbf{U}^t_{\beta t}(\mathbf{X}_\beta, s)\mathbf{R}(t)\mathbf{U}_\beta(\mathbf{X}_\beta, t)),\ \mathfrak{R}^t_{\beta\boldsymbol{\kappa}}(\mathbf{X}_\beta, s),\ GRAD\mathfrak{R}^t_{\beta\boldsymbol{\kappa}}(\mathbf{X}_\beta, s), \\
\bar{\bar{\theta}}^t_\beta(\mathbf{x}, s),\ \mathbf{R}(t)\mathbf{R}^t_i(s)\nabla\bar{\bar{\theta}}^t_\beta(\mathbf{x}, s),\ \mathbf{R}(t)\mathbf{R}^t_i(s)\bar{\mathbf{i}}^t_\beta(\mathbf{x}, s)\mathbf{R}^t_i(s)\mathbf{R}(t), \\
\mathbf{R}(t)\mathbf{R}^t_i(s)\nabla\bar{\mathbf{i}}^t_\beta(\mathbf{x}, s)\mathbf{R}^t_i(s)\mathbf{R}(t)\mathbf{R}^t_i(s)\mathbf{R}(t),\ \mathbf{R}(t)\mathbf{R}^t_i(s)\mathbf{b}^{(\beta)t}(\mathbf{x}, s)\mathbf{R}^t_i(s)\mathbf{R}(t), \\
\mathbf{R}(t)\mathbf{R}^t_i(s)\nabla\boldsymbol{\nu}^{(\beta)t}(\mathbf{x}, s)\mathbf{R}^t_i(s)\mathbf{R}(t)\mathbf{R}^t_i(s)\mathbf{R}(t)]
\end{gathered}
\tag{6.8.18}
$$

where use is made of the fact that $\mathbf{Q}(t) = (\mathbf{R}^t_i(0)\mathbf{R}(t))^T = \mathbf{R}(t)^T$, which follows from (6.8.15) at $s = 0$. The result (6.8.18) implies that

$$
\begin{gathered}
G_{\alpha\boldsymbol{\kappa}}[\mathbf{F}^t_\beta(\mathbf{X}_\beta, s),\ GRAD\mathbf{F}^t_\beta(\mathbf{X}_\beta, s),\ \mathfrak{R}^t_{\beta\boldsymbol{\kappa}}(\mathbf{X}_\beta, s),\ GRAD\mathfrak{R}^t_{\beta\boldsymbol{\kappa}}(\mathbf{X}_\beta, s), \\
\bar{\bar{\theta}}^t_\beta(\mathbf{x}, s),\ \nabla\bar{\bar{\theta}}^t_\beta(\mathbf{x}, s),\ \bar{\mathbf{i}}^t_\beta(\mathbf{x}, s),\ \nabla\bar{\mathbf{i}}^t_\beta(\mathbf{x}, s),\ \mathbf{b}^{(\beta)t}(\mathbf{x}, s),\ \nabla\boldsymbol{\nu}^{(\beta)t}(\mathbf{x}, s)] \\
= \mathbf{R}(t) G_{\alpha\boldsymbol{\kappa}}[\mathbf{R}(t)^T \mathbf{U}^t_{\beta t}(\mathbf{X}_\beta, s)\mathbf{R}(t),\ \mathbf{U}_\beta(\mathbf{X}_\beta, t), \\
GRAD(\mathbf{R}(t)^T \mathbf{U}^t_{\beta t}(\mathbf{X}_\beta, s)\mathbf{R}(t)),\ GRAD\mathbf{U}_\beta(\mathbf{X}_\beta, t),\ \mathfrak{R}^t_{\beta\boldsymbol{\kappa}}(\mathbf{X}_\beta, s), \\
GRAD\mathfrak{R}^t_{\beta\boldsymbol{\kappa}}(\mathbf{X}_\beta, s),\ \bar{\bar{\theta}}^t_\beta(\mathbf{x}, s),\ \mathbf{R}(t)\mathbf{R}^t_i(s)\nabla\bar{\bar{\theta}}^t_\beta(\mathbf{x}, s), \\
\mathbf{R}(t)\mathbf{R}^t_i(s)\bar{\mathbf{i}}^t_\beta(\mathbf{x}, s)\mathbf{R}^t_i(s)\mathbf{R}(t),\ \mathbf{R}(t)\mathbf{R}^t_i(s)\nabla\bar{\mathbf{i}}^t_\beta(\mathbf{x}, s)\mathbf{R}^t_i(s)\mathbf{R}(t)\mathbf{R}^t_i(s)\mathbf{R}(t), \\
\mathbf{R}(t)\mathbf{R}^t_i(s)\mathbf{b}^{(\beta)t}(\mathbf{x}, s)\mathbf{R}^t_i(s)\mathbf{R}(t),\ \mathbf{R}(t)\mathbf{R}^t_i(s)\nabla\boldsymbol{\nu}^{(\beta)t}(\mathbf{x}, s)\mathbf{R}^t_i(s)\mathbf{R}(t)\mathbf{R}^t_i(s)\mathbf{R}(t)]
\end{gathered}
\tag{6.8.19}
$$

or in terms of the right CAUCHY-GREEN tensor $\mathbf{C}_\beta$ defined by (6.8.4), this result can also be written as

$$
\begin{gathered}
G_{\alpha\boldsymbol{\kappa}}[\mathbf{F}^t_\beta(\mathbf{X}_\beta, s),\ GRAD\mathbf{F}^t_\beta(\mathbf{X}_\beta, s),\ \mathfrak{R}^t_{\beta\boldsymbol{\kappa}}(\mathbf{X}_\beta, s),\ GRAD\mathfrak{R}^t_{\beta\boldsymbol{\kappa}}(\mathbf{X}_\beta, s), \\
\bar{\bar{\theta}}^t_\beta(\mathbf{x}, s),\ \nabla\bar{\bar{\theta}}^t_\beta(\mathbf{x}, s),\ \bar{\mathbf{i}}^t_\beta(\mathbf{x}, s),\ \nabla\bar{\mathbf{i}}^t_\beta(\mathbf{x}, s),\ \mathbf{b}^{(\beta)t}(\mathbf{x}, s),\ \nabla\boldsymbol{\nu}^{(\beta)t}(\mathbf{x}, s)] \\
= \mathbf{R}(t) G_{\alpha\boldsymbol{\kappa}}[\mathbf{R}(t)^T \mathbf{C}^t_{\beta t}(\mathbf{X}_\beta, s)\mathbf{R}(t),\ \mathbf{C}_\beta(\mathbf{X}_\beta, t), \\
GRAD(\mathbf{R}(t)^T \mathbf{C}^t_{\beta t}(\mathbf{X}_\beta, s)\mathbf{R}(t)),\ GRAD\mathbf{C}_\beta(\mathbf{X}_\beta, t),\ \mathfrak{R}^t_{\beta\boldsymbol{\kappa}}(\mathbf{X}_\beta, s), \\
GRAD\mathfrak{R}^t_{\beta\boldsymbol{\kappa}}(\mathbf{X}_\beta, s),\ \bar{\bar{\theta}}^t_\beta(\mathbf{x}, s),\ \mathbf{R}(t)\mathbf{R}^t_i(s)\nabla\bar{\bar{\theta}}^t_\beta(\mathbf{x}, s), \\
\mathbf{R}(t)\mathbf{R}^t_i(s)\bar{\mathbf{i}}^t_\beta(\mathbf{x}, s)\mathbf{R}^t_i(s)\mathbf{R}(t),\ \mathbf{R}(t)\mathbf{R}^t_i(s)\nabla\bar{\mathbf{i}}^t_\beta(\mathbf{x}, s)\mathbf{R}^t_i(s)\mathbf{R}(t)\mathbf{R}^t_i(s)\mathbf{R}(t), \\
\mathbf{R}(t)\mathbf{R}^t_i(s)\mathbf{b}^{(\beta)t}(\mathbf{x}, s)\mathbf{R}^t_i(s)\mathbf{R}(t),\ \mathbf{R}(t)\mathbf{R}^t_i(s)\nabla\boldsymbol{\nu}^{(\beta)t}(\mathbf{x}, s)\mathbf{R}^t_i(s)\mathbf{R}(t)\mathbf{R}^t_i(s)\mathbf{R}(t)]
\end{gathered}
\tag{6.8.20}
$$

Equation (6.8.19) or (6.8.20) shows that it is not possible to express the effect of deformation, phase change, temperature, and structural properties of the mixture entirely by the present state of the mixture and that a fixed reference placement is required, in general.

For a multiphase mixture with *all* phases isotropic, equation (6.8.20) can be reduced further by taking in this equation $\mathbf{F}^t_\beta(\mathbf{X}_\beta, s) = \mathbf{F}^t_\beta(\mathbf{X}_\beta, s)\mathbf{R}^t(s)^T$ without changing the value of $G_{\alpha\boldsymbol{\kappa}}$, since isotropic materials have properties independent of direction. Thus, with the aid of (6.8.2) we have

$$\mathbf{F}^t_\beta(\mathbf{X}_\beta, s) = \mathbf{F}^t_\beta(\mathbf{X}_\beta, s)\mathbf{R}(t)^T = \mathbf{R}^t(s)\mathbf{U}^t_\beta(\mathbf{X}_\beta, s)\mathbf{R}^t(s)^T \qquad (6.8.21)$$

showing that $\mathbf{F}^t_\beta(\mathbf{X}_\beta, s)$ is positive and symmetric. By the polar decomposition theorem it follows then that $\mathbf{R}^t(s)$ can be replaced by $\mathbf{I}$, and $\mathbf{U}^t_\beta(\mathbf{X}_\beta, s)$ by $\mathbf{R}^t(s)\mathbf{U}^t_\beta(\mathbf{X}_\beta, s)\mathbf{R}^t(s)^T$. Moreover, $\mathbf{R}(t)^T\mathbf{C}^t_{\beta t}(\mathbf{X}_\beta, s)\mathbf{R}(t)$ is now replaced by $\mathbf{C}^t_{\beta t}(\mathbf{X}_\beta, s)$, and $\mathbf{C}_\beta(\mathbf{X}_\beta, t)$ is replaced by $\mathbf{R}(t)\mathbf{C}_\beta(\mathbf{X}_\beta, t)\mathbf{R}(t)^T$, which upon combining (6.8.2)-(6.8.5) is $\mathbf{B}_\beta(\mathbf{x}, t)$, *i.e.*

$$\begin{aligned}\mathbf{R}(t)\mathbf{C}_\beta(\mathbf{X}_\beta, t)\mathbf{R}(t)^T &= \mathbf{R}(t)\mathbf{F}_\beta(\mathbf{X}_\beta, t)^T\mathbf{F}_\beta(\mathbf{X}_\beta, t)\mathbf{R}(t)^T \\ &= \mathbf{R}(t)\mathbf{R}(t)^T\mathbf{V}_\beta(\mathbf{x}, t)\mathbf{V}_\beta(\mathbf{x}, t)\mathbf{R}(t)\mathbf{R}(t)^T = \mathbf{V}_\beta(\mathbf{x}, t)^2 = \mathbf{B}_\beta(\mathbf{x}, t)\end{aligned} \qquad (6.8.22)$$

Thus, *for an isotropic mixture* (6.8.20) is reduced to the following form:

$$\begin{aligned}&G_{\alpha\boldsymbol{\kappa}}[\mathbf{F}^t_\beta(\mathbf{X}_\beta, s),\ GRAD\mathbf{F}^t_\beta(\mathbf{X}_\beta, s),\ \mathfrak{R}^t_{\beta\boldsymbol{\kappa}}(\mathbf{X}_\beta, s),\ GRAD\mathfrak{R}^t_{\beta\boldsymbol{\kappa}}(\mathbf{X}_\beta, s),\\ &\quad \bar{\bar{\theta}}^t_\beta(\mathbf{x}, s),\ \nabla\bar{\bar{\theta}}^t_\beta(\mathbf{x}, s),\ \bar{\mathbf{i}}^t_\beta(\mathbf{x}, s),\ \nabla\bar{\mathbf{i}}^t_\beta(\mathbf{x}, s),\ \mathbf{b}^{(\beta)t}(\mathbf{x}, s),\ \nabla\boldsymbol{\nu}^{(\beta)t}(\mathbf{x}, s)]\\ &\quad = G_{\alpha\boldsymbol{\kappa}}[\mathbf{C}^t_{\beta t}(\mathbf{X}_\beta, s),\ \mathbf{B}_\beta(\mathbf{x}, t),\ GRAD\mathbf{C}^t_{\beta t}(\mathbf{X}_\beta, s),\ \nabla\mathbf{B}_\beta(\mathbf{x}, t),\\ &\quad \mathfrak{R}^t_{\beta\boldsymbol{\kappa}}(\mathbf{X}_\beta, s),\ GRAD\mathfrak{R}^t_{\beta\boldsymbol{\kappa}}(\mathbf{X}_\beta, s),\ \bar{\bar{\theta}}^t_\beta(\mathbf{x}, s),\ \nabla\bar{\bar{\theta}}^t_\beta(\mathbf{x}, s),\ \bar{\mathbf{i}}^t_\beta(\mathbf{x}, s),\\ &\quad \nabla\bar{\mathbf{i}}^t_\beta(\mathbf{x}, s),\ \mathbf{b}^{(\beta)t}(\mathbf{x}, s),\ \nabla\boldsymbol{\nu}^{(\beta)t}(\mathbf{x}, s)]\end{aligned} \qquad (6.8.23)$$

where the effect of rotation has been suppressed.

The transformation properties of (6.8.23) under the principle of the material frame-indifference can be determined by using (6.7.24) with all phases isotropic. In (6.8.23) we can therefore replace:

$$\mathbf{F}^t_\beta(\mathbf{X}_\beta, s) \ \ by \ \ \mathbf{Q}(t)\mathbf{F}^t_\beta(\mathbf{X}_\beta, s)\mathbf{Q}(t)^T$$

$$GRAD\mathbf{F}^t_\beta(\mathbf{X}_\beta, s) \ \ by \ \ \mathbf{Q}(t)(GRAD\mathbf{F}^t_\beta(\mathbf{X}_\beta, s))\mathbf{Q}(t)^T\mathbf{Q}(t)^T$$

$$\nabla\bar{\bar{\theta}}^t_\beta(\mathbf{x}, s) \ \ by \ \ \mathbf{Q}(t)\nabla\bar{\bar{\theta}}^t_\beta(\mathbf{x}, s)$$

$$\bar{\mathbf{i}}^t_\beta(\mathbf{x}, s) \ \ by \ \ \mathbf{Q}(t)\bar{\mathbf{i}}^t_\beta(\mathbf{x}, s)\mathbf{Q}(t)^T$$

$$\nabla \bar{\mathbf{i}}_\beta^t(\mathbf{x}, s) \quad by \quad \mathbf{Q}(t)\nabla \bar{\mathbf{i}}_\beta^t(\mathbf{x}, s)\mathbf{Q}(t)^T\mathbf{Q}(t)^T$$

$$\mathbf{b}^{(\beta)t}(\mathbf{x}, s) \quad by \quad \mathbf{Q}(t)\mathbf{b}^{(\beta)t}(\mathbf{x}, s)\mathbf{Q}(t)^T$$

$$\nabla \boldsymbol{\nu}^{(\beta)t}(\mathbf{x}, s) \quad by \quad \mathbf{Q}(t)\nabla \boldsymbol{\nu}^{(\beta)t}(\mathbf{x}, s)\mathbf{Q}(t)^T\mathbf{Q}(t)^T$$

and $\Upsilon_\alpha(\mathbf{x}, t)$ by $\Upsilon_\alpha(\mathbf{x}, t)$ if this is a scalar, by $\mathbf{Q}(t)\Upsilon_\alpha(\mathbf{x}, t)$ if this is a vector, and by $\mathbf{Q}(t)\Upsilon_\alpha(\mathbf{x}, t)\mathbf{Q}(t)^T$ if this is a second order tensor. Using (6.8.5), $\mathbf{B}_\beta(\mathbf{x}, t)$ and $\nabla\mathbf{B}_\beta(\mathbf{x}, t)$ are replaced by the following relations:

$$\begin{aligned} \mathbf{B}_\beta(\mathbf{x}, t) &= \mathbf{Q}(t)\mathbf{F}_\beta(\mathbf{X}_\beta, t)\mathbf{Q}(t)^T(\mathbf{Q}(t)\mathbf{F}_\beta(\mathbf{X}_\beta, t)\mathbf{Q}(t)^T)^T \\ &= \mathbf{Q}(t)\mathbf{F}_\beta(\mathbf{X}_\beta, t)\mathbf{F}_\beta(\mathbf{X}_\beta, t)^T\mathbf{Q}(t)^T = \mathbf{Q}(t)\mathbf{B}_\beta(\mathbf{x}, t)\mathbf{Q}(t)^T \end{aligned} \tag{6.8.24}$$

$$\nabla\mathbf{B}_\beta(\mathbf{x}, t) = \mathbf{Q}(t)\nabla\mathbf{B}_\beta(\mathbf{x}, t)\mathbf{Q}(t)^T\mathbf{Q}(t)^T \tag{6.8.25}$$

Using now (6.8.10) in (6.8.4) results in

$$\begin{aligned} \mathbf{C}_\beta(\mathbf{X}_\beta, \tau) &= \mathbf{F}_\beta(\mathbf{X}_\beta, \tau)^T\mathbf{F}_\beta(\mathbf{X}_\beta, \tau) \\ &= (\mathbf{F}_{\beta t}(\mathbf{x}, \tau)\mathbf{F}_\beta(\mathbf{X}_\beta, t))^T\mathbf{F}_{\beta t}(\mathbf{x}, \tau)\mathbf{F}_\beta(\mathbf{X}_\beta, t) \\ &= \mathbf{F}_\beta(\mathbf{X}_\beta, t)^T\mathbf{C}_{\beta t}(\mathbf{X}_\beta, \tau)\mathbf{F}_\beta(\mathbf{X}_\beta, t) \end{aligned} \tag{6.8.26}$$

from where

$$\mathbf{C}_{\beta t}(\mathbf{X}_\beta, \tau) = (\mathbf{F}_\beta(\mathbf{X}_\beta, t)^T)^{-1}\mathbf{F}_\beta(\mathbf{X}_\beta, \tau)^T\mathbf{F}_\beta(\mathbf{X}_\beta, \tau)\mathbf{F}_\beta(\mathbf{X}_\beta, t)^{-1} \tag{6.8.27}$$

Upon taking in this equation $\tau = t - s$ and using the above-mentioned substitution for

$$\mathbf{F}_\beta^t(\mathbf{X}_\beta, s) = \mathbf{Q}(t)\mathbf{F}_\beta^t(\mathbf{X}_\beta, s)\mathbf{Q}(t)^T$$

it follows that $\mathbf{C}_{\beta t}^t(\mathbf{X}_\beta, s)$ in (6.8.23) is replaced by $\mathbf{Q}(t)\mathbf{C}_{\beta t}^t(\mathbf{X}_\beta, s)\mathbf{Q}(t)^T$, and $GRAD\mathbf{C}_{\beta t}^t(\mathbf{X}_\beta, s)$ by

$$\mathbf{Q}(t)(GRAD\mathbf{C}_{\beta t}^t(\mathbf{X}_\beta, s))\mathbf{Q}(t)^T\mathbf{Q}(t)^T$$

Equation (6.8.23) is thus reduced to

$$\begin{aligned} &G_{\alpha\mathbf{K}}[\mathbf{Q}(t)\mathbf{C}_{\beta t}^t(\mathbf{X}_\beta, s)\mathbf{Q}(t)^T,\ \mathbf{Q}(t)\mathbf{B}_\beta(\mathbf{x}, t)\mathbf{Q}(t)^T, \\ &\mathbf{Q}(t)(GRAD\mathbf{C}_{\beta t}^t(\mathbf{X}_\beta, s))\mathbf{Q}(t)^T\mathbf{Q}(t)^T,\ \mathbf{Q}(t)\nabla\mathbf{B}_\beta(\mathbf{x}, t)\mathbf{Q}(t)^T\mathbf{Q}(t)^T, \\ &\mathfrak{R}_{\beta\mathbf{K}}^t(\mathbf{X}_\beta, s),\ GRAD\mathfrak{R}_{\beta\mathbf{K}}^t(\mathbf{X}_\beta, s),\ \bar{\bar{\theta}}_\beta^t(\mathbf{x}, s),\ \mathbf{Q}(t)\nabla\bar{\bar{\theta}}_\beta^t(\mathbf{x}, s), \\ &\mathbf{Q}(t)\bar{\mathbf{i}}_\beta^t(\mathbf{x}, s)\mathbf{Q}(t)^T,\ \mathbf{Q}(t)\nabla\bar{\mathbf{i}}_\beta^t(\mathbf{x}, s)\mathbf{Q}(t)^T\mathbf{Q}(t)^T, \\ &\mathbf{Q}(t)\mathbf{b}^{(\beta)t}(\mathbf{x}, s)\mathbf{Q}(t)^T,\ \mathbf{Q}(t)\nabla\boldsymbol{\nu}^{(\beta)t}(\mathbf{x}, s)\mathbf{Q}(t)^T\mathbf{Q}(t)^T] \\ &= \mathbf{Q}(t)G_{\alpha\mathbf{K}}[\mathbf{C}_{\beta t}^t(\mathbf{X}_\beta, s),\ \mathbf{B}_\beta(\mathbf{x}, t),\ GRAD\mathbf{C}_{\beta t}^t(\mathbf{X}_\beta, s),\ \nabla\mathbf{B}_\beta(\mathbf{x}, t), \\ &\mathfrak{R}_{\beta\mathbf{K}}^t(\mathbf{X}_\beta, s),\ GRAD\mathfrak{R}_{\beta\mathbf{K}}^t(\mathbf{X}_\beta, s),\ \bar{\bar{\theta}}_\beta^t(\mathbf{x}, s),\ \nabla\bar{\bar{\theta}}_\beta^t(\mathbf{x}, s),\ \bar{\mathbf{i}}_\beta^t(\mathbf{x}, s), \\ &\nabla\bar{\mathbf{i}}_\beta^t(\mathbf{x}, s),\ \mathbf{b}^{(\beta)t}(\mathbf{x}, s),\ \nabla\boldsymbol{\nu}^{(\beta)t}(\mathbf{x}, s)] \end{aligned} \tag{6.8.28}$$

***This equation represents a combined result of the principle of local action, material frame-indifference, and material symmetry, satisfying the isotropicity requirement of each phase in the mixture.***

## 6.9 Solidlike and Fluidlike Mixtures

The physical behavior of "solids" is different from "fluids" in that solids have preferred configurations and their thermomechanical response is different from different configurations. This difference forms a basis for categorizing solidlike and fluidlike multiphase mixtures.

### 6.9.1 Solidlike Mixtures

A phase $\delta$ can be said to be *solidlike* if there exists a reference configuration $\kappa_\delta$ such that the isotropy group $g_{\kappa\delta} = (\mathbf{H}_\delta, \mathbf{J}_\delta) = (\mathbf{H}_\delta, GRAD\mathbf{H}_\delta)$ is a linear group $(\mathbf{H}_\delta, \mathbf{O})$ and a subgroup of the orthogonal group $O_\delta$. Formally,

$$g_{\kappa\delta} = (\mathbf{H}_\delta, \mathbf{O}) \quad \rightarrow \quad g_{\kappa\delta} = \mathbf{H}_\delta \subset O_\delta \tag{6.9.1}$$

where the reference configurations with this property are called the undisturbed states of the solid. If the phase $\delta$ possesses a reference configuration $\kappa_\delta$ such that the isotropy group $g_{\kappa\delta}$ is a proper subgroup of the full orthogonal group $O_\delta$ then such a material can be categorized as being *anisotropic or aelotropic*. The isotropy group of the solid may be any subgroup of the orthogonal group that contains the inversion transformation $-\mathbf{I}$ and it can be, therefore, generated by $-\mathbf{I}$ and a subgroup of rotations $g_{R\kappa\delta}$ (consisting of proper orthogonal transformations). TRUESDELL & NOLL (1965) in section 33 of their treatise discuss 11 such rotation groups corresponding to 32 crystal classes. Here it is only noted that the *transverse isotropy* is characterized by a group of material symmetries generated by $g_{R\kappa\delta}$ in an undistorted material state for which any rotations about an axis for a given direction is a symmetry operation (the materials with the laminated structure). Materials with $g_{R\kappa\delta}$ consisting of only $\mathbf{I}$ and $-\mathbf{I}$ are solids corresponding to the *triclinic* system, whereas an *orthotropic* solid is characterized by an isotropy group containing reflections on three mutually perpendicular planes.

A material that is both solidlike and isotropic is referred as an *isotropic solid.* In this situation the phase $\delta$ must possess a local reference configuration such that its linear isotropy group coincides with the orthogonal group, *i.e.*

$$g_{\kappa\delta} = (\mathbf{H}_\delta, \mathbf{O}) = O_\delta \tag{6.9.2}$$

with $\mathbf{H}_\delta$ an arbitrary orthogonal transformation. If, moreover, phase $\delta$ cannot undergo a phase change, then $\mathbf{H}_\delta$ is unimodular as demonstrated in section 7 by (6.7.10).

To determine the constitutive response functional $G_{\alpha\mathcal{K}}$ for isotropic solid phases, it is necessary to examine the combined effects of material symmetry and the requirement (6.9.2) imposed on this functional. For this purpose, we can start from (6.7.9) by choosing a necessary condition $\mathbf{H}_\delta = \mathbf{R}^t(s)^T = \mathbf{R}^{tT}$ and using (6.8.3) and (6.8.5) with $\beta = \delta$. For the $\delta$'th phase in (6.7.9) we thus have

$$\mathbf{F}^t_\delta \mathbf{H}_\delta = \mathbf{V}^t_\delta \mathbf{R}^t \mathbf{R}^{tT} = \mathbf{V}^t_\delta = (\mathbf{B}^t_\delta)^{1/2} \tag{6.9.3}$$

and

$$(GRAD\mathbf{F}^t_\delta)\mathbf{H}_\delta \mathbf{H}_\delta + \mathbf{F}^t_\delta GRAD\mathbf{H}_\delta = (GRAD\mathbf{V}^t_\delta \mathbf{R}^t)\mathbf{R}^{tT}\mathbf{R}^{tT} \tag{6.9.4}$$

which in component notation is written as

$$\begin{aligned} \frac{\partial^2 x_i}{\partial X_{\delta L}\partial X_{\delta M}} R^{tT}_{Lk} R^{tT}_{Mq} &= \frac{\partial^2 x_i}{\partial X_{\delta L}\partial X_{\delta M}} (F^t_{\delta Lj})^{-1} V^t_{\delta jk} (F^t_{\delta M\ell})^{-1} V^t_{\delta \ell q} \\ &= \frac{\partial^2 x_i}{\partial X_{\delta L}\partial X_{\delta M}} \frac{\partial X_{\delta L}}{\partial x_j} \frac{\partial X_{\delta M}}{\partial x_\ell} V^t_{\delta jk} V^t_{\delta \ell q} \end{aligned} \tag{6.9.5}$$

since $\mathbf{F}^t_\delta = \mathbf{V}^t_\delta \mathbf{R}^t$ and $(\mathbf{R}^t)^{-1} = \mathbf{R}^{tT} = (\mathbf{F}^t_\delta)^{-1}\mathbf{V}^t_\delta$. Using now (6.8.5) it is possible to show that (6.9.5) can be transformed into the following form:

$$2\frac{\partial^2 x_i}{\partial X_{\delta L}\partial X_{\delta M}} \frac{\partial X_{\delta L}}{\partial x_j} \frac{\partial X_{\delta M}}{\partial x_\ell} = (B^t_{\delta jq})^{-1} \frac{\partial B^t_{\delta qi}}{\partial x_\ell} + (B^t_{\delta \ell q})^{-1} \frac{\partial B^t_{\delta qi}}{\partial x_j} + B^t_{\delta qi} \frac{\partial (B^t_{\delta j\ell})^{-1}}{\partial x_q} \tag{6.9.6}$$

Moreover,

$$|det\mathbf{H}_\delta| = |det\mathbf{R}^{tT}| = 1 \tag{6.9.7}$$

Equations (6.9.3)-(6.9.7) illustrate that a function of

$$\mathbf{F}^t_\delta \mathbf{H}_\delta,\ (GRAD\mathbf{F}^t_\delta)\mathbf{H}_\delta\mathbf{H}_\delta + \mathbf{F}^t_\delta GRAD\mathbf{H}_\delta,\ det\mathbf{H}_\delta$$

can be replaced by one of $\mathbf{B}^t_\delta$ and $\nabla\mathbf{B}^t_\delta$. Consequently, for the $\delta$'th phase to be an isotropic solid the constitutive response functional (6.7.9) becomes

$$\begin{aligned} \Upsilon_\alpha(\mathbf{x},t) = G_{\alpha\mathcal{K}}[&\mathbf{F}^t_1, \ldots, \mathbf{B}^t_\delta, \ldots, \mathbf{F}^t_\gamma, GRAD\mathbf{F}^t_1, \ldots, \nabla\mathbf{B}^t_\delta, \ldots, \\ &GRAD\mathbf{F}^t_\gamma, \mathfrak{R}^t_{\beta\mathcal{K}}, GRAD\mathfrak{R}^t_{\beta\mathcal{K}}, \bar{\theta}^t_\beta, \nabla\bar{\theta}^t_\beta, \bar{\mathbf{i}}^t_\beta, \nabla\bar{\mathbf{i}}^t_\beta, \mathbf{b}^{(\beta)t}, \nabla\boldsymbol{\nu}^{(\beta)t}] \end{aligned} \tag{6.9.8}$$

where for each $\alpha$, $\alpha=1,\ldots,\gamma$, $\beta$ ranges from 1 to $\gamma$. Combining this result with the restriction imposed by the principle of material frame-indifference as expressed by (6.7.23) yields a condition on the functional $G_{\alpha\boldsymbol{\kappa}}$ with $\delta$'th *phase being an isotropic solid.* Thus, with $\Upsilon_\alpha(\mathbf{x},t)$ a vector it follows that

$$\begin{aligned}
&G_{\alpha\boldsymbol{\kappa}}[\mathbf{Q}(t)\mathbf{F}_1^t,\ldots,\ \mathbf{Q}(t)\mathbf{B}_\delta^t\mathbf{Q}(t)^T,\ldots,\ \mathbf{Q}(t)\mathbf{F}_\gamma^t,\ \mathbf{Q}(t)GRAD\mathbf{F}_1^t,\ldots,\\
&\quad \mathbf{Q}(t)\nabla\mathbf{B}_\delta^t\mathbf{Q}(t)^T\mathbf{Q}(t)^T,\ldots,\ \mathbf{Q}(t)GRAD\mathbf{F}_\gamma^t,\ \mathfrak{R}_{\beta\boldsymbol{\kappa}}^t,\ GRAD\mathfrak{R}_{\beta\boldsymbol{\kappa}}^t,\\
&\qquad \bar{\bar{\theta}}_\beta^t,\ \mathbf{Q}(t)\nabla\bar{\bar{\theta}}_\beta^t,\ \mathbf{Q}(t)\bar{\bar{\mathbf{i}}}_\beta^t\mathbf{Q}(t)^T,\ \mathbf{Q}(t)\nabla\bar{\bar{\mathbf{i}}}_\beta^t\mathbf{Q}(t)^T\mathbf{Q}(t)^T,\\
&\qquad\quad \mathbf{Q}(t)\mathbf{b}^{(\beta)t}\mathbf{Q}(t)^T,\ \mathbf{Q}(t)\nabla\boldsymbol{\nu}^{(\beta)t}\mathbf{Q}(t)^T\mathbf{Q}(t)^T]\\
&= \mathbf{Q}(t)G_{\alpha\boldsymbol{\kappa}}[\mathbf{F}_1^t,\ldots,\ \mathbf{B}_\delta^t,\ldots,\ \mathbf{F}_\gamma^t,\ GRAD\mathbf{F}_1^t,\ldots,\ \nabla\mathbf{B}_\delta^t,\ldots,\ GRAD\mathbf{F}_\gamma^t,\\
&\qquad \mathfrak{R}_{\beta\boldsymbol{\kappa}}^t,\ GRAD\mathfrak{R}_{\beta\boldsymbol{\kappa}}^t,\ \bar{\bar{\theta}}_\beta^t,\ \nabla\bar{\bar{\theta}}_\beta^t,\ \bar{\bar{\mathbf{i}}}_\beta^t,\ \nabla\bar{\bar{\mathbf{i}}}_\beta^t,\ \mathbf{b}^{(\beta)t},\ \nabla\boldsymbol{\nu}^{(\beta)t}]
\end{aligned}\tag{6.9.9}$$

When all phases in a mixture are isotropic solids, it is necessary to replace in the above equation $\mathbf{Q}(t)\mathbf{F}_\epsilon^t$, $\epsilon\neq\delta$, by $\mathbf{Q}(t)\mathbf{B}_\epsilon^t\mathbf{Q}(t)^T$, $\mathbf{F}_\epsilon^t$ by $\mathbf{B}_\epsilon^t$, $\mathbf{Q}(t)GRAD\mathbf{F}_\epsilon^t$ by $\mathbf{Q}(t)\nabla\mathbf{B}_\epsilon^t\mathbf{Q}(t)^T\mathbf{Q}(t)^T$, and $GRAD\mathbf{F}_\epsilon^t$ by $\nabla\mathbf{B}_\epsilon^t$. In this situation the result becomes equivalent to (6.8.28), since for isotropic materials a function of $\mathbf{B}_\beta^t(\mathbf{x},s)$ and $\nabla\mathbf{B}_\beta^t(\mathbf{x},s)$ can be replaced by the one of

$$\mathbf{C}_{\beta t}^t(\mathbf{X}_\beta,s),\ \mathbf{B}_\beta(\mathbf{x},t),\ GRAD\mathbf{C}_{\beta t}^t(\mathbf{X}_\beta,s),\ \nabla\mathbf{B}_\beta(\mathbf{x},t)$$

The proof of this assertion follows from (6.8.27) and (6.8.5), since

$$\begin{aligned}
&\mathbf{C}_{\beta t}(\mathbf{X}_\beta,\tau)\\
&= \mathbf{B}_\beta(\mathbf{x},t)^{-1}\mathbf{F}_\beta(\mathbf{X}_\beta,t)\mathbf{F}_\beta(\mathbf{X}_\beta,\tau)^T\mathbf{F}_\beta(\mathbf{X}_\beta,\tau)\mathbf{F}_\beta(\mathbf{X}_\beta,t)^T\mathbf{B}_\beta(\mathbf{x},t)^{-1}
\end{aligned}\tag{6.9.10}$$

and for isotropic materials a necessary condition for the deformation gradient is that

$$\mathbf{F}_\beta(\mathbf{X}_\beta,\tau) = \mathbf{V}_\beta(\mathbf{X}_\beta,\tau)\mathbf{R}(t)\tag{6.9.11}$$

where $\mathbf{R}(t)$ is the history-independent rotation tensor. Upon substitution of this equation into (6.9.10) gives

$$\mathbf{C}_{\beta t}(\mathbf{X}_\beta,\tau) = (\mathbf{V}_\beta(\mathbf{X}_\beta,t)^T)^{-1}\mathbf{V}_\beta(\mathbf{X}_\beta,\tau)^T\mathbf{V}_\beta(\mathbf{X}_\beta,\tau)\mathbf{V}_\beta(\mathbf{X}_\beta,t)^T\mathbf{B}_\beta(\mathbf{x},t)^{-1}\tag{6.9.12}$$

which when using (6.8.5) and putting $\tau=t-s$ proves the assertion.

### 6.9.2 Fluidlike Mixtures

When phase change occurs for the $\delta$'th phase, $det\mathbf{H}_\delta \neq 1$ necessarily, and $\mathbf{H}_\delta$ is not a unimodular tensor. Following CROSS (1973), we will define the $\delta$'th ***phase to be a fluid with phase change*** whenever the following condition is satisfied:

$$(\mathbf{H}_\delta, \mathbf{J}_\delta) \subset g_{\boldsymbol{\kappa}\delta} \tag{6.9.13}$$

that is, the group element $(\mathbf{H}_\delta, \mathbf{J}_\delta)$ belongs to the isotropy group $g_{\boldsymbol{\kappa}\delta}$. As a necessary condition we can take $(\mathbf{H}_\delta, \mathbf{J}_\delta) = ((\mathbf{F}^t_\delta)^{-1}, GRAD(\mathbf{F}^t_\delta)^{-1})$ from where it follows that

$$\mathbf{F}^t_\delta \mathbf{H}^t_\delta = \mathbf{F}^t_\delta (\mathbf{F}^t_\delta)^{-1} = \mathbf{I} \tag{6.9.14}$$

$$\begin{aligned}(GRAD\mathbf{F}^t_\delta)\mathbf{H}_\delta\mathbf{H}_\delta + \mathbf{F}^t_\delta GRAD\mathbf{H}_\delta &= \\ (GRAD\mathbf{F}^t_\delta)(\mathbf{F}^t_\delta)^{-1}(\mathbf{F}^t_\delta)^{-1} + \mathbf{F}^t_\delta \boldsymbol{\nabla}(\mathbf{F}^t_\delta)^{-1} &= \mathbf{0}\end{aligned} \tag{6.9.15}$$

The last result is proved by using the fact that $\mathbf{F}^t_\delta(\mathbf{F}^t_\delta)^{-1} = \mathbf{I}$. Moreover, upon using (6.2.3) and (6.9.14) we obtain

$$\mathfrak{R}^t_{\delta\boldsymbol{\kappa}}|det\mathbf{H}_\delta| = \bar{\rho}^t_\delta |det\mathbf{F}^t_\delta||det\mathbf{H}_\delta| = \bar{\rho}^t_\delta |det\mathbf{F}^t_\delta \mathbf{H}_\delta| = \bar{\rho}^t_\delta \tag{6.9.16}$$

and

$$GRAD(\mathfrak{R}^t_{\delta\boldsymbol{\kappa}}|det\mathbf{H}_\delta|) = \boldsymbol{\nabla}\bar{\rho}^t_\delta \tag{6.9.17}$$

These results show that for a fluid phase $\delta$, a function of

$$\mathbf{F}^t_\delta\mathbf{H}_\delta,\ (GRAD\mathbf{F}^t_\delta)\mathbf{H}_\delta\mathbf{H}_\delta + \mathbf{F}^t_\delta GRAD\mathbf{H}_\delta,\ \mathfrak{R}^t_{\delta\boldsymbol{\kappa}}|det\mathbf{H}_\delta|,\ GRAD(\mathfrak{R}^t_{\delta\boldsymbol{\kappa}}|det\mathbf{H}_\delta|)$$

can be replaced by one of $(\bar{\rho}^t_\delta, \boldsymbol{\nabla}\bar{\rho}^t_\delta)$, thereby reducing (6.7.9) to the following form:

$$\begin{aligned}\Upsilon_\alpha(\mathbf{x},t) = G_{\alpha\boldsymbol{\kappa}}[&\mathbf{F}^t_1,\ldots,\ \mathbf{F}^t_{\delta-1},\ \mathbf{F}^t_{\delta+1},\ldots,\ \mathbf{F}^t_\gamma,\ GRAD\mathbf{F}^t_1,\ldots, \\ &GRAD\mathbf{F}^t_{\delta-1},\ GRAD\mathbf{F}^t_{\delta+1},\ldots,\ GRAD\mathbf{F}^t_\gamma,\ \mathfrak{R}^t_{1\boldsymbol{\kappa}},\ldots,\ \mathfrak{R}^t_{\delta-1\boldsymbol{\kappa}},\ \bar{\rho}^t_\delta, \\ &\mathfrak{R}^t_{\delta+1\boldsymbol{\kappa}},\ldots,\ \mathfrak{R}^t_{\gamma\boldsymbol{\kappa}},\ GRAD\mathfrak{R}^t_{1\boldsymbol{\kappa}},\ldots,\ GRAD\mathfrak{R}^t_{\delta-1\boldsymbol{\kappa}},\ \boldsymbol{\nabla}\bar{\rho}^t_\delta, \\ &GRAD\mathfrak{R}^t_{\delta+1\boldsymbol{\kappa}},\ldots,\ GRAD\mathfrak{R}^t_{\gamma\boldsymbol{\kappa}},\ \bar{\bar{\theta}}^t_\beta,\ \boldsymbol{\nabla}\bar{\bar{\theta}}^t_\beta,\ \bar{\mathbf{i}}^t_\beta,\ \boldsymbol{\nabla}\bar{\mathbf{i}}^t_\beta,\ \mathbf{b}^{(\beta)t},\ \boldsymbol{\nabla}\boldsymbol{\nu}^{(\beta)t}]\end{aligned} \tag{6.9.18}$$

where, as before, for each $\alpha$, $\alpha = 1,\ldots,\gamma$, $\beta$ ranges from 1 to $\gamma$. This result is a necessary consequence of the material symmetry requirement for the $\delta$'th phase to be a fluid, and when it is combined with (6.7.23), a combined restriction of material symmetry and principle of material frame-indifference is obtained. Hence,

$$
\begin{aligned}
&G_{\alpha\boldsymbol{\kappa}}[\mathbf{Q}(t)\mathbf{F}_1^t,\ldots,\ \mathbf{Q}(t)\mathbf{F}_{\delta-1}^t,\ \mathbf{Q}(t)\mathbf{F}_{\delta+1}^t,\ldots,\ \mathbf{Q}(t)\mathbf{F}_\gamma^t,\ \mathbf{Q}(t)GRAD\mathbf{F}_1^t,\ldots,\\
&\quad \mathbf{Q}(t)GRAD\mathbf{F}_{\delta-1}^t,\ \mathbf{Q}(t)GRAD\mathbf{F}_{\delta+1}^t,\ldots,\ \mathbf{Q}(t)GRAD\mathbf{F}_\gamma^t,\ \mathfrak{R}_{1\boldsymbol{\kappa}}^t,\ldots,\\
&\quad \mathfrak{R}_{\delta-1\boldsymbol{\kappa}}^t,\ \bar{\rho}_\delta^t,\ \mathfrak{R}_{\delta+1\boldsymbol{\kappa}}^t,\ldots,\ \mathfrak{R}_{\gamma\boldsymbol{\kappa}}^t,\ GRAD\mathfrak{R}_{1\boldsymbol{\kappa}}^t,\ldots,\ GRAD\mathfrak{R}_{\delta-1\boldsymbol{\kappa}}^t,\\
&\quad \mathbf{Q}(t)\boldsymbol{\nabla}\bar{\rho}_\delta^t,\ GRAD\mathfrak{R}_{\delta+1\boldsymbol{\kappa}}^t,\ldots,\ GRAD\mathfrak{R}_{\gamma\boldsymbol{\kappa}}^t,\ \bar{\bar{\theta}}_\beta^t,\ \mathbf{Q}(t)\boldsymbol{\nabla}\bar{\bar{\theta}}_\beta^t,\ \mathbf{Q}(t)\bar{\mathbf{i}}_\beta^t\mathbf{Q}(t)^T,\\
&\quad \mathbf{Q}(t)\boldsymbol{\nabla}\bar{\mathbf{i}}_\beta^t\mathbf{Q}(t)^T\mathbf{Q}(t)^T,\ \mathbf{Q}(t)\mathbf{b}^{(\beta)t}\mathbf{Q}(t)^T,\ \mathbf{Q}(t)\boldsymbol{\nabla}\boldsymbol{\nu}^{(\beta)t}\mathbf{Q}(t)^T\mathbf{Q}(t)^T]\\
&= \mathbf{Q}(t)G_{\alpha\boldsymbol{\kappa}}[\mathbf{F}_1^t,\ldots,\ \mathbf{F}_{\delta-1}^t,\ \mathbf{F}_{\delta+1}^t,\ldots,\ \mathbf{F}_\gamma^t,\ GRAD\mathbf{F}_1^t,\ldots,\\
&\quad GRAD\mathbf{F}_{\delta-1}^t,\ GRAD\mathbf{F}_{\delta+1}^t,\ldots,\ GRAD\mathbf{F}_\gamma^t,\ \mathfrak{R}_{1\boldsymbol{\kappa}}^t,\ldots,\ \mathfrak{R}_{\delta-1\boldsymbol{\kappa}}^t,\ \bar{\rho}_\delta^t,\\
&\quad \mathfrak{R}_{\delta+1\boldsymbol{\kappa}}^t,\ldots,\ \mathfrak{R}_{\gamma\boldsymbol{\kappa}}^t,\ GRAD\mathfrak{R}_{1\boldsymbol{\kappa}}^t,\ldots,\ GRAD\mathfrak{R}_{\delta-1\boldsymbol{\kappa}}^t,\ \boldsymbol{\nabla}\bar{\rho}_\delta^t,\\
&\quad GRAD\mathfrak{R}_{\delta+1\boldsymbol{\kappa}}^t,\ldots,\ GRAD\mathfrak{R}_{\gamma\boldsymbol{\kappa}}^t,\ \bar{\bar{\theta}}_\beta^t,\ \boldsymbol{\nabla}\bar{\bar{\theta}}_\beta^t,\ \bar{\mathbf{i}}_\beta^t,\ \boldsymbol{\nabla}\bar{\mathbf{i}}_\beta^t,\ \mathbf{b}^{(\beta)t},\ \boldsymbol{\nabla}\boldsymbol{\nu}^{(\beta)t}]
\end{aligned}
\tag{6.9.19}
$$

When all phases in the mixture are fluids, it is necessary to put in the above equation $\mathbf{F}_\beta^t = \mathbf{0}$, $\mathfrak{R}_{\beta\boldsymbol{\kappa}}^t = \bar{\rho}_\beta^t$, $GRAD\mathfrak{R}_{\beta\boldsymbol{\kappa}}^t = \boldsymbol{\nabla}\bar{\rho}_\beta^t$, for $\beta = 1,\ldots,\delta-1,\delta+1,\ldots,\gamma$ and let $\delta$ range from 1 to $\gamma$.

For a multiphase mixture with the $\beta$'th phase a fluid, the constitutive relation (6.8.28) cannot depend on $\mathbf{B}_\beta(\mathbf{x},t)$ and $\boldsymbol{\nabla}\mathbf{B}_\beta(\mathbf{x},t)$, but on $\bar{\rho}_\beta(\mathbf{x},t)$ and $\boldsymbol{\nabla}\bar{\rho}_\beta(\mathbf{x},t)$. This proof is left to the reader.

### 6.9.3 Mixtures of Isotropic Solids and Fluids

For a mixture consisting of isotropic solids and fluids, the results of sections 6.9.1 and 6.9.2 can be combined. Thus, for a *multiphase mixture consisting of $M$ isotropic solid phases and $\gamma - M$ fluid phases*, the combination of (6.9.9) and (6.9.19) gives the following form of the constitutive response functional:

$$
\begin{aligned}
&G_{\alpha\boldsymbol{\kappa}}[\mathbf{Q}(t)\mathbf{B}_1^t\mathbf{Q}(t)^T,\ldots,\ \mathbf{Q}(t)\mathbf{B}_M^t\mathbf{Q}(t)^T,\ \bar{\rho}_{M+1}^t,\ldots,\ \bar{\rho}_\gamma^t,\\
&\quad \mathbf{Q}(t)\boldsymbol{\nabla}\mathbf{B}_1^t\mathbf{Q}(t)^T\mathbf{Q}(t)^T,\ldots,\ \mathbf{Q}(t)\boldsymbol{\nabla}\mathbf{B}_M^t\mathbf{Q}(t)^T\mathbf{Q}(t)^T,\\
&\quad \mathbf{Q}(t)\boldsymbol{\nabla}\bar{\rho}_{M+1}^t,\ldots,\ \mathbf{Q}(t)\boldsymbol{\nabla}\bar{\rho}_\gamma^t,\ \mathfrak{R}_{1\boldsymbol{\kappa}}^t,\ldots,\ \mathfrak{R}_{M\boldsymbol{\kappa}}^t,\ GRAD\mathfrak{R}_{1\boldsymbol{\kappa}}^t,\ldots,\\
&\quad GRAD\mathfrak{R}_{M\boldsymbol{\kappa}}^t,\ \bar{\bar{\theta}}_\beta^t,\ \mathbf{Q}(t)\boldsymbol{\nabla}\bar{\bar{\theta}}_\beta^t,\ \mathbf{Q}(t)\bar{\mathbf{i}}_\beta^t\mathbf{Q}(t)^T,\\
&\quad \mathbf{Q}(t)\boldsymbol{\nabla}\bar{\mathbf{i}}_\beta^t\mathbf{Q}(t)^T\mathbf{Q}(t)^T,\ \mathbf{Q}(t)\mathbf{b}^{(\beta)t}\mathbf{Q}(t)^T,\ \mathbf{Q}(t)\boldsymbol{\nabla}\boldsymbol{\nu}^{(\beta)t}\mathbf{Q}(t)^T\mathbf{Q}(t)^T]\\
&= \mathbf{Q}(t)G_{\alpha\boldsymbol{\kappa}}[\mathbf{B}_1^t,\ldots,\ \mathbf{B}_M^t,\ \bar{\rho}_{M+1}^t,\ldots,\ \bar{\rho}_\gamma^t,\ \boldsymbol{\nabla}\mathbf{B}_1^t,\ldots,\ \boldsymbol{\nabla}\mathbf{B}_M^t,\\
&\quad \boldsymbol{\nabla}\bar{\rho}_{M+1}^t,\ldots,\ \boldsymbol{\nabla}\bar{\rho}_\gamma^t,\ \mathfrak{R}_{1\boldsymbol{\kappa}}^t,\ldots,\ \mathfrak{R}_{M\boldsymbol{\kappa}}^t,\ GRAD\mathfrak{R}_{1\boldsymbol{\kappa}}^t,\ldots,\ GRAD\mathfrak{R}_{M\boldsymbol{\kappa}}^t,\\
&\quad \bar{\bar{\theta}}_\beta^t,\ \boldsymbol{\nabla}\bar{\bar{\theta}}_\beta^t,\ \bar{\mathbf{i}}_\beta^t,\ \boldsymbol{\nabla}\bar{\mathbf{i}}_\beta^t,\ \mathbf{b}^{(\beta)t},\ \boldsymbol{\nabla}\boldsymbol{\nu}^{(\beta)t}]
\end{aligned}
\tag{6.9.20}
$$

where for each $\alpha$, $\alpha = 1, \ldots, \gamma$, the range of $\beta$ is from 1 to $\gamma$, and the notation of $\phi_\beta^t = \phi_\beta(\mathbf{x}, t - s)$ is used for the independent variables $\phi$.

## 6.10 Combined Principles of Local Action and Smooth and Local Memory for a Mixture of Isotropic Solids and Fluids

In section (6.5) we discussed the combination of constitutive relations (6.3.8) and (6.4.6) expressing the combined effect of the principles of local action and smooth and local memory. This result, expressed by (6.5.1), may be reduced further by invoking the additional principle of the material frame-indifference and isotropicity assumption on solid and fluid phases. Towards this objective, the starting point can be choosen more conveniently as equation (6.6.28) satisfying the combined principles of local action, smooth and local memory, and material frame-indifference. For $M$ isotropic solids, we can simply put in (6.9.9) $s = 0$ and use the result to replace in (6.6.28): (1) $\mathbf{QF}_\beta$ by $\mathbf{QB}_\beta\mathbf{Q}^T$, and (2) $\mathbf{Q}GRAD\mathbf{F}_\beta$ by $\mathbf{Q}\nabla\mathbf{B}_\beta\mathbf{Q}^T\mathbf{Q}^T$. In a similar manner, use can be made of (6.6.19) with $s = 0$ and $\gamma - M$ fluids to replace in (6.6.28): (1) $\mathbf{QF}_\beta$ and $\mathfrak{R}_{\beta\boldsymbol{\kappa}}$ by $\bar{\rho}_\beta$, and (2) $\mathbf{Q}GRAD\mathbf{F}_\beta$ and $GRAD\mathfrak{R}_{\beta\boldsymbol{\kappa}}$ by $\nabla\bar{\rho}_\beta$. The result is then given by

$$
\begin{aligned}
&G_{\alpha\boldsymbol{\kappa}}[\mathbf{QB}_1\mathbf{Q}^T, \ldots, \mathbf{QB}_M\mathbf{Q}^T, \mathbf{Q}\nabla\mathbf{B}_1\mathbf{Q}^T\mathbf{Q}^T, \ldots, \mathbf{Q}\nabla\mathbf{B}_M\mathbf{Q}^T\mathbf{Q}^T, \mathfrak{R}_{1\boldsymbol{\kappa}}, \ldots,\\
&\mathfrak{R}_{M\boldsymbol{\kappa}}, GRAD\mathfrak{R}_{1\boldsymbol{\kappa}}, \ldots, GRAD\mathfrak{R}_{M\boldsymbol{\kappa}}, \bar{\rho}_{M+1}, \ldots, \bar{\rho}_\gamma, \mathbf{Q}\nabla\bar{\rho}_{M+1}, \ldots, \mathbf{Q}\nabla\bar{\rho}_\gamma,\\
&\mathbf{QD}_{M+1}\mathbf{Q}^T, \ldots, \mathbf{QD}_\gamma\mathbf{Q}^T, \mathbf{Q}(\mathbf{W}_{M+1} - \mathbf{W}_\gamma)\mathbf{Q}^T, \ldots, \mathbf{Q}(\mathbf{W}_{\gamma-1} - \mathbf{W}_\gamma)\mathbf{Q}^T,\\
&\grave{\mathfrak{R}}_{1\boldsymbol{\kappa}}, \ldots, \grave{\mathfrak{R}}_{M\boldsymbol{\kappa}}, \grave{\bar{\rho}}_{M+1}, \ldots, \grave{\bar{\rho}}_\gamma, \bar{\bar{\theta}}_1, \ldots, \bar{\bar{\theta}}_\gamma, \grave{\bar{\bar{\theta}}}_1, \ldots, \grave{\bar{\bar{\theta}}}_\gamma, \mathbf{Q}\nabla\bar{\bar{\theta}}_1, \ldots,\\
&\mathbf{Q}\nabla\bar{\bar{\theta}}_\gamma, \mathbf{Q}(\tilde{\mathbf{v}}_1 - \tilde{\mathbf{v}}_\gamma), \ldots, \mathbf{Q}(\tilde{\mathbf{v}}_{\gamma-1} - \tilde{\mathbf{v}}_\gamma), \mathbf{Q}(\grave{\tilde{\mathbf{i}}}_1 - \mathbf{W}_1\tilde{\mathbf{i}}_1 - \tilde{\mathbf{i}}_1\mathbf{W}_1^T)\mathbf{Q}^T, \ldots,\\
&\mathbf{Q}(\grave{\tilde{\mathbf{i}}}_\gamma - \mathbf{W}_\gamma\tilde{\mathbf{i}}_\gamma - \tilde{\mathbf{i}}_\gamma\mathbf{W}_\gamma^T)\mathbf{Q}^T, \mathbf{Q}\tilde{\mathbf{i}}_1\mathbf{Q}^T, \ldots, \mathbf{Q}\tilde{\mathbf{i}}_\gamma\mathbf{Q}^T, \mathbf{Q}\nabla\tilde{\mathbf{i}}_1\mathbf{Q}^T\mathbf{Q}^T, \ldots,\\
&\mathbf{Q}\nabla\tilde{\mathbf{i}}_\gamma\mathbf{Q}^T\mathbf{Q}^T, \mathbf{Q}\mathbf{b}^{(1)}\mathbf{Q}^T, \ldots, \mathbf{Q}\mathbf{b}^{(\gamma)}\mathbf{Q}^T, \mathbf{Q}\nabla\boldsymbol{\nu}^{(1)}\mathbf{Q}^T\mathbf{Q}^T, \ldots, \mathbf{Q}\nabla\boldsymbol{\nu}^{(\gamma)}\mathbf{Q}^T\mathbf{Q}^T]\\
&= \mathbf{Q}G_{\alpha\boldsymbol{\kappa}}[\mathbf{B}_1, \ldots, \mathbf{B}_M, \nabla\mathbf{B}_1, \ldots, \nabla\mathbf{B}_M, \mathfrak{R}_{1\boldsymbol{\kappa}}, \ldots, \mathfrak{R}_{M\boldsymbol{\kappa}}, GRAD\mathfrak{R}_{1\boldsymbol{\kappa}}, \ldots,\\
&\quad GRAD\mathfrak{R}_{M\boldsymbol{\kappa}}, \bar{\rho}_{M+1}, \ldots, \bar{\rho}_\gamma, \nabla\bar{\rho}_{M+1}, \ldots, \nabla\bar{\rho}_\gamma, \mathbf{D}_{M+1}, \ldots, \mathbf{D}_\gamma,\\
&\quad \mathbf{W}_{M+1} - \mathbf{W}_\gamma, \ldots, \mathbf{W}_{\gamma-1} - \mathbf{W}_\gamma, \grave{\mathfrak{R}}_{1\boldsymbol{\kappa}}, \ldots, \grave{\mathfrak{R}}_{M\boldsymbol{\kappa}}, \grave{\bar{\rho}}_{M+1}, \ldots, \grave{\bar{\rho}}_\gamma,\\
&\quad \bar{\bar{\theta}}_1, \ldots, \bar{\bar{\theta}}_\gamma, \grave{\bar{\bar{\theta}}}_1, \ldots, \grave{\bar{\bar{\theta}}}_\gamma, \nabla\bar{\bar{\theta}}_1, \ldots, \nabla\bar{\bar{\theta}}_\gamma, \tilde{\mathbf{v}}_1 - \tilde{\mathbf{v}}_\gamma, \ldots, \tilde{\mathbf{v}}_{\gamma-1} - \tilde{\mathbf{v}}_\gamma,\\
&\quad \grave{\tilde{\mathbf{i}}}_1 - \mathbf{W}_1\tilde{\mathbf{i}}_1 - \tilde{\mathbf{i}}_1\mathbf{W}_1^T, \ldots, \grave{\tilde{\mathbf{i}}}_\gamma - \mathbf{W}_\gamma\tilde{\mathbf{i}}_\gamma - \tilde{\mathbf{i}}_\gamma\mathbf{W}_\gamma^T, \tilde{\mathbf{i}}_1, \ldots, \tilde{\mathbf{i}}_\gamma,\\
&\quad \nabla\tilde{\mathbf{i}}_1, \ldots, \nabla\tilde{\mathbf{i}}_\gamma, \ldots, \mathbf{b}^{(1)}, \ldots, \mathbf{b}^{(\gamma)}, \nabla\boldsymbol{\nu}^{(1)}, \ldots, \nabla\boldsymbol{\nu}^{(\gamma)}]
\end{aligned}
$$

(6.10.1)

where $\mathbf{Q} = \mathbf{Q}(t)$, $\Re_{\beta K} = \Re_{\beta K}(\mathbf{X}_\beta, t)$, and all other independent variables depend explicitly on $(\mathbf{x}, t)$. Notice that this result contains a number of important assumptions which can significantly limit the usefulness of the constitutive equations.

## 6.11 The Role of Entropy Inequality or Axiom of Dissipation in the Constitutive Theory

The exact role of the entropy inequality in the constitutive theory of multiphase mixtures is a matter of some controversy. As discussed in section 2.5, the derivation of the multiphase mixture field equations using the averaging approach leads naturally to an entropy inequality for *each phase*, as long as it is assumed that the local axiom of dissipation or entropy inequality within each sub-body $\delta$ of phase $\alpha$ is assumed valid, as common in modeling the single phase continua. In the postulatory theories of multiphase mixtures, however, only a global axiom of dissipation or entropy inequality for the *mixture as a whole* is postulated, as common in modeling the single phase multicomponent mixtures (BOWEN,1976).

The role of entropy inequality in the constitutive theory is to restrict the forms of constitutive assumptions. That is, a thermodynamic process that is compatible with an entropy inequality is called an *admissible thermodynamic process* and the inequality can be used to place restrictions on the constitutive response functionals. The resulting constitutive equations will thus depend not only on other principles of the constitutive theory as discussed in previous sections, but also on whether or not use is made of the mixture or phasic entropy inequality to restrict the forms of the constitutive equations. In the studies of constitutive equations in the following chapters, use will be made of the phasic axiom of dissipation (2.4.45) rather than of the mixture as a whole (equation (2.4.50)) for the reasons as discussed above and in chapter 2.

## 6.12 Internal Constraints

### 6.12.1 General Considerations

The internal constraints of a multiphase mixture are *a priori* assigned conditions or constitutive relations of material deformation, energy transfer, or phase change. As such, these constraints require a modification of the principle of determinism discussed at the beginning of the chapter. Constrained multiphase mixtures are more easily modeled than unconstrained mixtures, since one or more phases in the mixture may be incompressible, not undergoing a phase change, or the mixture as a whole may be saturated. The *saturation condition* refers to the distribution of phasic volumetric fractions $\phi_\alpha$ in a mixture which, in general, satisfy the following inequality:

$$\nu = \sum_\alpha \phi_\alpha \leq 1, \quad \phi_\alpha = \frac{U_\alpha}{U} \tag{6.12.1}$$

expressing the notion that a mixture may also consist of voids, as physically possible, for example, in granular mixtures. When in (6.12.1) the equality sign prevails, the mixture is called *saturated*; otherwise it is an *unsaturated* mixture.

Being *a priori* constitutive relations, the internal constraints must be maintained by mechanical, thermal, chemical, and other forces of the mixture, and cannot be determined by the history of the thermokinetic process. The principle of determinism discussed in section 6.2 should be, therefore, revised such that the constitutive relations (6.2.5) account for any possible *a priori* restriction on the material behavior due to mechanical, thermal, or other forces in the mixture. Formally, we may express this requirement by the following expression:

$$\hat{\Upsilon}_\alpha = \Upsilon_\alpha + \Upsilon''_\alpha \tag{6.12.2}$$

where $\Upsilon''_\alpha$ *are the internal constraint relations*, whereas $\Upsilon_\alpha$ are the constitutive relations (6.2.5) which are determined by the history of the thermokinetic process (6.2.6). Depending on the nature of internal constraints it should be clear from (6.12.2) and the above discussion that many $\Upsilon''_\alpha$ will, in general, be equal to zero.

From the definition of internal constraints, it may be argued that they cannot produce any energy or entropy dissipation. If they did, they would violate the original premise that internal constraints are maintained by forces which are *independent* of the history of the thermokinetic process. TRUESDELL & NOLL (1965) point out that constraints are maintained by reaction forces and that there are possibly infinitely many systems of reaction forces which suffice to maintain a given constraint. They also assert that the simplest forces are the ones which do no work. For a multiphase mixture, a great care is needed in asserting that the internal constarints produce no work or entropy generation, since the following problems need to be reconciled:

- Does the no work production imply or is implied by the no entropy production?

- Does the assertion pertain to each phase or to the mixture as a whole?

Before returning to the discussion concerned with the possible resolution of these questions, it will prove useful to summarize the past works on internal constraints.

### 6.12.2 Previous Works

GURTIN & GUIDUGLI (1973) developed a thermodynamic theory of constraints for single phase materials where the internal constraints do not produce *entropy*. With this assumption in view, NUNZIATO & WALSH (1980) developed constitutive equations for multiphase mixtures by utilizing the field equations of PASSMAN (1977) and used the entropy inequality for the mixture as a whole to include within it the "no-entropy production constraint"

$$\lambda \sum_{\alpha} \dot{\phi}_\alpha = 0 \tag{6.12.3}$$

with $\lambda$ being the Lagrange multiplier corresponding to the saturation constraint $\nu = 1$. PASSMAN *et al.* (1984) criticized the use of the above constraint in an entropy inequality on the basis that it may lead to the nonvanishing of the momentum source (5.4.9) when each phase possesses a different temperature, and, instead, they used the assumption of *workless constraints*. In particular, when treating the saturation constraint on phase $\alpha$, PASSMAN and co-workers introduced the reaction force $\pi$ for this constraint and let the power expended per unit volume of phase $\alpha$ be given by

$$\check{W}_\alpha = \pi \left(\frac{\partial \phi_\alpha}{\partial t}\right)_{\mathbf{x}} = \phi(\dot{\phi}_\alpha - \tilde{\mathbf{v}}_\alpha \cdot \nabla \phi_\alpha) \tag{6.12.4}$$

so that if this is added to the energy source $\check{e}_\alpha$ in (5.4.6) and summed over all phases in the mixture as in (5.4.13), then there would be no net power expended for the *mixture as a whole*. The reaction force $\pi$ in (6.12.4) is identified as the interfacial pressure and argued that such a quantity also appears in the works of ISHII (1975) (who used the time averaging approach to derive the field equations) and BEDFORD & DRUMHELLER (1978). This argument does not, of course, prove that $\pi$ is indeed the interfacial pressure, since the interphase pressure is the *pressure between two phases of the mixture* which by the above definition all such pressures in a mixture with three or more phases should be identical and equal to $\pi$ - a condition that is clearly too restrictive and superficial.

PASSMAN *et al.* (1984) also discuss the handling of the incompressiblity constraint within their theory of mixtures. To illustrate the approach, we first note that

$$\bar{\rho}_\alpha = \phi_\alpha \bar{\bar{\rho}}_\alpha \tag{6.12.5}$$

which follows from (2.3.7), where $\bar{\bar{\rho}}_\alpha$ is the (true) mass density of phase $\alpha$ and assumed to be constant, since the phase $\alpha$ is, by definition, incompressible. Using this equation, it follows from (2.4.14) that

$$\grave{\phi}_\alpha + \phi_\alpha \nabla \cdot \bar{\mathbf{v}}_\alpha = \frac{\hat{c}_\alpha}{\bar{\bar{\rho}}_\alpha} \tag{6.12.6}$$

from where it is argued that the work expended per unit volume due to the incompressibility constraint is

$$\check{W}_\alpha = p_\alpha(\grave{\phi}_\alpha + \phi_\alpha \nabla \cdot \bar{\mathbf{v}}_\alpha - \frac{\hat{c}_\alpha}{\bar{\bar{\rho}}_\alpha}) \tag{6.12.7}$$

In this expression $p_\alpha$ is the reaction stress associated with the constraint and identified as the ***hydrostatic pressure.*** Since $\check{W}_\alpha = 0$ in (6.12.7), the condition that the incompressibility constraint produces no work for the mixture as a whole is again satisfied as for the saturation constraint discussed above. But the question is: why is the power expended of phase $\alpha$ by the saturation constraint (6.12.4) different from zero, whereas that due to the incompressibility constraint (6.12.7) it is identically equal to zero? Is it because the saturation constraint applies to the mixture as a whole and hence no net power can be expended in the mixture as a whole, whereas the incompressibility constraint applies to a phase and, consequently, it cannot produce any power expenditure for that particular phase. If the answer to this is affirmative then we have an *internal constraint rule*; otherwise we have an artifact in the theory that may reside in the field equations or constitutive principles.

### 6.12.3 Restrictions Imposed on Internal Constraints Relations by No Entropy and Work Production Principles

The handling of internal constraints within the theory of multiphase mixtures is an unsolved problem. Towards the resolution of this problem it is useful to derive a set of conditions which the internal constraints should satisfy in the situations when they do not produce entropy, work, and both entropy and work for each phase in the mixture.

Starting from the phasic entropy inequality (2.4.45) it is possible to eliminate $\bar{\rho}_\alpha \tilde{r}_\alpha$ by using the energy equation (2.4.35) in which $\hat{\epsilon}_\alpha$ is found from

(3.4.17) and (3.4.15), $\hat{p}_{\alpha i}$ from (3.3.3), and $\hat{I}_{\alpha imn}$ from (4.3.47). Carrying out these substitutions and using the definition of the Helmholtz potential

$$\bar{\psi}_\alpha = \bar{\epsilon}_\alpha - \bar{\bar{\theta}}_\alpha \bar{s}_\alpha \tag{6.12.8}$$

the inequality (2.4.45) is reduced to the following form:

$$\begin{aligned}
&-\bar{\rho}_\alpha(\bar{s}_\alpha \grave{\bar{\bar{\theta}}}_\alpha + \grave{\bar{\psi}}_\alpha) - \frac{1}{\bar{\bar{\theta}}_\alpha}\bar{\mathbf{q}}_\alpha \cdot \nabla \bar{\bar{\theta}}_\alpha + \hat{s}_\alpha \bar{\bar{\theta}}_\alpha + \bar{T}_{\alpha ij}\bar{v}_{\alpha i,j} + \hat{c}_\alpha(-\bar{\psi}_\alpha + \hat{\epsilon}_\alpha) \\
&\quad -\bar{q}_{s\alpha} + \nu_{jm}^{(\alpha)}(\bar{S}_{\alpha mj} + \bar{\rho}_\alpha \bar{\ell}_{\alpha mj}) - \bar{v}_{\alpha j,i}\nu_{jm}^{(\alpha)}\nu_{iq}^{(\alpha)}\bar{\rho}_\alpha \bar{\imath}_{\alpha mq} \\
&\quad -\frac{1}{2}\bar{\rho}_\alpha(\overline{\nu_{im}^{(\alpha)}\nu_{iq}^{(\alpha)}\bar{\imath}_{\alpha mq}})^{\grave{}} + \frac{1}{2}\hat{c}_\alpha \nu_{im}^{(\alpha)}\nu_{iq}^{(\alpha)}(\hat{\imath}_{\alpha mq} - \bar{\imath}_{\alpha mq}) \\
&\quad +(\nu_{jm}^{(\alpha)}\bar{\lambda}_{\alpha mji} - \nu_{im}^{(\alpha)}\bar{\rho}_\alpha \bar{\ell\epsilon}_{\alpha m} - \frac{1}{2}(\nu_{jm}^{(\alpha)}\nu_{jn}^{(\alpha)}\nu_{i\ell}^{(\alpha)}U_{\alpha mn\ell}))_{,i} \geq 0
\end{aligned} \tag{6.12.9}$$

For each of the constitutive variables

$$\Upsilon_\alpha = \Upsilon_\alpha[\bar{s}_\alpha,\ \bar{\psi}_\alpha,\ \bar{\mathbf{q}}_\alpha,\ \hat{s}_\alpha,\ \bar{\mathbf{T}}_\alpha,\ \bar{\boldsymbol{\lambda}}_\alpha,\ \hat{c}_\alpha,\ \bar{\mathbf{S}}_\alpha,\ \hat{\mathbf{i}}_\alpha,\ \hat{\epsilon}_\alpha,\ \bar{q}_{s\alpha},\ \bar{\ell\epsilon}_\alpha,\ \mathbf{U}_\alpha]$$

in the above inequality, use is now made of (6.12.2) with the calorodynamic variables $\Upsilon_\alpha$ satisfying (6.12.9) and the internal constraint variables $\Upsilon''_\alpha$ requiring not to produce entropy, *i.e.*

$$\begin{aligned}
&-\bar{\rho}_\alpha(\bar{s}''_\alpha \grave{\bar{\bar{\theta}}}_\alpha + \grave{\bar{\psi}}''_\alpha) - \frac{1}{\bar{\bar{\theta}}_\alpha}\bar{\mathbf{q}}''_\alpha \cdot \nabla \bar{\bar{\theta}}_\alpha + \hat{s}''_\alpha \bar{\bar{\theta}}_\alpha + \bar{T}''_{\alpha ij}\bar{v}_{\alpha i,j} + \hat{c}''_\alpha(-\bar{\psi}_\alpha + \hat{\epsilon}_\alpha) \\
&\quad -\bar{q}''_{s\alpha} + (\hat{c}_\alpha + \hat{c}''_\alpha)(-\bar{\psi}''_\alpha + \hat{\epsilon}''_\alpha) - (\nu_{im}^{(\alpha)}\bar{\rho}_\alpha \bar{\ell\epsilon}''_{\alpha m})_{,i} + \nu_{jm}^{(\alpha)}\bar{S}''_{\alpha mj} \\
&\quad +(\nu_{jm}^{(\alpha)}\bar{\lambda}''_{\alpha mji})_{,i} - \frac{1}{2}(\nu_{jm}^{(\alpha)}\nu_{jn}^{(\alpha)}\nu_{ia}^{(\alpha)}U''_{\alpha mna})_{,i} \\
&\quad +\frac{1}{2}(\hat{c}_\alpha + \hat{c}''_\alpha)\nu_{im}^{(\alpha)}\nu_{iq}^{(\alpha)}\hat{\imath}''_{\alpha mq} + \frac{1}{2}\hat{c}''_\alpha \nu_{im}^{(\alpha)}\nu_{iq}^{(\alpha)}(\hat{\imath}_{\alpha mq} - \bar{\imath}_{\alpha mq}) = 0
\end{aligned} \tag{6.12.10}$$

But, by the definition, the internal constraint relations cannot depend on the thermokinetic process and, consequently, it follows from the above equation that

$$\hat{c}''_\alpha = 0 \tag{6.12.11}$$

$$-\bar{\psi}''_\alpha + \hat{\epsilon}''_\alpha = 0, \quad \hat{\imath}''_{\alpha mq} = 0 \tag{6.12.12}$$

*if* $\hat{c}_\alpha \neq 0$, for otherwise the condition (6.12.12) need not be satisfied. Equation (6.12.10) is, therefore, reduced to the following form:

$$-\bar{\rho}_\alpha(\bar{s}''_\alpha \grave{\bar{\bar{\theta}}}_\alpha + \grave{\bar{\psi}}''_\alpha) - \frac{1}{\bar{\bar{\theta}}_\alpha}\bar{\mathbf{q}}''_\alpha \cdot \nabla \bar{\bar{\theta}}_\alpha + \hat{s}''_\alpha \bar{\bar{\theta}}_\alpha + \bar{T}''_{\alpha ij}\bar{v}_{\alpha i,j} - \bar{q}''_{s\alpha} - (\nu_{im}^{(\alpha)}\bar{\rho}_\alpha \bar{\ell\epsilon}''_{\alpha m})_{,i}$$

$$+\nu_{jm}^{(\alpha)}\bar{S}''_{amj}+(\nu_{jm}^{(\alpha)}\bar{\lambda}''_{amji})_{,i}-\frac{1}{2}(\nu_{jm}^{(\alpha)}\nu_{jn}^{(\alpha)}\nu_{ia}^{(\alpha)}U''_{amna})_{,i}=0$$

(6.12.13)

which together with equations (6.12.11) and (6.12.12) form sufficient and necessary conditions for the internal constraints

$$\hat{c}''_{\alpha},\ \tilde{\psi}''_{\alpha},\ \hat{\hat{\epsilon}}''_{\alpha},\ \tilde{s}''_{\alpha},\ \bar{\mathbf{q}}''_{\alpha},\ \hat{s}''_{\alpha},\ \bar{\mathbf{T}}''_{\alpha},\ \bar{\boldsymbol{\lambda}}''_{\alpha},\ \bar{q}''_{s\alpha},\ \tilde{\ell\epsilon}''_{\alpha},\ \bar{\mathbf{S}}''_{\alpha},\ \hat{\mathbf{i}}''_{\alpha},\ \mathbf{U}''_{\alpha}$$

not to produce entropy in any phase of the mixture.

Before discussing the above results, we will determine the requirements imposed on the internal constraint relations when they are not allowed to produce any work. These results follow from the energy equation (2.4.35) with the above-mentioned substitutions for $\hat{\epsilon}_{\alpha}$, $\hat{p}_{\alpha i}$, and $\hat{I}_{\alpha imn}$, *i.e.*

$$\begin{aligned}\bar{\rho}_{\alpha}\grave{\tilde{\epsilon}}_{\alpha}=&-\nabla\cdot\bar{\mathbf{q}}_{\alpha}+\bar{\rho}_{\alpha}\tilde{r}_{\alpha}+\bar{T}_{\alpha ij}\tilde{v}_{\alpha i,j}-\hat{c}_{\alpha}(\tilde{\epsilon}_{\alpha}-\hat{\hat{\epsilon}}_{\alpha})-\bar{q}_{s\alpha}\\&+\nu_{jm}^{(\alpha)}(\bar{S}_{amj}+\bar{\rho}_{\alpha}\tilde{\ell}_{amj})-\tilde{v}_{\alpha j,i}(\nu_{jm}^{(\alpha)}\nu_{iq}^{(\alpha)}\bar{\rho}_{\alpha}\tilde{i}_{amq})-\frac{1}{2}\bar{\rho}_{\alpha}\overline{(\nu_{im}^{(\alpha)}\nu_{iq}^{(\alpha)}\tilde{i}_{amq})}^{\grave{}}\\&+(\nu_{jm}^{(\alpha)}\bar{\lambda}_{amji}-\nu_{im}^{(\alpha)}\bar{\rho}_{\alpha}\tilde{\ell\epsilon}_{am}-\frac{1}{2}\nu_{jm}^{(\alpha)}\nu_{jn}^{(\alpha)}\nu_{il}^{(\alpha)}U_{amnl})_{,i}\\&+\frac{1}{2}\hat{c}_{\alpha}\nu_{im}^{(\alpha)}\nu_{iq}^{(\alpha)}(\hat{i}_{amq}-\tilde{i}_{amq})\end{aligned}\qquad(6.12.14)$$

Following the same procedure as in the derivation of the sufficient and necessary conditions for internal constraints not to produce entropy in the entropy inequality (6.12.9), we may also obtain the sufficient and necessary conditions for internal constraint relations

$$\hat{c}''_{\alpha},\ \tilde{\epsilon}''_{\alpha},\ \hat{\hat{\epsilon}}''_{\alpha},\ \tilde{s}''_{\alpha},\ \bar{\mathbf{q}}''_{\alpha},\ \hat{s}''_{\alpha},\ \bar{\mathbf{T}}''_{\alpha},\ \bar{\boldsymbol{\lambda}}''_{\alpha},\ \bar{q}''_{s\alpha},\ \tilde{\ell\epsilon}''_{\alpha},\ \bar{\mathbf{S}}''_{\alpha},\ \hat{\mathbf{i}}''_{\alpha},\ \mathbf{U}''_{\alpha}$$

not to produce the work from (6.12.14). The results are:

$$\hat{c}''_{\alpha}=0\qquad(6.12.15)$$

$$-\tilde{\epsilon}''_{\alpha}+\hat{\hat{\epsilon}}''_{\alpha}=0,\qquad \hat{i}''_{amq}=0\qquad(6.12.16)$$

$$\begin{aligned}&-\bar{\rho}_{\alpha}\grave{\tilde{\epsilon}}''_{\alpha}-\nabla\cdot\bar{\mathbf{q}}''_{\alpha}+\bar{T}''_{\alpha ij}\tilde{v}_{\alpha i,j}-\bar{q}''_{s\alpha}+\nu_{jm}^{(\alpha)}\bar{S}''_{amj}+(\nu_{jm}^{(\alpha)}\bar{\lambda}''_{amji})_{,i}\\&\qquad-(\nu_{im}^{(\alpha)}\bar{\rho}_{\alpha}\tilde{\ell\epsilon}''_{am})_{,i}-\frac{1}{2}(\nu_{jm}^{(\alpha)}\nu_{jn}^{(\alpha)}\nu_{ia}^{(\alpha)}U''_{amna})_{,i}=0\end{aligned}\qquad(6.12.17)$$

where the condition (6.12.16) is not required *if* $\hat{c}_{\alpha}=0$.

The sufficient and necessary conditions for internal constraints relations not to produce *entropy and work* are obtained by combining (6.12.11)-(6.12.13) and (6.12.15)-(6.12.17). These conditions consist of the no-work-production relations expressed by (6.12.15)-(6.12.17) *and the following additional requirements*:

$$\bar{s}''_\alpha = 0 \quad (if\ \hat{c}_\alpha \neq 0) \tag{6.12.18}$$

$$\bar{\rho}_\alpha \dot{\bar{s}}''_\alpha + \nabla \cdot \left(\frac{\bar{\mathbf{q}}''_\alpha}{\bar{\bar{\theta}}_\alpha}\right) + \hat{s}''_\alpha = 0 \tag{6.12.19}$$

### 6.12.4 Special Results

The above results demonstrate that the constraints which do not produce work are, in general, different from the constraints which do not produce entropy, even in the absence of the internal constraints on $\bar{\mathbf{S}}''_\alpha$, $\hat{\mathbf{i}}''_\alpha$, $\bar{\boldsymbol{\lambda}}''_\alpha$, $\tilde{\ell\epsilon}''_\alpha$, and $\mathbf{U}''_\alpha$. With the latter constraints absent, the entropy conditions (6.12.11)-(6.12.13) give

$$\hat{c}''_\alpha = 0, \quad -\tilde{\psi}''_\alpha + \hat{\hat{\epsilon}}''_\alpha = 0 \quad (if\ \hat{c}''_\alpha \neq 0) \tag{6.12.20}$$

$$-\bar{\rho}_\alpha(\bar{s}''_\alpha \dot{\bar{\bar{\theta}}}_\alpha + \dot{\tilde{\psi}}''_\alpha) - \frac{1}{\bar{\bar{\theta}}_\alpha}\bar{\mathbf{q}}''_\alpha \cdot \nabla \bar{\bar{\theta}}_\alpha + \hat{s}''_\alpha \bar{\bar{\theta}}_\alpha + \bar{T}''_{\alpha ij}\tilde{v}_{\alpha i,j} - \bar{q}''_{s\alpha} = 0 \tag{6.12.21}$$

whereas the energy conditions (6.12.15)-(6.12.17) are reduced to the form

$$\hat{c}''_\alpha = 0, \quad -\tilde{\epsilon}''_\alpha + \hat{\hat{\epsilon}}''_\alpha = 0 \quad (if\ \hat{c}''_\alpha \neq 0) \tag{6.12.22}$$

$$-\bar{\rho}_\alpha \dot{\tilde{\epsilon}}''_\alpha - \nabla \cdot \bar{\mathbf{q}}''_\alpha + \bar{T}''_{\alpha ij}\tilde{v}_{\alpha i,j} - \bar{q}''_{s\alpha} = 0 \tag{6.12.23}$$

For a single phase mixture, the entropy condition (6.12.21) is further reduced to

$$-\bar{\rho}_\alpha(\bar{s}''_\alpha \dot{\bar{\bar{\theta}}}_\alpha + \dot{\tilde{\psi}}''_\alpha) - \frac{1}{\bar{\bar{\theta}}_\alpha}\bar{\mathbf{q}}''_\alpha \cdot \nabla \bar{\bar{\theta}}_\alpha + \bar{T}''_{\alpha ij}\tilde{v}_{\alpha i,j} = 0 \tag{6.12.24}$$

which upon setting $\dot{\tilde{\psi}}''_\alpha = 0$ is reduced to the result of GURTIN & GUIDUGLI (1973) discussed above. For single phase simple materials exhibiting only mechanical effects, it follows from (6.12.21) and (6.12.23) that

$$-\bar{\rho}_\alpha \dot{\tilde{\psi}}''_\alpha + \bar{T}''_{\alpha ij}\tilde{v}_{\alpha i,j} = 0 \tag{6.12.25}$$

$$-\bar{\rho}_\alpha \overset{\searrow''}{\epsilon}_\alpha + \bar{T}''_{\alpha ij}\tilde{v}_{\alpha i,j} = 0 \tag{6.12.26}$$

These results are in accord with those of TRUESDELL & NOLL (1965) when $\overset{\searrow''}{\psi}_\alpha = \overset{\searrow''}{\epsilon}_\alpha = 0$, since for $\bar{\mathbf{T}}_\alpha$ symmetric they can be written as

$$tr(\bar{\mathbf{T}}''_\alpha \mathbf{D}_\alpha) = 0 \tag{6.12.27}$$

Choosing

$$\bar{\mathbf{T}}''_\alpha = -p_\alpha \mathbf{I} \tag{6.12.28}$$

it then follows that

$$tr\mathbf{D}_\alpha = \nabla \cdot \tilde{v}_\alpha = 0 \tag{6.12.29}$$

which is the required form of single phase continuity equation for incompressible flow. The constitutive equation for the stress tensor has then the following form:

$$\bar{\mathbf{T}}_\alpha = -p_\alpha \mathbf{I} + \bar{\tau} \tag{6.12.30}$$

where the stress $\bar{\tau}$ is determined by the thermokinetic process through the constitutive equation.

### 6.12.5 Discussion

The theory of multiphase mixtures which is simplified by the constitutive asumption (5.1.1) does not appear to recover the incompressibility constraint (6.12.7) in (6.12.11)-(6.12.13) or (6.12.15)-(6.12.17). This may be seen from (6.12.17) when $\overset{\searrow''}{\epsilon}_\alpha = \bar{q}''_\alpha = \bar{\ell\epsilon}''_{\alpha m} = U''_{\alpha mna} = 0$, *i.e.*

$$\bar{T}''_{\alpha ij}\tilde{v}_{\alpha i,j} - \bar{q}''_{s\alpha} + \nu^{(\alpha)}_{jm}\bar{S}''_{\alpha mj} + (\nu^{(\alpha)}_{jm}\bar{\lambda}''_{\alpha mji})_{,i} = 0 \tag{6.12.31}$$

In the above equation, the intrinsic stress moment constraint $\bar{\lambda}''_{\alpha imq}$ may be ignored by noting from (3.2.11) and (3.1.13) (for an incompressible phase) that

$$\bar{\lambda}''_{\alpha miq} = \frac{1}{U}\sum_\delta \int_{U(\alpha\delta)} \xi^{(\alpha\delta)}_m T''^{(\alpha\delta)}_{iq}\, dU = -p_\alpha \delta_{iq}\frac{1}{U}\sum_\delta \int_{U(\alpha\delta)} \xi^{(\alpha\delta)}_m\, dU = 0 \tag{6.12.32}$$

where $p_\alpha$ is an average value of the reaction stress due to the incompressibility constraint in phase $\alpha$. Similarly, substituting $T^{(\alpha\delta)}_{km} = -p_\alpha \delta_{km}$ into (3.2.12), the surface traction moment becomes

$$\bar{S}_{\alpha mj} = -p_\alpha \delta_{mj}\phi_\alpha \tag{6.12.33}$$

Using these results, $\nu_{ij}^{(\alpha)} = \grave{\phi}_\alpha \delta_{ij}$, and

$$\bar{T}''_{\alpha ij} = -p_\alpha \phi_\alpha \delta_{ij} \tag{6.12.34}$$

in (6.12.31) (and for the moment concentrating only on the mechanical effects where $\bar{q}''_{s\alpha} = 0$), we obtain

$$p_\alpha(\tilde{v}_{\alpha i,i} + \grave{\phi}_\alpha) = 0 \tag{6.12.35}$$

a result which is clearly *not equivalent* to the result (6.12.7) when $\hat{c}_\alpha = 0$. This is perhaps unfortunate for it signifies that either the assumption of the material deformation relative to the center of mass (5.1.1) or the theoretical framework of constraints laid down in section (6.12.3) is incomplete or incorrect.

Towards the resolution of the above problem it may be recalled that the assumption (5.1.1) *is somewhat arbitrary* in representing the phasic dilatational effects because *it is also possible to assume that:*

$$\nu_{ij}^{(\alpha)} = \hat{\nu}_{ij}^{(\alpha)} + \frac{\grave{\phi}_\alpha}{\phi_\alpha}\delta_{ij} \tag{6.12.36}$$

where $\grave{\phi}_\alpha/\phi_\alpha$ now represents the effect of dilatation, and where the result (5.1.7) is modified as follows

$$\xi_k^{(\alpha\delta)} = \frac{\phi_\alpha}{\phi_{\alpha 0}}\,\xi_k^{(\alpha\delta)}(\phi_{\alpha 0}) \tag{6.12.37}$$

Ignoring again the thermal effects, using (6.12.32)-(6.12.34) and (6.12.36) with $\hat{\nu}_{ij}^{(\alpha)} = 0$ in (6.12.31), we obtain

$$p_\alpha(\grave{\phi}_\alpha + \phi_\alpha \tilde{v}_{\alpha i,i}) = 0 \tag{6.12.38}$$

a result that is now *consistent* with the incompressibility constraint (6.12.7) when $\hat{c}_\alpha = 0$.

The phase change process within the incompressibility constraint may be accounted by retaining $\bar{q}''_{s\alpha}$ in (6.12.31) and by arguing that[3]

$$\bar{q}''_{s\alpha} = -p_\alpha \frac{\hat{c}_\alpha}{\bar{\bar{\rho}}_\alpha} \tag{6.12.39}$$

Using this equation, (6.12.32)-(6.12.34), and (6.12.36) with $\hat{\nu}_{ij}^{(\alpha)} = 0$ in (6.12.31) gives now a result that is consistent with the incompressible form of the continuity equation (6.12.7) with phase change, *i.e.*

$$p_\alpha(\dot{\phi}_\alpha + \phi_\alpha \bar{v}_{\alpha i,i} - \frac{\hat{c}_\alpha}{\bar{\bar{\rho}}_\alpha}) = 0 \tag{6.12.40}$$

The recovery of saturation constraint (6.12.3) from (6.12.11)-(6.12.13) or (6.12.15)-(6.12.17) is not feasible. For this reason, the condition (6.12.3) in the theory of multiphase mixtures in this book will be taken as *constitutive* and not as a constraining relation. Taking the condition (6.12.3) as the constraining relation allows for the work or entropy production within each phase or within the mixture as a whole, as discussed in section (6.12.2) for the postulatory theories of mixtures. The postulatory theories of multiphase mixtures also assume a single reaction force for the mixture as a whole and identify this force as the interface pressure. Except for a mixture of two phases, the concept of single interface pressure for many phases in a mixture cannot be physically justified, in general. As will be seen in the following chapter, the theory of multiphase mixtures of this book produces an interface pressure for each phase in the mixture as physically required.

## 6.13 Principle of Phase Separation

In formulating the constitutive assumption or calorodynamic process (6.2.6), use was made of the *principle of equipresence*. This principle has been used extensively in single phase multicomponent mixture theories (BOWEN,1976)

[3]The plausibility of (6.12.39) may be established from (2.2.6) by noting that the constrained energy transfer rate across the interface of phase $\alpha$ is equal to

$$\sum_\delta \int_{a^{(\wedge\delta)}} [m^{(\alpha\delta)}\epsilon''^{(\alpha\delta)} + (\mathbf{q}''^{(\alpha\delta)} - \mathbf{T}''^{(\alpha\delta)}\mathbf{v}''^{(\alpha\delta)})\cdot\mathbf{n}^{(\alpha\delta)}]\, da$$

from where, if it is assumed that such an energy is balanced by the work of the constrained mechanical forces, it follows that

$$\sum_\delta \int_{a^{(\wedge\delta)}} [m^{(\alpha\delta)}\epsilon''^{(\alpha\delta)} + \mathbf{q}''^{(\alpha\delta)}\cdot\mathbf{n}^{(\alpha\delta)}]\, da = 0$$

Using now the definitions (3.3.9) and (2.4.15), and assuming that $\epsilon''^{(\alpha\delta)} = -p_\alpha/\bar{\bar{\rho}}_\alpha$, we obtain from the above equation the result expressed by (6.12.39). The assumed form of the constrained energy function $\epsilon''^{(\alpha\delta)}$ is plausible owing to the requirement of $s''^{(\alpha\delta)} = 0$ (*cit.* (6.12.18)).

which states that there should be no *a priori* reason for discriminating any thermomechanical variable in the list (6.2.7) from entering into each of the response functionals (6.2.5). For multiphase mixtures, it is argued by some that the phases are physically separated and that the principle of equipresence should be replaced by the *principle of phase separation*, or "Hypothesis of phase separation" as proposed by DREW & SIEGEL (1971). According to this principle, the constitutive relations (6.2.5) should be divided into two groups:

1. The bulk phase variables:

$$\bar{\mathbf{T}}_\alpha,\ \hat{\mathbf{M}}_\alpha,\ \tilde{\epsilon}_\alpha,\ \bar{\mathbf{q}}_\alpha,\ \tilde{s}_\alpha,\ \mathbf{U}_\alpha,\ \bar{\boldsymbol{\lambda}}_\alpha,\ \tilde{\boldsymbol{\ell}}\epsilon_\alpha \tag{6.13.1}$$

2. The interphase variables:

$$\hat{c}_\alpha,\ \hat{s}_\alpha,\ \hat{\mathbf{i}}_\alpha,\ \bar{\mathbf{S}}_\alpha,\ \bar{\mathbf{t}}_\alpha,\ \hat{\hat{\epsilon}}_\alpha,\ \bar{q}_{s\alpha} \tag{6.13.2}$$

The bulk phase variables are required to depend only on the independent variables from the same phase, whereas the interphase variables are required to depend on *all* independent variables of the mixture.

The principle of phase separation has been extensively used by PASSMAN and co-workers (PASSMAN *et al.* (1984)), whereas the principle of equipresence was used by DOBRAN (1984b) in the studies of constitutive equations for multiphase mixtures. The reason for using the principle of equipresence in the constitutive assumption (6.2.6) is that it is the correct choice for a theory of multiphase mixtures based on the volume averaging approach, or for any theory based on the superimposed continua model. The constitutive variables (6.2.5) are volume- and area-averaged quantities defined at the same point of space of the superimposed continua as discussed in chapter 2. They depend not only on the volume averaging region $U$, but also on the distribution of all other phases in this volume. Consequently, there should be no *a priori* reason for discriminating any thermokinetic variable in the constitutive relations (6.2.6). The most convincing argument for using the principle of phase separation in the constitutive relations is to reduce substantially the number of independent variables entering into the constitutive assumption and thereby producing a simpler model. Such a reason cannot, however, by justified *a priori* and the use of the principle of phase separation on *a priori* grounds appears too artificial and will not be used in the subsequent study of constitutive equations.

## 6.14 Concluding Remarks

The concepts and principles of constitutive theory of multiphase mixtures are fundamental to the studies of constitutive equations. These concepts and principles were discussed at length in order to provide a basis for studies of constitutive relations in the following chapters and elsewhere, and to identify the unsolved problems in the theory. By utilizing the ideas of this chapter and field equations of previous chapters, it is possible to develop a wide variety of models of structured multiphase mixtures. In the subsequent chapters, the studies of constitutive equations will necessarily be limited to mixtures with simple structural properties, primarily to determine the usefulness of the theory to model the simplest physical phenomena.

CHAPTER 7

# CONSTITUTIVE EQUATIONS

## 7.1 Purpose of the Chapter

The concepts and principles of the constitutive theory are fundamental to the study of constitutive equations. In this chapter, use will be made of the balance equations and principles of the constitutive theory discussed in previous chapters to develop constitutive equations for special multiphase mixtures. The constitutive equations are models or assumptions of the material behavior under the imposed mechanical and thermal loadings. With an established set of field equations of multiphase mixtures it may be possible to determine the relative merit of different constitutive assumptions and thereby establish the range of applicability of various models. The evaluation of models should consider, however, that different studies of constitutive equations may be employing different constitutive principles, such as the principle of equipresence vs. the principle of phase separation, an entropy inequality or axiom of dissipation for the mixture as a whole, vs. an entropy inequality or axiom of dissipation for each phase, the constraints which do not produce entropy in the mixture or in each phase, *etc.* The constitutive equations may also conveniently consist of approximations of the full tensor representation of the independent variables, thus further complicating the evaluation of relative merits of different models.

The multiphase mixture models are difficult to evaluate quantitatively due to their complexity and lack of information on the phenomenological or transport coefficients in the constitutive equations. The transport coefficients may be determined from fundamental experiments where the physical variables should be consistent with averaged variables used in the construction of the theory. In a theory of multiphase mixtures which is constructed by the volume averaging approach, the physical variables are instantaneous quantities determined on a sample of volume of the mixture and may be determined experimentally in some situations from a holographic image analysis of the flow field. At the present, however, this imaging technique is not sufficiently developed, and the current experimental methods produce essentially time-averaged variables which may be consistently used with a time-averaged set of field equations or time-averaged form of the volume-averaged equations. It should be noted, however, that in the discussion of kinematics of multiphase mixtures in section 2.4.1, the particles $\mathbf{X}_\alpha$ and $\mathbf{X}^{(\alpha)}$ have different meanings, or the material particles of phase $\alpha$ should not be confused with molecules of phase $\alpha$. For this reason, as in the postulatory theories, the above "measurement" problem does not really appear in the mathematical theory of mixtures.

To account for the incompressibility constraint within the theory of constraints of multiphase mixtures discussed in section 6.12, it was necessary to redefine the affine deformation assumption (5.1.1) to the form of (6.12.36), in order for this constraint not to produce entropy and work in each phase of the mixture. This procedure is reasonable, since the incompressibility constraint is then not maintained by the phasic or mixture's thermokinetic processes. The penalty for this construction is clearly a further divorce of the volume-averaged field equations from the postulatory theories of mixtures as discussed in chapter 5. For compressible phases, the theory of constraints may not require the modification (6.12.36). The difficulty of accounting for the saturation constraint in the theory of constraints is discussed in section 6.12.5, where it is concluded that this constraint within the volume averaging theory should be treated as a constitutive equation rather than a constraint. In the postulatory theories of multiphase mixtures, the saturation constraint produces the work and entropy in each phase of the mixture.

To continue with the development of a theory of structured multiphase mixtures based on the volume averaging approach, it is reasonable, therefore, to develop the constitutive equations based on the affine deformation assumption (6.12.36) rather than (5.1.1). In the following, the assumption (6.12.36) will be adopted to derive first an equivalent set of field equations of sections 5.2 and 5.3. These equations will then be used, together with the methods and principles of the constitutive theory of chapter 6, to investigate constitutive equations for mixtures of compressible and incompressible phases in order to ascertain the utility of results to model structured multiphase mixtures.

## 7.2 Balance Equations of Multiphase Mixtures with Rotation and Dilatation

Repeating the procedure of section 5.2, but with the affine deformation assumption

$$\nu_{m\ell}^{(\alpha)} = \hat{\nu}_{m\ell}^{(\alpha)} + \frac{\grave{\phi}_\alpha}{\phi_\alpha}\delta_{m\ell} \qquad (6.12.36)$$

instead of (5.1.1), the equivalent equations to (5.2.2) and (5.2.3) are, respectively, as follows

$$\bar{\rho}_\alpha(\grave{\bar{i}}_\alpha - 2\bar{i}_\alpha \frac{\grave{\phi}_\alpha}{\phi_\alpha}) = -\hat{c}_\alpha \bar{i}_\alpha + \hat{k}_{\alpha jj}; \quad j = n \tag{7.2.1}$$

$$\hat{k}_{\alpha jn} = 0; \quad j \neq n \tag{7.2.2}$$

Equation (7.2.1) may also be written in the form of (5.2.4), *i.e.*

$$\bar{\rho}_\alpha \grave{\bar{i}}_\alpha = -\hat{c}_\alpha \bar{i}_\alpha + K_{\alpha jj} + k_{\alpha jji,i} + \check{k}_{\alpha jj} \tag{7.2.3}$$

with the conditions (5.2.5)-(5.2.7) modified as follows

$$\hat{k}_{\alpha jj} = \hat{c}_\alpha \hat{i}_{\alpha jj} - (\bar{\rho}_\alpha \hat{I}_{\alpha mjj})_{,m} = \hat{c}_\alpha \hat{i}_{\alpha jj} - \hat{\nu}^{(\alpha)}_{mn,m} U_{\alpha jjn} - (\frac{\grave{\phi}_\alpha}{\phi_\alpha} U_{\alpha jjm})_{,m} \tag{7.2.4}$$

$$k_{\alpha jji,i} = (-\bar{\rho}_\alpha \hat{I}_{\alpha ijj})_{,i} = -\hat{\nu}^{(\alpha)}_{in,i} U_{\alpha jjn} - (\frac{\grave{\phi}_\alpha}{\phi_\alpha} U_{\alpha jji})_{,i} \tag{7.2.5}$$

$$K_{\alpha jj} = 2\bar{\rho}_\alpha \frac{\grave{\phi}_\alpha}{\phi_\alpha} \bar{i}_\alpha \tag{7.2.6}$$

$$\check{k}_{\alpha jj} = \hat{c}_\alpha \hat{i}_{\alpha jj} \tag{5.2.8}$$

$$K_{\alpha jn} = 0, \quad \check{k}_{\alpha jn} + k_{\alpha jni,i} = 0, \quad for \quad j \neq n \tag{5.2.9}$$

where use is made of (5.2.11). The balance equations equivalent to (5.2.13)-(5.2.18) are

$$\begin{aligned} \bar{\rho}_\alpha \bar{i}_\alpha [(\frac{d}{dt}(\frac{\grave{\phi}_\alpha}{\phi_\alpha}) + (\frac{\grave{\phi}_\alpha}{\phi_\alpha})^2 - \mu_i^{(\alpha)} \mu_i^{(\alpha)}) \delta_{jk} + \mu_j^{(\alpha)} \mu_k^{(\alpha)}] \\ = \bar{S}_{\alpha(jk)} + \bar{\rho}_\alpha \bar{\ell}_{\alpha(jk)} + \bar{\lambda}_{\alpha(jk)m,m} + \hat{\bar{T}}_{\alpha(jk)} \end{aligned} \tag{7.2.7}$$

$$\bar{\rho}_\alpha \bar{i}_\alpha (\grave{\mu}_\ell^{(\alpha)} + 2\frac{\grave{\phi}_\alpha}{\phi_\alpha} \mu_\ell^{(\alpha)}) = \frac{1}{2} \epsilon_{\ell jk} (\bar{S}_{\alpha[jk]} + \bar{\rho}_\alpha \bar{\ell}_{\alpha[jk]} + \bar{\lambda}_{\alpha[jk]m,m} + \hat{\bar{T}}_{\alpha[jk]}) \tag{7.2.8}$$

where

$$\hat{\bar{T}}_{\alpha jk} = \hat{g}_{\alpha jk} - \hat{\nu}^{(\alpha)}_{kn}\hat{k}_{\alpha jn} - \frac{\grave{\phi}_\alpha}{\phi_\alpha}\hat{k}_{\alpha jk} - \bar{\rho}_\alpha \hat{\nu}^{(\alpha)}_{mj}\bar{v}_{\alpha k,m}\bar{i}_\alpha - \bar{\rho}_\alpha \frac{\grave{\phi}_\alpha}{\phi_\alpha}\bar{v}_{\alpha k,j}\bar{i}_\alpha \tag{7.2.9}$$

$$\hat{\bar{T}}_{\alpha(jk)} = -\bar{\rho}_\alpha \bar{i}_\alpha \frac{\grave{\phi}_\alpha}{\phi_\alpha} D_{\alpha jk} - \frac{\grave{\phi}_\alpha}{\phi_\alpha}\frac{1}{2}(\nu^{(\alpha)}_{k\ell,i}U_{\alpha j\ell i} + \nu^{(\alpha)}_{j\ell,i}U_{\alpha k\ell i})$$
$$+\frac{1}{2}\mu^{(\alpha)}_i[\epsilon_{m\ell i}(\nu^{(\alpha)}_{kn,m}U_{\alpha jn\ell} + \nu^{(\alpha)}_{jn,m}U_{\alpha kn\ell}) + \bar{\rho}_\alpha \bar{i}_\alpha(\bar{v}_{\alpha k,m}\epsilon_{mji} + \bar{v}_{\alpha j,m}\epsilon_{mki})]$$

$$\hat{\bar{T}}_{\alpha[jk]} = \bar{\rho}_\alpha \bar{i}_\alpha \frac{\grave{\phi}_\alpha}{\phi_\alpha} W_{\alpha jk} - \frac{\grave{\phi}_\alpha}{\phi_\alpha}\frac{1}{2}(\nu^{(\alpha)}_{k\ell,i}U_{\alpha j\ell i} - \nu^{(\alpha)}_{j\ell,i}U_{\alpha k\ell i})$$
$$+\frac{1}{2}\mu^{(\alpha)}_i[\epsilon_{m\ell i}(\nu^{(\alpha)}_{kn,m}U_{\alpha jn\ell} - \nu^{(\alpha)}_{jn,m}U_{\alpha kn\ell}) + \bar{\rho}_\alpha \bar{i}_\alpha(\bar{v}_{\alpha k,m}\epsilon_{mji} - \bar{v}_{\alpha j,m}\epsilon_{mki})]$$

$$\hat{\bar{T}}_{\alpha jj} = -\frac{\grave{\phi}_\alpha}{\phi_\alpha}(\nu^{(\alpha)}_{jn,\ell}U_{\alpha jn\ell} + \bar{\rho}_\alpha \bar{i}_\alpha \bar{v}_{\alpha j,j}) + \mu^{(\alpha)}_i \epsilon_{m\ell i}(\nu^{(\alpha)}_{jn,m}U_{\alpha jn\ell} + \bar{\rho}_\alpha \bar{i}_\alpha \bar{v}_{\alpha \ell,m})$$

$$\bar{\rho}_\alpha \bar{\ell}_{\alpha(jk)} = \bar{\rho}_\alpha \bar{i}_\alpha(\mu^{(\alpha)}_j \mu^{(\alpha)}_k - \mu^{(\alpha)}_i \mu^{(\alpha)}_i \delta_{jk})$$
$$-\frac{1}{2}\mu^{(\alpha)}_i[\epsilon_{m\ell i}(\nu^{(\alpha)}_{kn,m}U_{\alpha jn\ell} + \nu^{(\alpha)}_{jn,m}U_{\alpha kn\ell}) + \bar{\rho}_\alpha \bar{i}_\alpha(\bar{v}_{\alpha k,m}\epsilon_{mji} + \bar{v}_{\alpha j,m}\epsilon_{mki})] \tag{7.2.10}$$

Equation (7.2.7) thus becomes

$$\bar{\rho}_\alpha \bar{i}_\alpha[\frac{d}{dt}(\frac{\grave{\phi}_\alpha}{\phi_\alpha}) + (\frac{\grave{\phi}_\alpha}{\phi_\alpha})^2]\delta_{jk} = \bar{S}_{\alpha(jk)} + \bar{\lambda}_{\alpha(jk)m,m} - \bar{\rho}_\alpha \bar{i}_\alpha \frac{\grave{\phi}_\alpha}{\phi_\alpha} D_{\alpha jk}$$
$$-\frac{1}{2}\frac{\grave{\phi}_\alpha}{\phi_\alpha}(\nu^{(\alpha)}_{k\ell,i}U_{\alpha j\ell i} + \nu^{(\alpha)}_{j\ell,i}U_{\alpha k\ell i}) \tag{7.2.11}$$

Multiplying (7.2.1) by $\grave{\phi}_\alpha/\phi_\alpha$ and adding the result to (7.2.11) gives

$$\bar{\rho}_\alpha[\grave{\overline{\bar{i}_\alpha \frac{\grave{\phi}_\alpha}{\phi_\alpha}}} - \bar{i}_\alpha(\frac{\grave{\phi}_\alpha}{\phi_\alpha})^2]\delta_{jk} = \bar{S}_{\alpha(jk)} + \bar{\lambda}_{\alpha(jk)m,m} - \hat{c}_\alpha \bar{i}_\alpha \frac{\grave{\phi}_\alpha}{\phi_\alpha}\delta_{jk} + \hat{k}_{\alpha ii}\frac{\grave{\phi}_\alpha}{\phi_\alpha}\delta_{jk}$$
$$-\bar{\rho}_\alpha \bar{i}_\alpha \frac{\grave{\phi}_\alpha}{\phi_\alpha} D_{\alpha jk} - \frac{1}{2}\frac{\grave{\phi}_\alpha}{\phi_\alpha}(\nu^{(\alpha)}_{k\ell,i}U_{\alpha j\ell i} + \nu^{(\alpha)}_{j\ell,i}U_{\alpha k\ell i}) \tag{7.2.12}$$

Notice that equation (7.2.7) is equivalent to (5.2.13), (7.2.9) to (5.2.15), (7.2.10) to (5.2.16), (7.2.8) to (5.2.19), and (7.2.12) to (5.2.18). The energy sources (3.4.15)-(3.4.17), and energy equations (3.4.18) and (3.4.19), become:

$$\check{e}_\alpha = \hat{e}_\alpha + \hat{c}_\alpha \frac{1}{2}\left(\tilde{v}_{\alpha i}\tilde{v}_{\alpha i} + \hat{\nu}^{(\alpha)}_{im}\hat{\nu}^{(\alpha)}_{im}\bar{i}_\alpha + \frac{\grave{\phi}_\alpha}{\phi_\alpha}\frac{\grave{\phi}_\alpha}{\phi_\alpha}\bar{i}_\alpha\right) + \left(\hat{\nu}^{(\alpha)}_{im} + \frac{\grave{\phi}_\alpha}{\phi_\alpha}\delta_{im}\right)\bar{S}_{\alpha mi} \tag{7.2.13}$$

$$\begin{aligned}\hat{\epsilon}_\alpha = &-\hat{p}_{\alpha i}\tilde{v}_{\alpha i} - \hat{c}_\alpha\tilde{\epsilon}_\alpha - \epsilon_{jmn}[\mu^{(\alpha)}_n(\bar{S}_{\alpha mj} + \bar{\rho}_\alpha\bar{\ell}_{\alpha mj}) + (\mu^{(\alpha)}_n\bar{\lambda}_{\alpha mji} \\ &+\epsilon_{nik}\mu^{(\alpha)}_m\mu^{(\alpha)}_k\bar{i}_\alpha\tilde{v}_{\alpha j}\bar{\rho}_\alpha)_{,i}] + \left[\frac{\grave{\phi}_\alpha}{\phi_\alpha}(\bar{S}_{\alpha ii} + \bar{\rho}_\alpha\bar{\ell}_{\alpha ii}) + \left(\frac{\grave{\phi}_\alpha}{\phi_\alpha}\bar{\lambda}_{\alpha mmi} - \frac{\grave{\phi}_\alpha}{\phi_\alpha}\frac{\grave{\phi}_\alpha}{\phi_\alpha}\tilde{v}_{\alpha i}\bar{\rho}_\alpha\bar{i}_\alpha\right)_{,i}\right] \\ &-\bar{\rho}_\alpha\left(\overline{\grave{\bar{i}_\alpha\mu^{(\alpha)}_m\mu^{(\alpha)}_m}}\right) - \frac{1}{2}\bar{\rho}_\alpha\left(\overline{\grave{\bar{i}_\alpha\frac{\grave{\phi}_\alpha}{\phi_\alpha}\frac{\grave{\phi}_\alpha}{\phi_\alpha}}}\right) + \hat{e}_\alpha\end{aligned} \tag{7.2.14}$$

$$\begin{aligned}\bar{\rho}_\alpha\left(\grave{\tilde{\epsilon}}_\alpha + \frac{1}{2}\overline{\grave{\tilde{v}_{\alpha i}\tilde{v}_{\alpha i}}} + \overline{\grave{\bar{i}_\alpha\mu^{(\alpha)}_n\mu^{(\alpha)}_n}} + \frac{1}{2}\overline{\grave{\bar{i}_\alpha\frac{\grave{\phi}_\alpha}{\phi_\alpha}\frac{\grave{\phi}_\alpha}{\phi_\alpha}}}\right) = &(\bar{T}_{\alpha ij}\tilde{v}_{\alpha i})_{,j} - \bar{q}_{\alpha i,i} + \bar{\rho}_\alpha\bar{b}_{\alpha i}\tilde{v}_{\alpha i} \\ -\hat{c}_\alpha\tilde{\epsilon}_\alpha - \epsilon_{jmn}[\mu^{(\alpha)}_n(\bar{S}_{\alpha mj} + \bar{\rho}_\alpha\bar{\ell}_{\alpha mj}) + &(\mu^{(\alpha)}_n\bar{\lambda}_{\alpha mji} + \epsilon_{nik}\mu^{(\alpha)}_m\mu^{(\alpha)}_k\bar{i}_\alpha\tilde{v}_{\alpha j}\bar{\rho}_\alpha)_{,i}] \\ +\bar{\rho}_\alpha\tilde{r}_\alpha + \frac{\grave{\phi}_\alpha}{\phi_\alpha}(\bar{S}_{\alpha ii} + \bar{\rho}_\alpha\bar{\ell}_{\alpha ii}) + &\left(\frac{\grave{\phi}_\alpha}{\phi_\alpha}\bar{\lambda}_{\alpha mmi} - \frac{\grave{\phi}_\alpha}{\phi_\alpha}\frac{\grave{\phi}_\alpha}{\phi_\alpha}\tilde{v}_{\alpha i}\bar{\rho}_\alpha\bar{i}_\alpha\right)_{,i} + \hat{e}_\alpha\end{aligned} \tag{7.2.15}$$

$$\begin{aligned}\bar{\rho}_\alpha\grave{\tilde{\epsilon}}_\alpha = &\bar{T}_{\alpha ij}\tilde{v}_{\alpha i,j} - \bar{q}_{\alpha i,i} + \bar{\rho}_\alpha\tilde{r}_\alpha + \left(\hat{\nu}^{(\alpha)}_{im,q} + \left(\frac{\grave{\phi}_\alpha}{\phi_\alpha}\right)_{,q}\delta_{im}\right)\bar{h}_{\alpha miq} \\ &-\check{m}_{\alpha i}\tilde{v}_{\alpha i} - \left(\hat{\nu}^{(\alpha)}_{im} + \frac{\grave{\phi}_\alpha}{\phi_\alpha}\delta_{im}\right)(\check{g}_{\alpha mi} + \bar{S}_{\alpha mi}) - \hat{c}_\alpha\left[\tilde{\epsilon}_\alpha - \frac{1}{2}\tilde{v}_{\alpha i}\tilde{v}_{\alpha i}\right. \\ &\left.-\frac{1}{2}\bar{i}_\alpha\left(\hat{\nu}^{(\alpha)}_{im}\hat{\nu}^{(\alpha)}_{im} + \frac{\grave{\phi}_\alpha}{\phi_\alpha}\frac{\grave{\phi}_\alpha}{\phi_\alpha}\right)\right] + \frac{1}{2}\bar{\rho}_\alpha\left(\hat{\nu}^{(\alpha)}_{im}\hat{\nu}^{(\alpha)}_{im} + \frac{\grave{\phi}_\alpha}{\phi_\alpha}\frac{\grave{\phi}_\alpha}{\phi_\alpha}\right)\grave{\bar{i}}_\alpha + \check{e}_\alpha\end{aligned} \tag{7.2.16}$$

Using (5.1.9), (6.12.36), and (4.3.47) in (3.3.3), (3.4.10), (3.4.11), and (3.4.13), gives

$$\hat{p}_{\alpha i} = \bar{t}_{\alpha i} - \left(\hat{\nu}^{(\alpha)}_{im} + \frac{\grave{\phi}_\alpha}{\phi_\alpha}\delta_{im}\right)_{,j}\bar{\rho}_\alpha\bar{i}_\alpha\left(\hat{\nu}^{(\alpha)}_{jm} + \frac{\grave{\phi}_\alpha}{\phi_\alpha}\delta_{jm}\right) \tag{7.2.17}$$

$$\hat{g}_{\alpha jk} = \hat{c}_\alpha \hat{i}_{\alpha jn}(\hat{\nu}^{(\alpha)}_{kn} + \frac{\grave{\phi}_\alpha}{\phi_\alpha}\delta_{kn})$$

$$-[(\hat{\nu}^{(\alpha)}_{kn}\hat{\nu}^{(\alpha)}_{m\ell} + \frac{\grave{\phi}_\alpha}{\phi_\alpha}(\delta_{kn}\hat{\nu}^{(\alpha)}_{m\ell} + \hat{\nu}^{(\alpha)}_{kn}\delta_{m\ell}))U_{\alpha jn\ell} + (\frac{\grave{\phi}_\alpha}{\phi_\alpha})^2 U_{\alpha jkm}]_{,m} \quad (7.2.18)$$

$$\bar{h}_{\alpha jkm} = \bar{\lambda}_{\alpha jkm} - \hat{\nu}^{(\alpha)}_{mj}\bar{\rho}_\alpha \tilde{v}_{\alpha k}\bar{i}_\alpha - \bar{\rho}_\alpha \frac{\grave{\phi}_\alpha}{\phi_\alpha}\tilde{v}_{\alpha k}\bar{i}_\alpha \delta_{mj} \quad (7.2.19)$$

$$\breve{g}_{\alpha jk} = \bar{\rho}_\alpha \bar{i}_\alpha(\hat{\nu}^{(\alpha)}_{jn}\hat{\nu}^{(\alpha)}_{kn} + (\frac{\grave{\phi}_\alpha}{\phi_\alpha})^2\delta_{jk}) + \hat{g}_{\alpha jk} + \tilde{v}_{\alpha k}[\bar{\rho}_\alpha \bar{i}_\alpha(\hat{\nu}^{(\alpha)}_{mj} + \frac{\grave{\phi}_\alpha}{\phi_\alpha}\delta_{mj})]_{,m} \quad (7.2.20)$$

## 7.3 Balance Equations of Multiphase Mixtures with Dilatation

The balance equations of multiphase mixtures involving only the dilatation are obtained by ignoring the rotational effect ($\hat{\boldsymbol{\nu}}^{(\alpha)} = \mathbf{0}$ or $\boldsymbol{\mu}^{(\alpha)} = \mathbf{0}$) in the equations of section 7.2. The complete set of these equations (which is equivalent to equations of section 5.3), are:

$$\grave{\bar{\rho}}_\alpha + \bar{\rho}_\alpha \tilde{v}_{\alpha i,i} = \hat{c}_\alpha \quad (2.4.14)$$

$$\bar{\rho}_\alpha \grave{\tilde{v}}_{\alpha i} = \bar{T}_{\alpha ij,j} + \bar{\rho}_\alpha \tilde{b}_{\alpha i} + \check{m}_{\alpha i} - \hat{c}_\alpha \tilde{v}_{\alpha i} \quad (3.4.2)$$

$$\hat{M}_{\alpha ij} = \bar{T}_{\alpha ij} - \bar{T}_{\alpha ji} \quad (2.4.31)$$

$$\bar{\rho}_\alpha \grave{\bar{i}}_\alpha = -\hat{c}_\alpha \bar{i}_\alpha + k_{\alpha jji,i} + K_{\alpha jj} + \check{k}_{\alpha jj} \quad (7.2.3)$$

$$\bar{\rho}_\alpha(\overline{\bar{i}_\alpha \frac{\grave{\phi}_\alpha}{\phi_\alpha}})^{\grave{}}\delta_{jk} = -\hat{c}_\alpha \bar{i}_\alpha \frac{\grave{\phi}_\alpha}{\phi_\alpha}\delta_{jk} + \bar{S}_{\alpha(jk)} + \bar{h}_{\alpha(jk)m,m} + \breve{g}_{\alpha(jk)} \quad (7.3.1)$$

$$\bar{S}_{\alpha[jk]} + \bar{\rho}_\alpha \bar{\ell}_{\alpha[jk]} + \bar{\lambda}_{\alpha[jk]m,m} + \bar{\rho}_\alpha \bar{i}_\alpha \frac{\grave{\phi}_\alpha}{\phi_\alpha} W_{\alpha jk} - \frac{\grave{\phi}_\alpha}{\phi_\alpha}(\frac{\grave{\phi}_\alpha}{\phi_\alpha})_{,i} U_{\alpha[jk]i} = 0 \quad (7.3.2)$$

$$\bar{\rho}_\alpha \grave{\tilde{\epsilon}}_\alpha = \bar{T}_{\alpha ij}\tilde{v}_{\alpha i,j} - \bar{q}_{\alpha k,k} + \bar{\rho}_\alpha \tilde{r}_\alpha + \hat{\epsilon}_\alpha \quad (7.3.3)$$

$$\bar{\rho}_\alpha \grave{\bar{s}}_\alpha + \nabla\cdot(\frac{\bar{\mathbf{q}}_\alpha}{\bar{\bar{\theta}}_\alpha}) - \frac{\bar{\rho}_\alpha \tilde{r}_\alpha}{\bar{\bar{\theta}}_\alpha} + \hat{c}_\alpha \bar{s}_\alpha + \hat{s}_\alpha \geq 0 \tag{2.4.45}$$

where

$$\check{m}_{\alpha i} = \hat{p}_{\alpha i} + \hat{c}_\alpha \tilde{v}_{\alpha i} \tag{3.4.1}$$

$$\hat{p}_{\alpha i} = \bar{t}_{\alpha i} - (\frac{\grave{\phi}_\alpha}{\phi_\alpha})_{,i} \bar{\rho}_\alpha \frac{\grave{\phi}_\alpha}{\phi_\alpha} \bar{i}_\alpha \tag{7.3.4}$$

$$k_{\alpha jji} = -\frac{\grave{\phi}_\alpha}{\phi_\alpha} U_{\alpha jji} \tag{7.3.5}$$

$$K_{\alpha jj} = 2\bar{\rho}_\alpha \frac{\grave{\phi}_\alpha}{\phi_\alpha} \bar{i}_\alpha \tag{7.2.6}$$

$$\check{k}_{\alpha jj} = \hat{c}_\alpha \hat{i}_{\alpha jj} \tag{5.2.8}$$

$$\bar{h}_{\alpha(jk)m} = \bar{\lambda}_{\alpha(jk)m} - \frac{1}{2} \bar{\rho}_\alpha \bar{i}_\alpha \frac{\grave{\phi}_\alpha}{\phi_\alpha} (\tilde{v}_{\alpha k} \delta_{mj} + \tilde{v}_{\alpha j} \delta_{mk}) \tag{7.3.6}$$

$$\check{g}_{\alpha(jk)} = \hat{g}_{\alpha(jk)} + \bar{\rho}_\alpha \bar{i}_\alpha (\frac{\grave{\phi}_\alpha}{\phi_\alpha})^2 \delta_{jk} + \frac{1}{2} [\tilde{v}_{\alpha k} (\bar{\rho}_\alpha \bar{i}_\alpha \frac{\grave{\phi}_\alpha}{\phi_\alpha})_{,j} + \tilde{v}_{\alpha j} (\bar{\rho}_\alpha \bar{i}_\alpha \frac{\grave{\phi}_\alpha}{\phi_\alpha})_{,k}] \tag{7.3.7}$$

$$\hat{g}_{\alpha(jk)} = \frac{1}{2} \hat{c}_\alpha \frac{\grave{\phi}_\alpha}{\phi_\alpha} (\hat{i}_{\alpha jk} + \hat{i}_{\alpha kj}) - \frac{1}{2} [(\frac{\grave{\phi}_\alpha}{\phi_\alpha} \frac{\grave{\phi}_\alpha}{\phi_\alpha} U_{\alpha jkm})_{,m} + (\frac{\grave{\phi}_\alpha}{\phi_\alpha} \frac{\grave{\phi}_\alpha}{\phi_\alpha} U_{\alpha kjm})_{,m}] \tag{7.3.8}$$

$$\hat{\epsilon}_\alpha = -\hat{c}_\alpha (\bar{\epsilon}_\alpha - \hat{\bar{\epsilon}}_\alpha) - \bar{q}_{s\alpha} + (\frac{\grave{\phi}_\alpha}{\phi_\alpha})_{,i} \bar{\lambda}_{\alpha jji} - (\frac{\grave{\phi}_\alpha}{\phi_\alpha} \bar{\rho}_\alpha \tilde{\ell} \epsilon_{\alpha i})_{,i} \tag{7.3.9}$$

where (7.3.9) follows from (3.4.15), (3.4.17), (7.2.3), and (7.3.1). From (7.2.10), it follows that $\tilde{\ell}_{\alpha(jk)} = 0$.

The entropy inequality (2.4.45) may be written in an alternate form by eliminating in this inequality $\bar{\rho}_\alpha \tilde{r}_\alpha$ by the aid of the energy equation (7.3.3) and Helmholtz potential (6.12.8), *i.e.*

$$
\begin{aligned}
&-\bar{\rho}_\alpha(\tilde{s}_\alpha \grave{\bar{\bar{\theta}}}_\alpha + \grave{\tilde{\psi}}_\alpha) - \frac{1}{\bar{\bar{\theta}}_\alpha}\bar{\mathbf{q}}_\alpha \cdot \nabla \bar{\bar{\theta}}_\alpha + \hat{s}_\alpha \bar{\bar{\theta}}_\alpha + \bar{T}_{\alpha ij}\tilde{v}_{\alpha i,j} + (\frac{\grave{\phi}_\alpha}{\phi_\alpha})_{,m}\bar{\lambda}_{\alpha kkm} \\
&-\bar{q}_{s\alpha} + \hat{c}_\alpha(\hat{\bar{\epsilon}}_\alpha - \tilde{\psi}_\alpha) - (\frac{\grave{\phi}_\alpha}{\phi_\alpha}\bar{\rho}_\alpha \tilde{\ell}\epsilon_{\alpha i})_{,i} + p_\alpha(\grave{\phi}_\alpha + \phi_\alpha \tilde{v}_{\alpha i,i} - \frac{\hat{c}_\alpha}{\bar{\bar{\rho}}_\alpha}) \geq 0
\end{aligned}
\tag{7.3.10}
$$

where the incompressibility constraint (6.12.40) has been introduced with $p_\alpha$ being the reaction force due to this constraint. For the purpose of developing constitutive equations of compressible and incompressible constituents of the mixture it is always desirable to include in the entropy inequality the incompressibility constraint, since in the former case one simply sets in the final results $p_\alpha = 0$ for the compressible phase $\alpha$.

Equations (2.4.14), (3.4.2), (7.2.3), and (7.3.1)-(7.3.3) provide a set of $13\gamma$ algebraic equations for $13\gamma$ unknowns

$$\bar{\rho}_\alpha,\ \tilde{v}_{\alpha i},\ \tilde{b}_{\alpha i},\ \phi_\alpha,\ \bar{\bar{\theta}}_\alpha,\ \tilde{i}_\alpha,\ \tilde{\ell}_{\alpha[jk]},\ \tilde{r}_\alpha$$

It is then sufficient that the remaining variables

$$
\begin{gathered}
\Upsilon_\alpha = (\hat{c}_\alpha,\ \bar{\mathbf{T}}_\alpha,\ \hat{\mathbf{M}}_\alpha,\ \bar{\mathbf{q}}_\alpha,\ \tilde{s}_\alpha,\ \tilde{\psi}_\alpha,\ \hat{s}_\alpha,\ \hat{\mathbf{i}}_\alpha,\ \mathbf{U}_\alpha,\ \bar{\boldsymbol{\lambda}}_\alpha,\ \bar{\mathbf{S}}_\alpha,\ \bar{\mathbf{t}}_\alpha,\ \tilde{\ell}\epsilon_\alpha,\ \hat{\bar{\epsilon}}_\alpha,\ \bar{q}_{s\alpha}); \\
\alpha = 1, \ldots, \gamma
\end{gathered}
\tag{7.3.11}
$$

be determined by the constitutive equations which are restricted by the entropy inequality (7.3.10). Notice that all the constitutive variables in (7.3.11) are frame-indifferent, as discussed in chapter 4. The body force moment is not frame-indifferent, and from (7.3.2) it is necessary, but not sufficient, that

$$\tilde{\ell}_{\alpha[jk]} = -\tilde{i}_\alpha \frac{\grave{\phi}_\alpha}{\phi_\alpha} W_{\alpha jk} \tag{7.3.12}$$

## 7.4 Constitutive Equations for Mixtures of Isotropic Fluids

As an illustration of the method to determine constitutive equations for multiphase mixtures, a somewhat detailed consideration will be given in this section to a multiphase mixture composed of the fluidlike phases. It will also be assumed that the mixture properties satisfy the principles of local action and smooth and local memory, and that the independent variables in the constitutive equations form a subset of the set of section 6.10, *i.e.*

$$\Upsilon_\alpha = G_{\alpha\boldsymbol{\kappa}}[\bar{\rho}_\beta, \nabla\bar{\rho}_\beta, \mathbf{D}_\beta, \mathbf{W}_\beta - \mathbf{W}_\gamma, \bar{\bar{\theta}}_\beta, \nabla\bar{\bar{\theta}}_\beta, \tilde{\mathbf{v}}_\beta - \tilde{\mathbf{v}}_\gamma, \bar{\mathbf{i}}_\beta, \frac{\grave{\phi}_\beta}{\phi_\beta}, \nabla(\frac{\grave{\phi}_\beta}{\phi_\beta})] \tag{7.4.1}$$

or since

$$\bar{\rho}_\beta = \phi_\beta \bar{\bar{\rho}}_\beta \tag{7.4.2}$$

and ignoring the independent variation of $\nabla\grave{\phi}_\beta$, gives

$$\begin{aligned}\Upsilon_\alpha = G_{\alpha\boldsymbol{\kappa}}[\bar{\bar{\rho}}_\beta, \nabla\bar{\bar{\rho}}_\beta, \mathbf{D}_\beta, \mathbf{W}_\beta - \mathbf{W}_\gamma, \bar{\bar{\theta}}_\beta, \nabla\bar{\bar{\theta}}_\beta, \tilde{\mathbf{v}}_\beta - \tilde{\mathbf{v}}_\gamma, \bar{\mathbf{i}}_\beta, \phi_\beta, \nabla\phi_\beta, \grave{\phi}_\beta] \\ = G_{\alpha\boldsymbol{\kappa}}[\{S_\beta\}]\end{aligned} \tag{7.4.3}$$

where for each $\alpha$, $\alpha=1,\ldots,\gamma$, $\beta=1,\ldots,\gamma$.

For the substitution into the entropy inequality (7.3.10), the time derivative of the Helmholtz potential is

$$\begin{aligned}\grave{\tilde{\psi}}_\alpha = \sum_{\beta=1}^{\gamma}[&\frac{\partial\tilde{\psi}_\alpha}{\partial\bar{\bar{\rho}}_\beta}(\grave{\bar{\bar{\rho}}}_\beta) + \frac{\partial\tilde{\psi}_\alpha}{\partial\bar{\bar{\rho}}_{\beta,i}}(\grave{\bar{\bar{\rho}}}_{\beta,i}) + \frac{\partial\tilde{\psi}_\alpha}{\partial D_{\beta ij}}(\grave{D}_{\beta ij}) + \frac{\partial\tilde{\psi}_\alpha}{\partial\bar{\bar{\theta}}_\beta}(\grave{\bar{\bar{\theta}}}_\beta) + \frac{\partial\tilde{\psi}_\alpha}{\partial\bar{\bar{\theta}}_{\beta,i}}(\grave{\bar{\bar{\theta}}}_{\beta,i}) \\ &+\frac{\partial\tilde{\psi}_\alpha}{\partial(W_{\beta ij} - W_{\gamma ij})}((W_{\beta ij} \grave{-} W_{\gamma ij})) + \frac{\partial\tilde{\psi}_\alpha}{\partial(\tilde{v}_{\beta i} - \tilde{v}_{\gamma i})}((\tilde{v}_{\beta i} \grave{-} \tilde{v}_{\gamma i})) \\ &+\frac{\partial\tilde{\psi}_\alpha}{\partial\bar{i}_\beta}(\grave{\bar{i}}_\beta) + \frac{\partial\tilde{\psi}_\alpha}{\partial\phi_\beta}(\grave{\phi}_\beta) + \frac{\partial\tilde{\psi}_\alpha}{\partial\phi_{\beta,i}}(\grave{\phi}_{\beta,i}) + \frac{\partial\tilde{\psi}_\alpha}{\partial\grave{\phi}_\beta}(\grave{\grave{\phi}}_\beta)]\end{aligned} \tag{7.4.4}$$

where, for example, $(\grave{\bar{\bar{\rho}}}_\beta)$ is the material derivative of $\bar{\bar{\rho}}_\beta$ following the motion of the *α'th phase* and it is defined as

$$\begin{aligned}(\grave{\bar{\bar{\rho}}}_\beta) = \frac{\partial\bar{\bar{\rho}}_\beta}{\partial t} + \nabla\bar{\bar{\rho}}_\beta\cdot\tilde{\mathbf{v}}_\alpha = \frac{\partial\bar{\bar{\rho}}_\beta}{\partial t} + \nabla\bar{\bar{\rho}}_\beta\cdot\tilde{\mathbf{v}}_\beta + \nabla\cdot\bar{\bar{\rho}}_\beta(\tilde{\mathbf{v}}_\alpha - \tilde{\mathbf{v}}_\beta) \\ = \grave{\bar{\bar{\rho}}}_\beta + \bar{\bar{\rho}}_{\beta,i}(\tilde{v}_{\alpha i} - \tilde{v}_{\beta i})\end{aligned} \tag{7.4.5}$$

with $\grave{\bar{\bar{\rho}}}_\beta$ having the usual meaning as the material derivative of $\bar{\bar{\rho}}_\beta$. Equation (7.4.4) can be, therefore, written in the following alternate form:

$$\grave{\tilde{\psi}}_\alpha = \sum_{\beta=1}^{\gamma}[\frac{\partial\tilde{\psi}_\alpha}{\partial\bar{\bar{\rho}}_\beta}(\grave{\bar{\bar{\rho}}}_\beta + \bar{\bar{\rho}}_{\beta,i}(\tilde{v}_{\alpha i} - \tilde{v}_{\beta i})) + \frac{\partial\tilde{\psi}_\alpha}{\partial\bar{\bar{\rho}}_{\beta,i}}(\{\grave{\bar{\bar{\rho}}}_{\beta,i}\} + \bar{\bar{\rho}}_{\beta,ij}(\tilde{v}_{\alpha j} - \tilde{v}_{\beta j}))$$

$$+\frac{\partial\tilde{\psi}_\alpha}{\partial D_{\beta ij}}(\grave{D}_{\beta ij}+D_{\beta ij,k}(\tilde{v}_{\alpha k}-\tilde{v}_{\beta k}))+\frac{\partial\tilde{\psi}_\alpha}{\partial\bar{\bar{\theta}}_\beta}(\grave{\bar{\bar{\theta}}}_\beta+\bar{\bar{\theta}}_{\beta,i}(\tilde{v}_{\alpha i}-\tilde{v}_{\beta i}))$$

$$+\frac{\partial\tilde{\psi}_\alpha}{\partial(W_{\beta ij}-W_{\gamma ij})}(\grave{W}_{\beta ij}-\grave{W}_{\gamma ij}+(W_{\beta ij,k}-W_{\gamma ij,k})(\tilde{v}_{\alpha k}-\tilde{v}_{\beta k})$$

$$-W_{\gamma ij,k}(\tilde{v}_{\beta k}-\tilde{v}_{\gamma k}))+\frac{\partial\tilde{\psi}_\alpha}{\partial\grave{\phi}_\beta}((\grave{\grave{\phi}}_\beta)+\{\grave{\phi}_\beta\}_{,i}(\tilde{v}_{\alpha i}-\tilde{v}_{\beta i}))$$

$$+\frac{\partial\tilde{\psi}_\alpha}{\partial\bar{\bar{\theta}}_{\beta,i}}(\{\grave{\bar{\bar{\theta}}_{\beta,i}}\}+\bar{\bar{\theta}}_{\beta,ij}(\tilde{v}_{\alpha j}-\tilde{v}_{\beta j}))+\frac{\partial\tilde{\psi}_\alpha}{\partial\bar{i}_\beta}(\grave{\bar{i}}_\beta+\bar{i}_{\beta,i}(\tilde{v}_{\alpha i}-\tilde{v}_{\beta i}))$$

$$+\frac{\partial\tilde{\psi}_\alpha}{\partial\phi_\beta}(\grave{\phi}_\beta+\phi_{\beta,i}(\tilde{v}_{\alpha i}-\tilde{v}_{\beta i}))+\frac{\partial\tilde{\psi}_\alpha}{\partial\phi_{\beta,i}}(\{\grave{\phi}_\beta\}_{,i}+\phi_{\beta,ij}(\tilde{v}_{\alpha j}-\tilde{v}_{\beta j})-\phi_{\beta,j}\tilde{v}_{\beta j,i})$$

$$+\frac{\partial\tilde{\psi}_\alpha}{\partial(\tilde{v}_{\beta i}-\tilde{v}_{\gamma i})}(\grave{\tilde{v}}_{\beta i}-\grave{\tilde{v}}_{\gamma i}+(\tilde{v}_{\beta i,j}-\tilde{v}_{\gamma i,j})(\tilde{v}_{\alpha j}-\tilde{v}_{\beta j})-\tilde{v}_{\gamma i,j}(\tilde{v}_{\beta j}-\tilde{v}_{\gamma j}))] \tag{7.4.6}$$

with the notation that the operation on the quantity enclosed by brackets $\{\ldots\}$ is performed first.

From the balance of mass equation (2.4.14) and equation (7.4.2) it is possible to solve for $\grave{\bar{\bar{\rho}}}_\beta$, *i.e.*

$$\grave{\bar{\bar{\rho}}}_\beta=\frac{\hat{c}_\beta}{\phi_\beta}-\bar{\bar{\rho}}_\beta\frac{\grave{\phi}_\beta}{\phi_\beta}-\bar{\bar{\rho}}_\beta\tilde{v}_{\beta i,i} \tag{7.4.7}$$

and substitute (7.4.6) and (7.4.7) into (7.3.10). Performing these operations and requiring that the resulting entropy inequality is satisfied for arbitarry values of

$$\{\grave{\bar{\bar{\rho}}_{\beta,i}}\},\ \bar{\bar{\rho}}_{\beta,ij},\ \grave{D}_{\beta ij},\ D_{\beta ij,k},\ \grave{W}_{\beta ij}-\grave{W}_{\gamma ij},\ W_{\beta ij,k}-W_{\gamma ij,k},$$
$$\grave{\bar{\bar{\theta}}}_\beta,\ \{\grave{\bar{\bar{\theta}}_{\beta,i}}\},\ \bar{\bar{\theta}}_{\beta,ij},\ \grave{\tilde{v}}_{\beta i}-\grave{\tilde{v}}_{\gamma i},\ \bar{i}_{\beta,i},\ \grave{\bar{i}}_\beta,\ (\grave{\grave{\phi}}_\beta),\ \{\grave{\phi}_\beta\}_{,i},\ \phi_{\beta,ij}$$

appearing in this inequality, it is sufficient and necessary that the coefficients which multiply these variables be set equal to zero. If this were not true, then it would be possible to find at least one thermokinetic process that would violate the entropy inequality. Thus:

*Coefficient of:*

$$\grave{\bar{\bar{\theta}}}_\alpha: \quad \bar{s}_\alpha=-\frac{\partial\tilde{\psi}_\alpha}{\partial\bar{\bar{\theta}}_\alpha} \tag{7.4.8}$$

$$\grave{\bar{\bar{\theta}}}_\beta: \quad \frac{\partial\tilde{\psi}_\alpha}{\partial\bar{\bar{\theta}}_\beta}=0;\quad \beta=1,\ldots,\alpha-1,\alpha+1,\ldots,\gamma \tag{7.4.9}$$

$$\bar{\bar{\theta}}_{\beta,ij},\ \{\grave{\bar{\bar{\theta}}}_{\beta,i}\}: \quad \frac{\partial\tilde{\psi}_\alpha}{\partial\bar{\bar{\theta}}_{\beta,i}} = 0; \quad \beta = 1,\ldots,\gamma \tag{7.4.10}$$

$$\{\grave{\bar{\bar{\rho}}}_{\beta,i}\},\ \bar{\bar{\rho}}_{\beta,ij}: \quad \frac{\partial\tilde{\psi}_\alpha}{\partial\bar{\bar{\rho}}_{\beta,i}} = 0; \quad \beta = 1,\ldots,\gamma \tag{7.4.11}$$

$$\grave{D}_{\beta ij}: \quad \frac{\partial\tilde{\psi}_\alpha}{\partial D_{\beta ij}} = 0; \quad \beta = 1,\ldots,\gamma \tag{7.4.12}$$

$$(\grave{W}_{\beta ij} - \grave{W}_{\gamma ij}): \quad \frac{\partial\tilde{\psi}_\alpha}{\partial(W_{\beta ij} - W_{\gamma ij})} = 0; \quad \beta = 1,\ldots,\gamma-1 \tag{7.4.13}$$

$$(\grave{\bar{v}}_{\beta i} - \grave{\bar{v}}_{\gamma i}): \quad \frac{\partial\tilde{\psi}_\alpha}{\partial(\bar{v}_{\beta i} - \bar{v}_{\gamma i})} = 0; \quad \beta = 1,\ldots,\gamma-1 \tag{7.4.14}$$

$$\{\grave{\phi}_\beta\}_{,i}: \quad \frac{\partial\tilde{\psi}_\alpha}{\partial\phi_{\beta,i}} = 0; \quad \beta = 1,\ldots,\alpha-1,\alpha+1,\ldots,\gamma \tag{7.4.15}$$

$$(\grave{\dot{\phi}}_\beta): \quad \frac{\partial\tilde{\psi}_\alpha}{\partial\dot{\phi}_\beta} = 0; \quad \beta = 1,\ldots,\gamma \tag{7.4.16}$$

$$\{\grave{\phi}_\alpha\}_{,i}: \quad \bar{\rho}_\alpha\frac{\partial\tilde{\psi}_\alpha}{\partial\phi_{\alpha,i}} = \frac{1}{\phi_\alpha}\bar{\lambda}_{\alpha kki} - \frac{1}{\phi_\alpha}\bar{\rho}_\alpha\bar{\ell}\epsilon_{\alpha i} \tag{7.4.17}$$

$$\grave{\bar{i}}_\beta,\ \bar{i}_{\beta,i}: \quad \frac{\partial\tilde{\psi}_\alpha}{\partial\bar{i}_\beta} = 0; \quad \beta = 1,\ldots,\gamma \tag{7.4.18}$$

and the entropy inequality (7.3.10) becomes

$$\begin{aligned}
&-\frac{1}{\bar{\bar{\theta}}_\alpha}\bar{q}_{\alpha i}\bar{\bar{\theta}}_{\alpha,i} - \bar{\rho}_\alpha\sum_{\beta\neq\alpha}^{\gamma}\Big[\frac{\partial\tilde{\psi}_\alpha}{\partial\bar{\bar{\rho}}_\beta}\frac{\hat{c}_\beta}{\phi_\beta} - \Big(\frac{\partial\tilde{\psi}_\alpha}{\partial\bar{\bar{\rho}}_\beta}\frac{\bar{\bar{\rho}}_\beta}{\phi_\beta} - \frac{\partial\tilde{\psi}_\alpha}{\partial\phi_\beta}\Big)\grave{\phi}_\beta \\
&+\Big(\frac{\partial\tilde{\psi}_\alpha}{\partial\bar{\bar{\rho}}_\beta}\bar{\bar{\rho}}_{\beta,i} + \frac{\partial\tilde{\psi}_\alpha}{\partial\phi_\beta}\phi_{\beta,i}\Big)(\bar{v}_{\alpha i} - \bar{v}_{\beta i}) - \frac{\partial\tilde{\psi}_\alpha}{\partial\bar{\bar{\rho}}_\beta}\bar{\bar{\rho}}_\beta\, tr\mathbf{D}_\beta\Big] \\
&+\hat{f}_\alpha + \frac{\grave{\phi}_\alpha}{\phi_\alpha}\big[\phi_\alpha(p_\alpha + \bar{\bar{\pi}}_\alpha) - \bar{\bar{\lambda}}^R_{\alpha kkm}\phi_{\alpha,m} - \phi_\alpha\beta_\alpha - (\bar{\bar{\rho}}_\alpha\phi_\alpha\bar{\ell}\epsilon_{\alpha i})_{,i}\big] \\
&+\hat{c}_\alpha\Big(-\tilde{\psi}_\alpha - \frac{p_\alpha}{\bar{\bar{\rho}}_\alpha} - \frac{\bar{\bar{\pi}}_\alpha}{\bar{\bar{\rho}}_\alpha} + \hat{\bar{\epsilon}}_\alpha\Big) + \bar{T}^e_{\alpha ij}\bar{v}_{\alpha i,j} - u_{\alpha i}\bar{t}_{\alpha i} = R_\alpha(\{S_\beta\}) \geq 0
\end{aligned} \tag{7.4.19}$$

where:

$\bar{T}^e_{\alpha ij}$ is the *extra stress:*

$$\bar{T}^e_{\alpha ij} = \bar{T}_{\alpha ij} + \phi_\alpha(p_\alpha + \bar{\bar{\pi}}_\alpha)\delta_{ij} + \bar{\bar{\rho}}_\alpha \phi_\alpha \frac{\partial \tilde{\psi}_\alpha}{\partial \phi_{\alpha,j}} \phi_{\alpha,i} \tag{7.4.20}$$

$\bar{\bar{\pi}}_\alpha$ is the *thermodynamic pressure:*

$$\bar{\bar{\pi}}_\alpha = \bar{\bar{\rho}}^2_\alpha \frac{\partial \tilde{\psi}_\alpha}{\partial \bar{\bar{\rho}}_\alpha} \tag{7.4.21}$$

$\beta_\alpha$ is the *configuration pressure* (PASSMAN *et al.*,1984):

$$\beta_\alpha = \bar{\bar{\rho}}_\alpha \phi_\alpha \frac{\partial \tilde{\psi}_\alpha}{\partial \phi_\alpha} \tag{7.4.22}$$

$\bar{\bar{\lambda}}^R_{\alpha k\ell m}$ is the *reduced intrinsic hyperstress:*

$$\bar{\bar{\lambda}}^R_{\alpha k\ell m} = \bar{\bar{\rho}}_\alpha \phi_\alpha \frac{\partial \tilde{\psi}_\alpha}{\partial \phi_{\alpha,m}} \delta_{k\ell} \tag{7.4.23}$$

and $\hat{f}_\alpha$ is a variable defined as

$$\hat{f}_\alpha = \hat{s}_\alpha \bar{\bar{\theta}}_\alpha - \bar{q}_{s\alpha} + \mathbf{u}_\alpha \cdot \bar{\mathbf{t}}_\alpha \tag{7.4.24}$$

whose properties will be discussed below. Notice in (7.4.20) that $p_\alpha = 0$ for a *compressible* phase $\alpha$ and that $\bar{\bar{\pi}}_\alpha = 0$ for an *incompressible* phase $\alpha$. From (7.4.17), (7.4.2), and (7.4.23), it follows that the reduced intrinsic hyperstress may be written as follows

$$\bar{\bar{\lambda}}^R_{\alpha kkm} = \frac{1}{\phi_\alpha} \bar{\lambda}_{\alpha kkm} - \bar{\bar{\rho}}_\alpha \tilde{\ell} \epsilon_{\alpha m} \tag{7.4.25}$$

Equations (7.4.8)-(7.4.18) show that the Helmholtz potential $\tilde{\psi}_\alpha$ has the following functional form:

$$\tilde{\psi}_\alpha = \tilde{\psi}_\alpha(\bar{\bar{\rho}}_\beta, \bar{\bar{\theta}}_\alpha, \phi_\beta, \nabla \phi_\alpha) \tag{7.4.26}$$

Thus the Helmholtz potential of phase $\alpha$ not only depends on the density and temperature (as in a single phase, single component mixture), but also on the structural properties of the mixture expressed by the volumetric fractions. Moreover, $\tilde{\phi}_\alpha$ does not satisfy the principle of phase separation, since for each $\alpha$, $\alpha = 1, \ldots, \gamma$, $\beta = 1, \ldots, \gamma$ in (7.4.26).

To obtain further restrictions on the constitutive variables appearing in the entropy inequality (7.4.19) it is useful to define an *equilibrium thermokinetic process* with the following characteristics:

1. The entropy inequality (7.4.19) becomes an equality.
2. $\nabla\bar{\bar{\theta}}_\beta = \mathbf{0}$ (no temperature gradient), $\tilde{\mathbf{v}}_\beta = \mathbf{v}$ (no diffusion), $\mathbf{D}_\beta = \mathbf{0}$ and $\mathbf{W}_\beta = \mathbf{0}$ (no viscous effects), and $\dot{\phi}_\beta = 0$ (no dilatational effects), for $\beta = 1, \ldots, \gamma$.

From (7.4.19) and (7.4.3), it then follows that

$$R_\alpha(\bar{\bar{\rho}}_\beta,\ \nabla\bar{\bar{\rho}}_\beta,\ \mathbf{D}_\beta = \mathbf{0},\ \mathbf{W}_\beta - \mathbf{W}_\gamma = \mathbf{0},\ \bar{\bar{\theta}}_\beta,\ \nabla\bar{\bar{\theta}}_\beta = \mathbf{0},\ \tilde{\mathbf{v}}_\beta - \tilde{\mathbf{v}}_\gamma = \mathbf{0},$$
$$\bar{i}_\beta,\ \phi_\beta,\ \nabla\phi_\beta,\ \dot{\phi}_\beta = 0) = R_\alpha(\{S_{\beta 0}\}) = 0; \quad \beta = 1, \ldots, \gamma \qquad (7.4.27)$$

This definition of the equilibrium state is consistent with the definition in the equilibrium thermodynamics, since in such a state no possibility is allowed for the temperature gradient, and viscous and dilatational effects. Clearly, a mixture with these properties is by no means an ideal mixture, since it is *not* assumed that $\nabla\bar{\bar{\rho}}_\beta = \nabla\phi_\beta = \mathbf{0}$.

The properties of the equilibrium thermokinetic process can be determined by using the method as in DOBRAN (1984b). For this purpose, let $\lambda$ be a parameter and set

$$\mathbf{D}_\beta = \lambda\mathbf{A}_\beta,\ \mathbf{W}_\beta = \lambda\mathbf{B}_\beta,\ \nabla\bar{\bar{\theta}}_\beta = \lambda\mathbf{a}_\beta,\ \tilde{\mathbf{v}}_\beta = \mathbf{v} + \lambda\mathbf{b}_\beta,$$
$$\nabla\tilde{\mathbf{v}}_\beta = \lambda\mathbf{C}_\beta,\ \dot{\phi}_\beta = \lambda\eta_\beta$$

Taking account of (7.4.25) and substituting these parametrized properties into the entropy inequality (7.4.19) results in

$$-\frac{1}{\bar{\bar{\theta}}_\alpha}\bar{q}_{\alpha i}(\{S_{\beta\lambda}\})\lambda a_{\alpha i} - \bar{\rho}_\alpha \sum_{\beta\neq\alpha}^{\gamma}\Big[\frac{\partial\tilde{\psi}_\alpha(\{S_{\beta\lambda}\})}{\partial\bar{\bar{\rho}}_\beta}\frac{\hat{c}_\beta(\{S_{\beta\lambda}\})}{\phi_\beta} - \Big(\frac{\partial\tilde{\psi}_\alpha(\{S_{\beta\lambda}\})}{\partial\bar{\bar{\rho}}_\beta}\frac{\bar{\bar{\rho}}_\beta}{\phi_\beta}$$
$$-\frac{\partial\tilde{\psi}_\alpha(\{S_{\beta\lambda}\})}{\partial\phi_\beta}\Big)\lambda\eta_\beta + \Big(\frac{\partial\tilde{\psi}_\alpha(\{S_{\beta\lambda}\})}{\partial\bar{\bar{\rho}}_\beta}\bar{\bar{\rho}}_{\beta,i} + \frac{\partial\tilde{\psi}_\alpha(\{S_{\beta\lambda}\})}{\partial\phi_\beta}\phi_{\beta,i}\Big)\lambda(b_{\alpha i} - b_{\beta i})$$
$$-\frac{\partial\tilde{\psi}_\alpha(\{S_{\beta\lambda}\})}{\partial\bar{\bar{\rho}}_\beta}\bar{\bar{\rho}}_\beta\lambda b_{\beta i,i}\Big] + \hat{f}_\alpha(\{S_{\beta\lambda}\}) + \frac{\lambda\eta_\alpha}{\phi_\alpha}\big[\phi_\alpha(p_\alpha + \bar{\bar{\pi}}_\alpha(\{S_{\beta\lambda}\}))$$
$$-\bar{\bar{\lambda}}^R_{\alpha kkm}(\{S_{\beta\lambda}\})\phi_{\alpha,m} - \phi_\alpha\beta_\alpha(\{S_{\beta\lambda}\}) - (\bar{\bar{\rho}}_\alpha\phi_\alpha\bar{\ell}\epsilon_{\alpha i}(\{S_{\beta\lambda}\}))_{,i}\big]$$
$$+\hat{c}_\alpha(\{S_{\beta\lambda}\})\Big[-\tilde{\psi}_\alpha(\{S_{\beta\lambda}\}) - \frac{p_\alpha}{\bar{\bar{\rho}}_\alpha} - \frac{\bar{\bar{\pi}}_\alpha(\{S_{\beta\lambda}\})}{\bar{\bar{\rho}}_\alpha} + \hat{\bar{\epsilon}}_\alpha(\{S_{\beta\lambda}\})\Big]$$
$$+\bar{T}^e_{\alpha ij}(\{S_{\beta\lambda}\})\lambda C_{\alpha ij} - \lambda b_{\alpha i}\bar{t}_{\alpha i}(\{S_{\beta\lambda}\}) = R_\alpha(\{S_{\beta\lambda}\}) \geq 0 \qquad (7.4.28)$$

where the set $\{S_{\beta\lambda}\}$ is comprised of the independent variables as in (7.4.3).

Since $R_\alpha(\{S_{\beta\lambda}\}) \geq 0$, and $R_\alpha(\{S_{\beta 0}\}) = 0$ (*cit.* (7.4.27)), it follows that *in the equilibrium state*

$$\left(\frac{dR_\alpha(\{S_{\beta\lambda}\})}{d\lambda}\right)_{\lambda=0} = 0 \qquad (7.4.29)$$

Performing the above operation on (7.4.28) and requiring the arbitrariness of $\mathbf{A}_\beta$, $\mathbf{B}_\beta$, $\mathbf{a}_\beta$, $\mathbf{b}_\beta$, $\mathbf{C}_\beta$, and $\eta_\beta$ for $\beta = 1,\ldots,\gamma$, results in the following equilibrium state properties:

$$\frac{\partial\tilde{\psi}_\alpha(\{S_{\beta 0}\})}{\partial\bar{\bar{\rho}}_\beta} = 0, \quad \frac{\partial\tilde{\psi}_\alpha(\{S_{\beta 0}\})}{\partial\phi_\beta} = 0; \quad \beta = 1,\ldots,\alpha-1,\alpha+1,\ldots,\gamma \tag{7.4.30}$$

$$\bar{q}_{\alpha i}(\{S_{\beta 0}\}) = 0, \quad \bar{T}^e_{\alpha ij}(\{S_{\beta 0}\}) = 0, \quad \bar{t}_{\alpha i}(\{S_{\beta 0}\}) = 0; \quad \alpha = 1,\ldots,\gamma \tag{7.4.31}$$

$$\begin{aligned}\phi_\alpha[p_\alpha + \bar{\bar{\pi}}_\alpha(\{S_{\beta 0}\})] - \bar{\bar{\lambda}}^R_{\alpha kkm}(\{S_{\beta 0}\})\phi_{\alpha,m} - \phi_\alpha\beta_\alpha(\{S_{\beta 0}\})\\ -[\bar{\bar{\rho}}_\alpha\phi_\alpha\bar{\ell}\epsilon_{\alpha i}(\{S_{\beta 0}\})]_{,i} = 0\end{aligned} \tag{7.4.32}$$

$$\hat{c}_\beta(\{S_{\beta 0}\}) = 0, \quad \frac{d\hat{c}_\beta(\{S_{\beta 0}\})}{d\lambda} = 0 \tag{7.4.33}$$

$$\hat{f}_\alpha(\{S_{\beta 0}\}) = 0, \quad \frac{d\hat{f}_\alpha(\{S_{\beta 0}\})}{d\lambda} = 0 \tag{7.4.34}$$

The results expressed by equation (7.4.30) can be used in (7.4.26) to show that

$$\tilde{\psi}_\alpha(\{S_{\beta 0}\}) = \tilde{\psi}^0_\alpha(\bar{\bar{\rho}}_\alpha, \bar{\bar{\theta}}_\alpha, \phi_\alpha, \nabla\phi_\alpha) \tag{7.4.35}$$

Moreover, from (7.4.20) and $(7.4.31)_2$ it follows that

$$\bar{T}_{\alpha ij}(\{S_{\beta 0}\}) = -\phi_\alpha[p_\alpha + \bar{\bar{\pi}}_\alpha(\{S_{\beta 0}\})]\delta_{ij} - \bar{\bar{\rho}}_\alpha\phi_\alpha\frac{\partial\tilde{\psi}_\alpha}{\partial\phi_{\alpha,j}}\phi_{\alpha,i} \tag{7.4.36}$$

whereas (7.3.1), (5.3.6), (7.3.6)-(7.3.8), $(7.4.33)_1$, and the definition of equilibrium state $\{S_{\beta 0}\}$ give

$$\bar{S}_{\alpha kk}(\{S_{\beta 0}\}) + \bar{\lambda}_{\alpha kkm,m}(\{S_{\beta 0}\}) = 0 \tag{7.4.37}$$

From (7.4.36) it is thus seen that even *in equilibrium the stress need not be a hydrostatic pressure.* This important result, which accounts for the structural properties of the mixture, is further discussed in section 7.8.

The results represented by equations (7.4.32) and (7.4.25) can be combined to produce a solution for the intrinsic stress $\bar{\lambda}_{\alpha kkm}$, *i.e.*

$$\begin{aligned}\bar{\lambda}_{\alpha kkm}(\{S_{\beta 0}\})\phi_{\alpha,m} + \phi^2_\alpha\beta_\alpha(\{S_{\beta 0}\}) - \phi^2_\alpha(p_\alpha + \bar{\bar{\pi}}_\alpha(\{S_{\beta 0}\}))\\ +\phi_\alpha(\bar{\bar{\rho}}_\alpha\bar{\ell}\epsilon_{\alpha i}(\{S_{\beta 0}\})_{,i} - \phi_{\alpha,m}\bar{\bar{\rho}}_\alpha\phi_\alpha\bar{\ell}\epsilon_{\alpha m}(\{S_{\beta 0}\}) = 0\end{aligned} \tag{7.4.38}$$

Taking the divergence of (7.4.25) and combining with (7.4.32) and (7.4.37) gives

$$\bar{\bar{\lambda}}^{R}_{\alpha kkm,m}(\{S_{\beta 0}\}) = \beta_\alpha(\{S_{\beta 0}\}) + \frac{1}{\phi_\alpha}\bar{\lambda}_{\alpha kkm,m} - (p_\alpha + \bar{\bar{\pi}}_\alpha(\{S_{\beta 0}\})$$
$$= \beta_\alpha(\{S_{\beta 0}\}) - (p_\alpha + \bar{\bar{\pi}}_\alpha(\{S_{\beta 0}\})) - \frac{1}{\phi_\alpha}\bar{S}_{\alpha kk}(\{S_{\beta 0}\}) \qquad (7.4.39)$$

showing that the equilibrium *surface traction pressure* $P_\alpha$ defined by

$$P_\alpha = -\frac{1}{\phi_\alpha}\bar{S}_{\alpha kk}(\{S_{\beta 0}\}) \qquad (7.4.40)$$

is maintanied by the configuration pressure $\beta_\alpha(\{S_{\beta 0}\})$, hydrostatic (thermodynamic) pressure $p_\alpha$ ($\bar{\bar{\pi}}_\alpha$), and inhomogeneous distribution of the reduced intrinsic stress $\bar{\bar{\lambda}}^{R}_{\alpha kkm}(\{S_{\beta 0}\})$. If phase $\alpha$ is characterized by an average interface pressure $P_\alpha$ and the adjoining phase $\beta$ by an average interface pressure $P_\beta$, then from (7.4.39) and (7.4.40) it follows that

$$-(P_\alpha - P_\beta) = \beta_\alpha(\{S_{\beta 0}\}) - \beta_\beta(\{S_{\beta 0}\})$$
$$-[(p_\alpha + \bar{\bar{\pi}}_\alpha(\{S_{\beta 0}\})) - (p_\beta + \bar{\bar{\pi}}_\beta(\{S_{\beta 0}\}))] - (\bar{\bar{\lambda}}^{R}_{\alpha kkm} - \bar{\bar{\lambda}}^{R}_{\beta kkm})_{,m} \qquad (7.4.41)$$

For a multiphase mixture which can be characterized by a *single interface pressure* $P$, the above equation is reduced to

$$(p_\alpha + \bar{\bar{\pi}}_\alpha) - (p_\beta + \bar{\bar{\pi}}_\beta) = (\beta_\alpha - \beta_\beta)(\{S_{\beta 0}\}) - (\bar{\bar{\lambda}}^{R}_{\alpha kkm} - \bar{\bar{\lambda}}^{R}_{\beta kkm})_{,m}(\{S_{\beta 0}\}) \qquad (7.4.42)$$

This result, which is similar to the one discussed by PASSMAN *et al.* (1984), shows that the difference between the phasic pressures in a homogeneous mixture is determined by the configuration pressure. But in a two-phase mixture in equilibrium, the difference between the phase pressures may be related to the surface tension and average interface curvature, as can be readily seen from equation (2.2.6) when specialized for the momentum equation in Table 2.1. Consequently, it may be permissible to state that the difference in configuration pressures in (7.4.42) can be related to an average surface tension force and curvature of the interface.

The function $\hat{f}_\alpha$ defined by (7.4.24) is positive semidefinite and vanishes in the equilibrium state $\{S_{\beta 0}\}$, as may be seen from (7.4.19) and (7.4.34). The entropy inequality (7.4.19) imposes restrictions on the material properties appearing in the constitutive equations, and in the next section a consideration will be given to the linearized analysis of constitutive equations to determine the simplest forms of these equations. The restrictions imposed by the entropy inequality on the material coefficients is discussed in section 7.6.

## 7.5  Linearized Constitutive Equations for Mixtures of Isotropic Fluids

The linear constitutive equations are usually not fully satisfactory to decribe multiphase flows and need nonlinear modifications to construct useful models. Nevertheless, the linear analysis of constitutive equations provides a starting point to the more complete physical modeling and it produces the simplest results which are often necessary for practical applications.

The linear analysis of constitutive equations (7.3.11) involves at first the selection of an equilibrium state for the mixture about which the linearization can be performed. This state will be selected as follows:

$$\{S_{\beta 0}'\} = \{\bar{\bar{\rho}}_\beta, \ \boldsymbol{\nabla}\bar{\bar{\rho}}_\beta = 0, \ \mathbf{D}_\beta = 0, \ \mathbf{W}_\beta - \mathbf{W}_\gamma = 0, \ \bar{\bar{\theta}}_\beta, \ \boldsymbol{\nabla}\bar{\bar{\theta}}_\beta = 0,$$

$$\tilde{\mathbf{v}}_\beta - \tilde{\mathbf{v}}_\gamma = 0, \ \tilde{\mathbf{i}}_\beta = 0, \ \phi_\beta, \ \boldsymbol{\nabla}\phi_\beta = 0, \ \dot{\phi}_\beta = 0\} \qquad (7.5.1)$$

such that upon comparison with (7.4.27) gives

$$\{S_{\beta 0}\} = \{S_{\beta 0}'\} \ \cup \ \boldsymbol{\nabla}\bar{\bar{\rho}}_\beta \ \cup \ \tilde{\mathbf{i}}_\beta \ \cup \ \boldsymbol{\nabla}\phi_\beta \qquad (7.5.2)$$

The selection of the equilibrium state $\{S_{\beta 0}'\}$ about which the linearization of constitutive equations can be performed implies that nonequilibrium effects and structural characteristics of the mixture are not very significant, if only the linear deviations of independent variables (7.4.3) in the constitutive equations (7.3.11) from this state are retained. A measure which relates the importance of these effects from the state $\{S_{\beta 0}'\}$ can be defined through a quantity which is defined as follows:

$$\varepsilon^2 = \sum_\beta^\gamma C_{1\beta} \boldsymbol{\nabla}\bar{\bar{\rho}}_\beta \cdot \boldsymbol{\nabla}\bar{\bar{\rho}}_\beta + \sum_\beta^\gamma C_{2\beta}\, tr(\mathbf{D}_\beta \mathbf{D}_\beta)$$

$$+ \sum_\beta^{\gamma-1} C_{3\beta}\, tr[(\mathbf{W}_\beta - \mathbf{W}_\gamma)(\mathbf{W}_\beta - \mathbf{W}_\gamma)]$$

$$+ \sum_\beta^\gamma C_{4\beta} \boldsymbol{\nabla}\bar{\bar{\theta}}_\beta \cdot \boldsymbol{\nabla}\bar{\bar{\theta}}_\beta + \sum_\beta^{\gamma-1} C_{5\beta}\, (\tilde{\mathbf{v}}_\beta - \tilde{\mathbf{v}}_\gamma)\cdot(\tilde{\mathbf{v}}_\beta - \tilde{\mathbf{v}}_\gamma) + \sum_\beta^\gamma C_{6\beta}\, tr(\tilde{\mathbf{i}}_\beta \tilde{\mathbf{i}}_\beta)$$

$$+ \sum_\beta^\gamma C_{7\beta} \boldsymbol{\nabla}\phi_\beta \cdot \boldsymbol{\nabla}\phi_\beta + \sum_\beta^\gamma C_{8\beta}\, \dot{\phi}_\beta\dot{\phi}_\beta \qquad (7.5.3)$$

When $\varepsilon = 0$, (7.5.3) implies that $\boldsymbol{\nabla}\bar{\bar{\rho}}_\beta = 0$, $\mathbf{D}_\beta = 0$, $\mathbf{W}_\beta = 0$, $\boldsymbol{\nabla}\bar{\bar{\theta}}_\beta = 0$, $\tilde{\mathbf{v}}_\beta = \mathbf{v}$, $\tilde{\mathbf{i}}_\beta = 0$, $\boldsymbol{\nabla}\phi_\beta = 0$, and $\dot{\phi}_\beta = 0$. Following the procedure as in DOBRAN (1984b), given an integer $n$, let the function of

$$\boldsymbol{\nabla}\bar{\bar{\rho}}_\beta, \ \mathbf{D}_\beta, \ \mathbf{W}_\beta - \mathbf{W}_\gamma, \ \boldsymbol{\nabla}\bar{\bar{\theta}}_\beta, \ \tilde{\mathbf{v}}_\beta - \tilde{\mathbf{v}}_\gamma, \ \tilde{\mathbf{i}}_\beta, \ \boldsymbol{\nabla}\phi_\beta, \ \dot{\phi}_\beta$$

be denoted by $O(\varepsilon^n)$ with the property that $\| O(\varepsilon^n) \| < M\varepsilon^n$ as $\varepsilon \to 0$, where $\| \ldots \|$ is the Euclidean norm, $M$ is a positive constant, and $0 < \varepsilon < 1$. The condition $\varepsilon < 1$ implies that the density gradients, viscous effects, temperature gradients, diffusion of phases, inertia, volumetric fraction gradients, and dilatational effects of the mixture are small. The coefficients $C_{1\beta}, \ldots, C_{8\beta}$ in (7.5.3) are suitable reference variables which allow $\varepsilon$ to be expressed in a nondimensional form.

With the above definitions, the constitutive equation for the heat flux vector $\bar{\mathbf{q}}_\alpha$ may be written as

$$\bar{q}_{\alpha i}(\{S_\beta\}) = \bar{q}_{\alpha i}(\{S'_{\beta 0}\}) + \sum_\beta^\gamma \frac{\partial \bar{q}_{\alpha i}(\{S'_{\beta 0}\})}{\partial \bar{\bar{\rho}}_{\beta,j}} \bar{\bar{\rho}}_{\beta,j} + \sum_\beta^\gamma \frac{\partial \bar{q}_{\alpha i}(\{S'_{\beta 0}\})}{\partial D_{\beta jk}} D_{\beta jk}$$
$$+ \sum_\beta^{\gamma-1} \frac{\partial \bar{q}_{\alpha i}(\{S'_{\beta 0}\})}{\partial (W_{\beta jk} - W_{\gamma jk})}(W_{\beta jk} - W_{\gamma jk}) + \sum_\beta^\gamma \frac{\partial \bar{q}_{\alpha i}(\{S'_{\beta 0}\})}{\partial \bar{\bar{\theta}}_{\beta,j}} \bar{\bar{\theta}}_{\beta,j}$$
$$+ \sum_\beta^\gamma \frac{\partial \bar{q}_{\alpha i}(\{S'_{\beta 0}\})}{\partial \bar{\bar{i}}_{\beta jk}} \bar{\bar{i}}_{\beta jk} + \sum_\beta^{\gamma-1} \frac{\partial \bar{q}_{\alpha i}(\{S'_{\beta 0}\})}{\partial (\tilde{v}_{\beta j} - \tilde{v}_{\gamma j})}(\tilde{v}_{\beta j} - \tilde{v}_{\gamma j})$$
$$+ \sum_\beta^\gamma \frac{\partial \bar{q}_{\alpha i}(\{S'_{\beta 0}\})}{\partial \phi_{\beta,j}} \phi_{\beta,j} + \sum_\beta^\gamma \frac{\partial \bar{q}_{\alpha i}(\{S'_{\beta 0}\})}{\partial \grave{\phi}_\beta} \grave{\phi}_\beta + O(\varepsilon^2) \qquad (7.5.4)$$

Notice in the above equation: (1) $\bar{q}_{\alpha i}(\{S'_{\beta 0}\}) = 0$ because of (7.5.2) and $(7.4.31)_1$, and (2) the third, fourth, sixth, and ninth term on the right of the equation are equal to zero, owing to the isotropicity of the constitutive function $\bar{\mathbf{q}}_\alpha$ and its derivatives. The odd order isotropic tensors change signs in an improper orthogonal transformation (SPENCER, 1971) and all odd order tensors in (7.5.4) are required to vanish, *i.e.*

$$\frac{\partial \bar{\mathbf{q}}_\alpha}{\partial \mathbf{D}_\alpha} = 0, \quad \frac{\partial \bar{\mathbf{q}}_\alpha}{\partial (\mathbf{W}_\beta - \mathbf{W}_\gamma)} = 0, \quad \frac{\partial \bar{\mathbf{q}}_\alpha}{\partial \bar{\bar{\mathbf{i}}}_\beta} = 0, \quad \frac{\bar{\mathbf{q}}_\alpha}{\partial \grave{\phi}_\beta} = 0$$

and the even order isotropic tensors have the following representation:

$$\frac{\partial \bar{q}_{\alpha i}(\{S'_{\beta 0}\})}{\partial \bar{\bar{\rho}}_{\beta,j}} = -\nu_{\alpha\beta}\delta_{ij}, \quad \frac{\partial \bar{q}_{\alpha i}(\{S'_{\beta 0}\})}{\partial \bar{\bar{\theta}}_{\beta,j}} = -\kappa_{\alpha\beta}\delta_{ij} \qquad (7.5.5)$$

$$\frac{\partial \bar{q}_{\alpha i}(\{S'_{\beta 0}\})}{\partial (\tilde{v}_{\beta j} - \tilde{v}_{\gamma j})} = -\zeta_{\alpha\beta}\delta_{ij}, \quad \frac{\partial \bar{q}_{\alpha i}(\{S'_{\beta 0}\})}{\partial \phi_{\beta,j}} = -\Gamma_{\alpha\beta}\delta_{ij}$$

where $\nu_{\alpha\beta}$, $\kappa_{\alpha\beta}$, $\zeta_{\alpha\beta}$, and $\Gamma_{\alpha\beta}$ are functions of the equilibrium state properties

$$\{S'_{\beta 0}\} = \{\bar{\bar{\rho}}_\beta,\ \bar{\bar{\theta}}_\beta,\ \phi_\beta\}, \quad \beta = 1, \ldots, \gamma$$

Equation (7.5.4) thus becomes

$$\bar{q}_{\alpha i}(\{S_\beta\}) = -\sum_\beta^\gamma \nu_{\alpha\beta}\,\bar{\bar{\rho}}_{\beta,i} - \sum_\beta^{\gamma-1} \zeta_{\alpha\beta}(\tilde{v}_{\beta i} - \tilde{v}_{\gamma i}) - \sum_\beta^\gamma \kappa_{\alpha\beta}\,\bar{\bar{\theta}}_{\beta,i}$$
$$-\sum_\beta^\gamma \Gamma_{\alpha\beta}\,\phi_{\beta,i} + O(\varepsilon^2) \qquad (7.5.6)$$

Following an analysis identical to that which produces (7.5.6) and using $(7.4.31)_3$ and $(7.4.33)_1$, the following forms of constitutive equations for the interaction force $\bar{\mathbf{t}}_\alpha$ and the internal energy density moment $\tilde{\boldsymbol{\ell}}\epsilon_\alpha$ of each phase can be established, *i.e.*

$$\bar{t}_{\alpha i}(\{S_\beta\}) = -\sum_\beta^\gamma \gamma_{\alpha\beta}\,\bar{\bar{\theta}}_{\beta,i} - \sum_\beta^{\gamma-1} \xi_{\alpha\beta}(\tilde{v}_{\beta i} - \tilde{v}_{\gamma i}) - \sum_\beta^\gamma \Delta_{\alpha\beta}\,\bar{\bar{\rho}}_{\beta,i}$$
$$-\sum_\beta^\gamma M_{\alpha\beta}\,\phi_{\beta,i} + O(\varepsilon^2) \qquad (7.5.7)$$

$$\bar{\bar{\rho}}_\alpha \phi_\alpha \tilde{\ell}\epsilon_{\alpha i}(\{S_\beta\}) = \sum_\beta^\gamma \epsilon^\alpha_{0,\beta}\,\bar{\bar{\theta}}_{\beta,i} + \sum_\beta^{\gamma-1} \epsilon^\alpha_{1,\beta}(\tilde{v}_{\beta i} - \tilde{v}_{\gamma i}) + \sum_\beta^\gamma \epsilon^\alpha_{2,\beta}\,\bar{\bar{\rho}}_{\beta,i}$$
$$+\sum_\beta^\gamma \epsilon^\alpha_{3,\beta}\,\phi_{\beta,i} + O(\varepsilon^2) \qquad (7.5.8)$$

Physically, it is expected that $T_{ij}^{(\alpha\delta)}(\{S'_{\beta 0}\}) = -p_\alpha \delta_{ij}$, where $p_\alpha$ is an average pressure in the equilibrium state $\{S'_{\beta 0}\}$, and upon substitution of this expression into (3.3.2) and making use of (2.3.3) and the equilibrium state $\{S'_{\beta 0}\}$ (where $\nabla\phi_\alpha = 0$), also yields the continuum mechanics result $(7.4.31)_3$, *i.e.*

$$\bar{t}_{\alpha i}(\{S'_{\beta 0}\}) = -p_\alpha \frac{1}{U} \sum_\delta \int_{a(\Lambda\delta)} n_i^{(\alpha\delta)}\, da = p_\alpha \phi_{\alpha,i} = 0 \qquad (7.5.9)$$

In deriving the constitutive equation for $\tilde{\boldsymbol{\ell}}\epsilon_\alpha$, use was made of another reasonable physical requirement that this energy density vanishes in the state $\{S'_{\beta 0}\}$, which may be seen by taking $\epsilon^{(\alpha\delta)}$=constant in (3.3.7) and using (3.1.13). A result from (7.5.8) which will become useful later is that

$$(\bar{\bar{\rho}}_\alpha \phi_\alpha \tilde{\ell}\epsilon_{\alpha i})_{,i} = \sum_\beta^{\gamma-1} \epsilon^\alpha_{1,\beta}(tr\mathbf{D}_\beta - tr\mathbf{D}_\gamma) + O(\varepsilon^2) \qquad (7.5.10)$$

The extra stress tensor $\bar{\mathbf{T}}^e_\alpha$ is a second order isotropic tensor and its representation is

$$\bar{T}^e_{\alpha ij}(\{S_\beta\}) = \bar{T}^e_{\alpha ij}(\{S'_{\beta 0}\}) + \sum_\beta^\gamma \frac{\partial \bar{T}^e_{\alpha ij}(\{S'_{\beta 0}\})}{\partial D_{\beta k\ell}} D_{\beta k\ell}$$

$$+ \sum_\beta^{\gamma-1} \frac{\partial \bar{T}^e_{\alpha ij}(\{S'_{\beta 0}\})}{\partial (W_{\beta k\ell} - W_{\gamma k\ell})}(W_{\beta k\ell} - W_{\gamma k\ell}) + \sum_\beta^\gamma \frac{\partial \bar{T}^e_{\alpha ij}(\{S'_{\beta 0}\})}{\partial \bar{\bar{i}}_\beta} \bar{\bar{i}}_\beta$$

$$+ \sum_\beta^\gamma \frac{\partial \bar{T}^e_{\alpha ij}(\{S'_{\beta 0}\})}{\partial \grave{\phi}_\beta} \grave{\phi}_\beta + O(\varepsilon^2) \tag{7.5.11}$$

with the following requirements (SPENCER,1971):

$$\frac{\partial \bar{T}^e_{\alpha ij}(\{S'_{\beta 0}\})}{\partial \bar{\bar{i}}_\beta} = \iota_{\alpha\beta}\delta_{ij}, \quad \frac{\partial \bar{T}^e_{\alpha ij}(\{S'_{\beta 0}\})}{\partial \grave{\phi}_\beta} = O_{\alpha\beta}\delta_{ij} \tag{7.5.12}$$

$$\begin{aligned} \frac{\partial \bar{T}^e_{\alpha ij}(\{S'_{\beta 0}\})}{\partial D_{\beta k\ell}} D_{\beta k\ell} &= (C_1\delta_{ij}\delta_{k\ell} + C_2\delta_{ik}\delta_{j\ell} + C_3\delta_{i\ell}\delta_{jk})D_{\beta k\ell} \\ &= \lambda_{\alpha\beta}(tr\mathbf{D}_\beta)\delta_{ij} + 2\mu_{\alpha\beta}D_{\beta ij} \end{aligned} \tag{7.5.13}$$

$$\frac{\partial \bar{T}^e_{\alpha ij}(\{S'_{\beta 0}\})}{\partial (W_{\beta k\ell} - W_{\gamma k\ell})}(W_{\beta k\ell} - W_{\gamma k\ell}) = 2e_{\alpha\beta}(W_{\beta ij} - W_{\gamma ij}) \tag{7.5.14}$$

where use is made of the properties of the symmetric tensor $\mathbf{D}_\beta$ and skew-symmetric tensor $\mathbf{W}_\beta$. With the above substitutions into (7.5.11) and using $(7.4.31)_2$, it follows that

$$\bar{T}^e_{\alpha ij}(\{S_\beta\}) = \sum_\beta^\gamma [\lambda_{\alpha\beta}(tr\mathbf{D}_\beta)\delta_{ij} + 2\mu_{\alpha\beta}D_{\beta ij}] + 2\sum_\beta^{\gamma-1} e_{\alpha\beta}(W_{\beta ij} - W_{\gamma ij})$$

$$+ \sum_\beta^\gamma \iota_{\alpha\beta}\, \bar{\bar{i}}_\beta \delta_{ij} + \sum_\beta^\gamma O_{\alpha\beta}\, \grave{\phi}_\beta \delta_{ij} + O(\varepsilon^2) \tag{7.5.15}$$

where $\lambda_{\alpha\beta}$, $\mu_{\alpha\beta}$, $e_{\alpha\beta}$, $\iota_{\alpha\beta}$, and $O_{\alpha\beta}$ depend on the state properties $\{S'_{\beta 0}\}$. Combining (7.4.20) with the above result yields the stress tensor

$$\bar{T}_{\alpha ij}(\{S_\beta\}) = -\phi_\alpha(p_\alpha + \bar{\bar{\pi}}_\alpha)\delta_{ij} - \bar{\bar{\rho}}_\alpha \phi_\alpha \frac{\partial \bar{\psi}_\alpha}{\partial \phi_{\alpha,j}} \phi_{\alpha,i} + 2\sum_\beta^{\gamma-1} e_{\alpha\beta}(W_{\beta ij} - W_{\gamma ij})$$

$$+ \sum_\beta^\gamma [\lambda_{\alpha\beta}(tr\mathbf{D}_\beta)\delta_{ij} + 2\mu_{\alpha\beta}D_{\beta ij}] + \sum_\beta^\gamma \iota_{\alpha\beta}\, \bar{\bar{i}}_\beta \delta_{ij} + \sum_\beta^\gamma O_{\alpha\beta}\, \grave{\phi}_\beta \delta_{ij} + O(\varepsilon^2)$$

(7.5.16)

and the angular momentum condition (2.4.31) is reduced to

$$\hat{M}_{\alpha ij} = \bar{T}_{\alpha ij} - \bar{T}_{\alpha ji} = 4\sum_{\beta}^{\gamma-1} e_{\alpha\beta}(W_{\beta ij} - W_{\gamma ij}) + O(\varepsilon^2) \qquad (7.5.17)$$

From equation (7.3.12) we can establish that

$$\bar{\ell}_{\alpha[jk]} = O(\varepsilon^3) \qquad (7.5.18)$$

Repeating the above procedure for the representation of second order isotropic tensors and using (7.4.40), the surface traction moment can be expressed as

$$\bar{S}_{\alpha ij}(\{S_\beta\}) = -\phi_\alpha P_\alpha(\{S'_{\beta 0}\})\delta_{ij} + \sum_{\beta}^{\gamma}[\Pi_{\alpha\beta}(tr\mathbf{D}_\beta)\delta_{ij} + \Phi_{\alpha\beta}D_{\beta ij}]$$

$$+\sum_{\beta}^{\gamma-1} T_{\alpha\beta}(W_{\beta ij} - W_{\gamma ij}) + \sum_{\beta}^{\gamma} K_{\alpha\beta}\bar{\dot{i}}_\beta\delta_{ij} + \sum_{\beta}^{\gamma} H_{\alpha\beta}\dot{\phi}_\beta\delta_{ij} + O(\varepsilon^2)$$

(7.5.19)

Similarly, using $(7.4.33)_1$, it may be shown that

$$\hat{c}_\alpha\hat{i}_{\alpha ij}(\{S_\beta\}) = \sum_{\beta}^{\gamma}[\Psi_{\alpha\beta}(tr\mathbf{D}_\beta)\delta_{ij} + \chi_{\alpha\beta}D_{\beta ij}] + \sum_{\beta}^{\gamma-1}\Omega_{\alpha\beta}(W_{\beta ij} - W_{\gamma ij})$$

$$+\sum_{\beta}^{\gamma} E_{\alpha\beta}\bar{\dot{i}}_\beta\delta_{ij} + \sum_{\beta}^{\gamma} Z_{\alpha\beta}\dot{\phi}_\beta\delta_{ij} + O(\varepsilon^2) \qquad (7.5.20)$$

where the material coefficients $\Pi_{\alpha\beta}$, $\Phi_{\alpha\beta}$, $T_{\alpha\beta}$, $K_{\alpha\beta}$, $H_{\alpha\beta}$, $\Psi_{\alpha\beta}$, $\chi_{\alpha\beta}$, $\Omega_{\alpha\beta}$, $E_{\alpha\beta}$, and $Z_{\alpha\beta}$ depend on the equilibrium state properties $\{S'_{\beta 0}\}$.

The third order tensors $\bar{\lambda}_{\alpha ijk}$ and $U_{\alpha ijk}$ transform with the change of frame according to (4.1.8). Thus, if we take $t^* = t$ and $\mathbf{Q} = -\mathbf{I}$, then according to (4.1.8) and (6.10.1) in the state $\{S'_{\beta 0}\}$

$$\bar{\boldsymbol{\lambda}}_\alpha(\{S'_{\beta 0}\}) = -\bar{\boldsymbol{\lambda}}_\alpha(\{S'_{\beta 0}\}), \quad \mathbf{U}_\alpha(\{S'_{\beta 0}\}) = -\mathbf{U}_\alpha(\{S'_{\beta 0}\})$$

from where it follows that

$$\bar{\boldsymbol{\lambda}}_\alpha(\{S'_{\beta 0}\}) = \mathbf{0}, \quad \mathbf{U}_\alpha(\{S'_{\beta 0}\}) = \mathbf{0} \qquad (7.5.21)$$

Notice that a special case of the result expressed by $(7.5.21)_1$ also follows from (7.4.25), (7.5.8), (7.4.23), and (7.5.26). The tensor representations for $\bar{\boldsymbol{\lambda}}_\alpha$ and $\mathbf{U}_\alpha$ may then be written as follows

$$\begin{aligned}\bar{\lambda}_{\alpha ijk}(\{S_\beta\}) = &\sum_\beta^\gamma \delta_{ij}[a^\alpha_{1,\beta}\bar{\bar{\rho}}_{\beta,k} + b^\alpha_{1,\beta}(\tilde{v}_{\beta k} - \tilde{v}_{\gamma k}) + c^\alpha_{1,\beta}\bar{\bar{\theta}}_{\beta,k} + d^\alpha_{1,\beta}\phi_{\beta,k}] \\ &+ \sum_\beta^\gamma \delta_{ik}[a^\alpha_{2,\beta}\bar{\bar{\rho}}_{\beta,j} + b^\alpha_{2,\beta}(\tilde{v}_{\beta j} - \tilde{v}_{\gamma j}) + c^\alpha_{2,\beta}\bar{\bar{\theta}}_{\beta,j} + d^\alpha_{2,\beta}\phi_{\beta,j}] \\ &+ \sum_\beta^\gamma \delta_{jk}[a^\alpha_{3,\beta}\bar{\bar{\rho}}_{\beta,i} + b^\alpha_{3,\beta}(\tilde{v}_{\beta i} - \tilde{v}_{\gamma i}) + c^\alpha_{3,\beta}\bar{\bar{\theta}}_{\beta,i} + d^\alpha_{3,\beta}\phi_{\beta,i}] + O(\varepsilon^2)\end{aligned} \tag{7.5.22}$$

$$\begin{aligned}U_{\alpha ijk}(\{S_\beta\}) = &\sum_\beta^\gamma \delta_{ij}[a^\alpha_{4,\beta}\bar{\bar{\rho}}_{\beta,k} + b^\alpha_{4,\beta}(\tilde{v}_{\beta k} - \tilde{v}_{\gamma k}) + c^\alpha_{4,\beta}\bar{\bar{\theta}}_{\beta,k} + d^\alpha_{4,\beta}\phi_{\beta,k}] \\ &+ \sum_\beta^\gamma \delta_{ik}[a^\alpha_{5,\beta}\bar{\bar{\rho}}_{\beta,j} + b^\alpha_{5,\beta}(\tilde{v}_{\beta j} - \tilde{v}_{\gamma j}) + c^\alpha_{5,\beta}\bar{\bar{\theta}}_{\beta,j} + d^\alpha_{5,\beta}\phi_{\beta,j}] \\ &+ \sum_\beta^\gamma \delta_{jk}[a^\alpha_{6,\beta}\bar{\bar{\rho}}_{\beta,i} + b^\alpha_{6,\beta}(\tilde{v}_{\beta i} - \tilde{v}_{\gamma i}) + c^\alpha_{6,\beta}\bar{\bar{\theta}}_{\beta,i} + d^\alpha_{6,\beta}\phi_{\beta,i}] + O(\varepsilon^2)\end{aligned} \tag{7.5.23}$$

An examination of the entropy inequality (7.4.19) and the above linearized constitutive equations indicates that the error in this inequality will be of $O(\varepsilon^3)$ if the constitutive variables $\hat{f}_\alpha$ and $\hat{c}_\alpha\hat{\hat{\varepsilon}}_\alpha$ are approximated within the same order of accuracy. For completeness and noting that $\hat{c}_\alpha\hat{\hat{\varepsilon}}_\alpha(\{S_{\beta 0}\})=0$ (*cit.* $(7.4.33)_1$) it may be shown that

$$\begin{aligned}\hat{c}_\alpha\hat{\hat{\varepsilon}}_\alpha(\{S_\beta\}) = &\sum_\beta^\gamma \tau^\alpha_{1,\beta}\dot{\phi}_\beta + \sum_\beta^\gamma \tau^\alpha_{2,\beta}(tr\mathbf{D}_\beta) + \sum_{\beta\le\delta,\delta}^\gamma \tau^\alpha_{3,\beta\delta}(tr\mathbf{D}_\beta\mathbf{D}_\delta) \\ &+ \sum_{\beta\le\delta,\delta}^\gamma \tau^\alpha_{4,\beta\delta}(tr\mathbf{D}_\beta)(tr\mathbf{D}_\delta) + \sum_{\beta\le\delta,\delta}^{\gamma-1} \tau^\alpha_{5,\beta\delta}\,tr[(\mathbf{W}_\beta - \mathbf{W}_\gamma)(\mathbf{W}_\delta - \mathbf{W}_\gamma)] \\ &+ \sum_{\beta\le\delta,\delta}^\gamma \tau^\alpha_{6,\beta\delta}\bar{\bar{\theta}}_{\beta,i}\bar{\bar{\theta}}_{\delta,i} + \sum_{\beta\le\delta,\delta}^{\gamma-1} \tau^\alpha_{7,\beta\delta}(\tilde{v}_{\beta i} - \tilde{v}_{\gamma i})(\tilde{v}_{\delta i} - \tilde{v}_{\gamma i}) \\ &+ \sum_\beta^\gamma\sum_\delta^{\gamma-1} \tau^\alpha_{8,\beta\delta}\bar{\bar{\theta}}_{\beta,i}(\tilde{v}_{\delta i} - \tilde{v}_{\gamma i}) + O(\varepsilon^3)\end{aligned} \tag{7.5.24}$$

An identical expression, but with different coefficients, may also be derived for $\hat{c}_\alpha$. Using $(7.4.34)_1$, we obtain

$$\hat{f}_\alpha(\{S_\beta\}) = \sum_\beta^\gamma \omega^\alpha_{0,\beta}\dot{i}_\beta + \sum_\beta^\gamma \omega^\alpha_{1,\beta}\dot{\phi}_\beta + \sum_\beta^\gamma \omega^\alpha_{2,\beta}(tr\mathbf{D}_\beta) + \sum_{\beta\le\delta,\delta}^\gamma \omega^\alpha_{3,\beta\delta}(tr\mathbf{D}_\beta\mathbf{D}_\delta)$$

$$+\sum_{\beta\le\delta,\delta}^{\gamma}\omega^{\alpha}_{4,\beta\delta}(tr\mathbf{D}_\beta)(tr\mathbf{D}_\delta)+\sum_{\beta\le\delta,\delta}^{\gamma}\omega^{\alpha}_{5,\beta\delta}\,tr[(\mathbf{W}_\beta-\mathbf{W}_\gamma)(\mathbf{W}_\delta-\mathbf{W}_\gamma)]$$

$$+\sum_{\beta\le\delta,\delta}^{\gamma}\omega^{\alpha}_{6,\beta\delta}\,\bar{\bar{\rho}}_{\beta,i}\bar{\bar{\rho}}_{\delta,i}+\sum_{\beta,\delta}^{\gamma}\omega^{\alpha}_{7,\beta\delta}\,\bar{\bar{\rho}}_{\beta,i}\bar{\bar{\theta}}_{\delta,i}+\sum_{\beta}^{\gamma}\sum_{\delta}^{\gamma-1}\omega^{\alpha}_{8,\beta\delta}\,\bar{\bar{\rho}}_{\beta,i}(\tilde{v}_{\delta i}-\tilde{v}_{\gamma i})$$

$$+\sum_{\beta,\delta}^{\gamma}\omega^{\alpha}_{9,\beta\delta}\,\bar{\bar{\rho}}_{\beta,i}\phi_{\delta,i}+\sum_{\beta\le\delta,\delta}^{\gamma}\omega^{\alpha}_{10,\beta\delta}\,\bar{\bar{\theta}}_{\beta,i}\bar{\bar{\theta}}_{\delta,i}+\sum_{\beta\le\delta,\delta}^{\gamma-1}\omega^{\alpha}_{11,\beta\delta}(\tilde{v}_{\beta i}-\tilde{v}_{\gamma i})(\tilde{v}_{\delta i}-\tilde{v}_{\gamma i})$$

$$\sum_{\beta\le\delta,\delta}^{\gamma}\omega^{\alpha}_{12,\beta\delta}\,\phi_{\beta,i}\phi_{\delta,i}+\sum_{\beta}^{\gamma}\sum_{\delta}^{\gamma-1}\omega^{\alpha}_{13,\beta\delta}\,\bar{\bar{\theta}}_{\beta,i}(\tilde{v}_{\delta i}-\tilde{v}_{\gamma i})+\sum_{\beta,\delta}^{\gamma}\omega^{\alpha}_{14,\beta\delta}\,\bar{\bar{\theta}}_{\beta,i}\phi_{\delta,i}$$

$$+\sum_{\beta}^{\gamma}\sum_{\delta}^{\gamma-1}\omega^{\alpha}_{15,\beta\delta}\,\phi_{\beta,i}(\tilde{v}_{\delta i}-\tilde{v}_{\gamma i})+O(\varepsilon^3) \qquad (7.5.25)$$

The coefficients $\tau$'s and $\omega$'s appearing in the above equations depend on the state properties $\{S'_{\beta 0}\}$.

The constitutive equation for the Helmholtz potential may also be expanded in series about the equilibrium state and use made of (7.4.26), giving the following result:

$$\tilde{\psi}_\alpha(\{S_\beta\})=\bar{\bar{\psi}}_\alpha(\bar{\bar{\rho}}_\beta,\bar{\bar{\theta}}_\alpha,\phi_\beta)+I_\alpha(\bar{\bar{\rho}}_\beta,\bar{\bar{\theta}}_\alpha,\phi_\beta)\phi_{\alpha,i}\phi_{\alpha,i}+O(\varepsilon^4) \qquad (7.5.26)$$

From (7.4.8), the entropy of phase $\alpha$ is

$$\begin{aligned}\tilde{s}_\alpha(\{S_\beta\}) &= -\frac{\partial\bar{\bar{\psi}}_\alpha}{\partial\bar{\bar{\theta}}_\alpha}-\frac{\partial I_\alpha}{\partial\bar{\bar{\theta}}_\alpha}\phi_{\alpha,i}\phi_{\alpha,i}+O(\varepsilon^4)\\ &= -\frac{\partial\bar{\bar{\psi}}_\alpha}{\partial\bar{\bar{\theta}}_\alpha}+O(\varepsilon^2)\end{aligned} \qquad (7.5.27)$$

The internal energy $\tilde{\epsilon}_\alpha$ is obtained from (6.12.8), (7.5.26), and (7.5.27), *i.e.*

$$\tilde{\epsilon}_\alpha(\{S_\beta\})=\bar{\bar{\psi}}_\alpha-\bar{\bar{\theta}}_\alpha\frac{\partial\bar{\bar{\psi}}_\alpha}{\partial\bar{\bar{\theta}}_\alpha}+I_\alpha\phi_{\alpha,i}\phi_{\alpha,i}-\bar{\bar{\theta}}_\alpha\frac{\partial I_\alpha}{\partial\bar{\bar{\theta}}_\alpha}\phi_{\alpha,i}\phi_{\alpha,i}+O(\varepsilon^4) \qquad (7.5.28)$$

or

$$\tilde{\epsilon}_\alpha(\{S_\beta\}=\bar{\bar{\psi}}_\alpha-\bar{\bar{\theta}}_\alpha\frac{\partial\bar{\bar{\psi}}_\alpha}{\partial\bar{\bar{\theta}}_\alpha}+O(\varepsilon^2) \qquad (7.5.29)$$

The equilibrium Helmholtz potential $\bar{\bar{\psi}}_\alpha(\bar{\bar{\rho}}_\beta,\bar{\bar{\theta}}_\alpha,\phi_\beta)$ is a property of the mixture and should be determined from experiments, similarly as for a single phase material where it only depends on the density and temperature and is referred as the "fundamental equation", because its knowledge allows for the determination of *all* other equilibrium thermodynamic variables (see CALLEN,1960, for example). Another result which will become useful in the next section is the combination of variables appearing in the entropy inequality (7.4.19). This can be derived through the following arguments:

$$\begin{aligned}
&\phi_\alpha(p_\alpha + \bar{\bar{\pi}}_\alpha) - \bar{\bar{\lambda}}^{R}_{\alpha kkm}\phi_{\alpha,m} - \phi_\alpha\beta_\alpha - (\bar{\bar{\rho}}_\alpha\phi_\alpha\tilde{\ell}\epsilon_{\alpha m})_{,m} \\
&\overset{(7.4.25)}{=\!=} \phi_\alpha(p_\alpha + \bar{\bar{\pi}}_\alpha) + \phi_\alpha\bar{\bar{\lambda}}^{R}_{\alpha kkm,m} - \bar{\lambda}_{\alpha kkm,m} - \phi_\alpha\beta_\alpha \\
&\overset{(7.5.19),(7.4.21)-(7.4.23),(7.5.26),(7.3.1)}{=\!=} -P_\alpha(\{S'_{\beta 0}\})\phi_\alpha \\
&+\sum_{\beta}^{\gamma}[(\Pi_{\alpha\beta} + \Phi_{\alpha\beta})tr\mathbf{D}_\beta + K_{\alpha\beta}\,\tilde{i}_\beta + H_{\alpha\beta}\,\grave{\phi}_\beta] + \phi_\alpha(p_\alpha + \bar{\bar{\pi}}_\alpha(\{S'_{\beta 0}\})) \\
&-\phi_\alpha\beta_\alpha(\{S'_{\beta 0}\}) + O(\varepsilon^2) \overset{(7.4.39),(7.4.40)\, in\ state\ \{S'_{\beta 0}\}}{=\!=} \\
&\sum_{\beta}^{\gamma}(\Pi_{\alpha\beta} + \Phi_{\alpha\beta})tr\mathbf{D}_\beta + \sum_{\beta}^{\gamma} K_{\alpha\beta}\,\tilde{i}_\beta + \sum_{\beta}^{\gamma} H_{\alpha\beta}\,\grave{\phi}_\beta + O(\varepsilon^2) \qquad (7.5.30)
\end{aligned}$$

The restriction of material coefficients appearing in the above constitutive equations may be determined by substituting these equations into the entropy inequality (7.4.19). To simplify the algebra, however, we will study the special case corresponding to no phase change and assume that $\hat{f}_\alpha = O(\varepsilon^3)$.

## 7.6 Restrictions of Material Coefficients of Linearized Constitutive Equations Without Phase Change

For the no phase change situation,

$$\hat{c}_\alpha = 0 \qquad (7.6.1)$$

and to reduce algebraic manipulations, it will be assumed that

$$\hat{f}_\alpha = O(\varepsilon^3) \qquad (7.6.2)$$

With these simplifications and making use of the constitutive equations of the previous section in the entropy inequality (7.4.19), yields

$$\mathbf{VAV}^T + \sum_{\beta\neq\alpha}^{\gamma}[\bar{\rho}_\alpha\frac{\partial\tilde{\psi}_\alpha}{\partial\bar{\bar{\rho}}_\beta}\bar{\bar{\rho}}_\beta + \frac{\grave{\phi}_\alpha}{\phi_\alpha}(\Pi_{\alpha\beta} + \Phi_{\alpha\beta})]tr\mathbf{D}_\beta + [\frac{\grave{\phi}_\alpha}{\phi_\alpha}(\Pi_{\alpha\alpha} + \Phi_{\alpha\alpha})$$

$$+\sum_{\beta}^{\gamma}(\iota_{\alpha\beta}\,\bar{i}_\beta+O_{\alpha\beta}\,\dot{\phi}_\beta)]tr\mathbf{D}_\alpha+\sum_{\beta}^{\gamma}\{\lambda_{\alpha\beta}(tr\mathbf{D}_\beta)(tr\mathbf{D}_\alpha)+2\mu_{\alpha\beta}\,tr(\mathbf{D}_\beta\mathbf{D}_\alpha)$$

$$+2e_{\alpha\beta}\,tr[(\mathbf{W}_\beta^T-\mathbf{W}_\gamma^T)\mathbf{W}_\alpha]\}+O(\epsilon^3)\geq 0 \tag{7.6.3}$$

where $\mathbf{V}A\mathbf{V}^T$ is defined by

$$\mathbf{V}A\mathbf{V}^T=\frac{\bar{\bar{\theta}}_{\alpha,i}}{\bar{\bar{\theta}}_\alpha}[\sum_{\beta}^{\gamma}\nu_{\alpha\beta}\,\bar{\bar{\rho}}_{\beta,i}+\sum_{\beta}^{\gamma-1}\zeta_{\alpha\beta}(\bar{v}_{\beta i}-\bar{v}_{\gamma i})+\sum_{\beta}^{\gamma}\kappa_{\alpha\beta}\,\bar{\bar{\theta}}_{\beta,i}+\sum_{\beta}^{\gamma}\Gamma_{\alpha\beta}\,\phi_{\beta,i}]$$

$$+\bar{\bar{\rho}}_\alpha\phi_\alpha\sum_{\beta\neq\alpha}^{\gamma}[(\frac{\partial\bar{\psi}_\alpha}{\partial\bar{\bar{\rho}}_\beta}\frac{\bar{\bar{\rho}}_\beta}{\phi_\beta}-\frac{\partial\bar{\psi}_\alpha}{\partial\phi_\beta})\dot{\phi}_\beta-(\frac{\partial\bar{\psi}_\alpha}{\partial\bar{\bar{\rho}}_\beta}\bar{\bar{\rho}}_{\beta,i}+\frac{\partial\bar{\psi}_\alpha}{\partial\phi_\beta}\phi_{\beta,i})(\bar{v}_{\alpha i}-\bar{v}_{\beta i})]$$

$$+\frac{\dot{\phi}_\alpha}{\phi_\alpha}\sum_{\beta}^{\gamma}(K_{\alpha\beta}\,\bar{i}_\beta+H_{\alpha\beta}\,\dot{\phi}_\beta)+\sum_{\beta}^{\gamma}u_{\alpha i}[\gamma_{\alpha\delta}\bar{\bar{\theta}}_{\beta,i}+\xi_{\alpha\beta}(\bar{v}_{\beta i}-\bar{v}_{\gamma i})$$

$$+\Delta_{\alpha\beta}\,\bar{\bar{\rho}}_{\beta,i}+M_{\alpha\beta}\,\phi_{\beta,i}] \tag{7.6.4}$$

The derivation of inequality (7.6.3) makes use of (4.3.16) and the fact that the trace of the product of symmetric and skew-symmetric matrices vanishes ($D_{\alpha ij}W_{\alpha ij}=0$). Moreover, noticing that

$$tr(\mathbf{D}_\beta\mathbf{D}_\alpha)=\frac{1}{3}(tr\mathbf{D}_\beta)(tr\mathbf{D}_\alpha)$$

$$+tr\{[\mathbf{D}_\beta-\frac{3+\sqrt{6}}{3}(tr\mathbf{D}_\beta)\mathbf{I}][\mathbf{D}_\alpha-\frac{3+\sqrt{6}}{3}(tr\mathbf{D}_\alpha)\mathbf{I}]\} \tag{7.6.5}$$

$$\sum_{\beta}^{\gamma-1}e_{\alpha\beta}\,tr[(\mathbf{W}_\beta^T-\mathbf{W}_\gamma^T)\mathbf{W}_\alpha]=\sum_{\beta}^{\gamma}\hat{e}_{\alpha\beta}\,tr(\mathbf{W}_\beta^T\mathbf{W}_\alpha) \tag{7.6.6}$$

where

$$\hat{e}_{\alpha\beta}=e_{\alpha\beta},\quad \beta=1,\ldots,\gamma-1$$

$$\hat{e}_{\alpha\gamma}=-\sum_{\beta=1}^{\gamma-1}e_{\alpha\beta} \tag{7.6.7}$$

and substituting these relations into (7.6.3), gives

$$\{\mathbf{V}A\mathbf{V}^T\}+\{\sum_{\beta\neq\alpha}^{\gamma}[\bar{\rho}_\alpha\frac{\partial\bar{\psi}_\alpha}{\partial\bar{\bar{\rho}}_\beta}\bar{\bar{\rho}}_\beta+\frac{\dot{\phi}_\alpha}{\phi_\alpha}(\Pi_{\alpha\beta}+\Phi_{\alpha\beta})]tr\mathbf{D}_\beta\}$$

$$+\{[\frac{\grave{\phi}_\alpha}{\phi_\alpha}(\Pi_{\alpha\alpha}+\Phi_{\alpha\alpha})+\sum_\beta^\gamma(\iota_{\alpha\beta}\bar{i}_\beta+O_{\alpha\beta}\grave{\phi}_\beta)]tr\mathbf{D}_\alpha\}$$

$$+\{\sum_{\beta,\delta}^\gamma\delta_{\alpha\delta}(\lambda_{\alpha\beta}+\frac{2}{3}\mu_{\alpha\beta})(tr\mathbf{D}_\beta)(tr\mathbf{D}_\delta)\}+\{\sum_{\beta,\delta}^\gamma 2\hat{e}_{\alpha\beta}\delta_{\alpha\delta}\,tr(\mathbf{W}_\beta^T\mathbf{W}_\delta)\}$$

$$+\{\sum_{\beta,\delta}^\gamma 2\mu_{\alpha\beta}\delta_{\alpha\delta}\,tr[(\mathbf{D}_\beta-\frac{3+\sqrt{6}}{3}(tr\mathbf{D}_\beta)\mathbf{I})(\mathbf{D}_\delta-\frac{3+\sqrt{6}}{3}(tr\mathbf{D}_\delta)\mathbf{I})]\}$$

$$+O(\varepsilon^3)=R_\alpha\geq 0 \tag{7.6.8}$$

where the Kronecker delta $\delta_{\alpha\delta}$ is defined as usual, *i.e.*

$$\delta_{\alpha\delta} = \begin{cases} 1, & \alpha=\delta \\ 0, & \alpha\neq\delta \end{cases} \tag{7.6.9}$$

The *necessary conditions* for the function $R_\alpha$ in (7.6.8) to be positive semidefinite are that each of the six terms which are enclosed by the brackets $\{\ldots\}$ be positive semidefinite. These conditions are also *sufficient* if in this inequality $O(\varepsilon^3)=0$.

The condition

$$[\frac{\grave{\phi}_\alpha}{\phi_\alpha}(\Pi_{\alpha\alpha}+\Phi_{\alpha\alpha})+\sum_\beta^\gamma(\iota_{\alpha\beta}\bar{i}_\beta+O_{\alpha\beta}\grave{\phi}_\beta)]tr\mathbf{D}_\alpha\geq 0 \tag{7.6.10}$$

for all $\mathbf{D}_\alpha$ requires

$$\frac{\grave{\phi}_\alpha}{\phi_\alpha}(\Pi_{\alpha\alpha}+\Phi_{\alpha\alpha}+\phi_\alpha O_{\alpha\alpha})+\sum_\beta^\gamma\iota_{\alpha\beta}\bar{i}_\beta+\sum_{\beta\neq\alpha}^\gamma O_{\alpha\beta}\grave{\phi}_\beta=0 \tag{7.6.11}$$

or, since $\bar{i}_\beta$ and $\grave{\phi}_\beta$ are independent variables in the constitutive equations, it follows that

$$\Pi_{\alpha\alpha}+\Phi_{\alpha\alpha}+\phi_\alpha O_{\alpha\alpha}=0 \tag{7.6.12}$$

$$\iota_{\alpha\beta}=0;\quad \beta=1,\ldots,\gamma \tag{7.6.13}$$

$$O_{\alpha\beta}=0;\quad \beta=1,\ldots,\alpha-1,\alpha+1,\ldots,\gamma \tag{7.6.14}$$

The condition

$$\sum_{\beta\neq\alpha}^\gamma[\bar{\rho}_\alpha\frac{\partial\bar{\psi}_\alpha}{\partial\bar{\bar{\rho}}_\beta}\bar{\bar{\rho}}_\beta+\frac{\grave{\phi}_\alpha}{\phi_\alpha}(\Pi_{\alpha\beta}+\Phi_{\alpha\beta})]tr\mathbf{D}_\beta\geq 0 \tag{7.6.15}$$

for all $\mathbf{D}_\beta$, $\beta \neq \alpha$, requires

$$\bar{\bar{\rho}}_\alpha \phi_\alpha \frac{\partial \bar{\psi}_\alpha}{\partial \bar{\bar{\rho}}_\beta} \bar{\bar{\rho}}_\beta + \frac{\grave{\phi}_\alpha}{\phi_\alpha}(\Pi_{\alpha\beta} + \Phi_{\alpha\beta}) = 0 \tag{7.6.16}$$

which upon using (7.5.26), gives

$$\bar{\bar{\rho}}_\alpha \phi_\alpha [\frac{\partial \bar{\bar{\psi}}_\alpha}{\partial \bar{\bar{\rho}}_\beta} + \frac{\partial I_\alpha}{\partial \bar{\bar{\rho}}_\beta} \phi_{\alpha,i}\phi_{\alpha,i}] \bar{\bar{\rho}}_\beta + \frac{\grave{\phi}_\alpha}{\phi_\alpha}(\Pi_{\alpha\beta} + \Phi_{\alpha\beta}) + O(\varepsilon^4) = 0 \tag{7.6.17}$$

so that (consistent with $(7.4.30)_1$) for independent variations of $\bar{\bar{\rho}}_\beta$, $\beta \neq \alpha$, $\grave{\phi}_\alpha$, and $\phi_{\alpha,i}$ it follows that (within $O(\varepsilon^4)$)

$$\Pi_{\alpha\beta} + \Phi_{\alpha\beta} = 0; \quad \beta = 1, \ldots, \alpha-1, \alpha+1, \ldots, \gamma \tag{7.6.18}$$

$$\bar{\bar{\psi}}_\alpha(\bar{\bar{\rho}}_\beta, \bar{\bar{\theta}}_\alpha, \phi_\beta) = \bar{\bar{\psi}}_\alpha(\bar{\bar{\rho}}_\alpha, \bar{\bar{\theta}}_\alpha, \phi_\alpha), \quad I_\alpha(\bar{\bar{\rho}}_\beta, \bar{\bar{\theta}}_\alpha, \phi_\beta) = I_\alpha(\bar{\bar{\rho}}_\alpha, \bar{\bar{\theta}}_\alpha, \phi_\alpha) \tag{7.6.19}$$

The condition

$$\sum_{\beta,\delta}^{\gamma} \delta_{\alpha\delta}(\lambda_{\alpha\beta} + \frac{2}{3}\mu_{\alpha\beta})(tr\mathbf{D}_\beta)(tr\mathbf{D}_\delta) = \sum_{\beta,\delta}^{\gamma} B^\alpha_{\beta\delta}(tr\mathbf{D}_\beta)(tr\mathbf{D}_\delta) \geq 0 \tag{7.6.20}$$

requires that all principal minors of $B^\alpha_{\beta\delta}$ be positive semidefinite (Sylvester's criterion). *For a two-phase mixture*, this requirement gives

$$B^\alpha_{\delta\delta} = \delta_{\alpha\delta}(\lambda_{\alpha\delta} + \frac{2}{3}\mu_{\alpha\delta}) \geq 0 \quad \lambda_{11} + \frac{2}{3}\mu_{11} \geq 0 \quad \lambda_{22} + \frac{2}{3}\mu_{22} \geq 0 \tag{7.6.21}$$

$$\begin{aligned} B^\alpha_{11}B^\alpha_{22} - \frac{1}{4}(B^\alpha_{12} + B^\alpha_{21})^2 &= [\delta_{\alpha 1}(\lambda_{\alpha 1} + \frac{2}{3}\mu_{\alpha 1})][\delta_{\alpha 2}(\lambda_{\alpha 2} + \frac{2}{3}\mu_{\alpha 2})] \\ &- \frac{1}{4}[\delta_{\alpha 2}(\lambda_{\alpha 1} + \frac{2}{3}\mu_{\alpha 1}) + \delta_{\alpha 1}(\lambda_{\alpha 2} + \frac{2}{3}\mu_{\alpha 2})]^2 \geq 0 \end{aligned}$$

giving for $\alpha = 1$ and $\alpha = 2$, respectively

$$\lambda_{12} + \frac{2}{3}\mu_{12} = 0, \quad \lambda_{21} + \frac{2}{3}\mu_{21} = 0 \tag{7.6.22}$$

Another condition in (7.6.8) requiring

$$\begin{aligned} &\sum_{\beta,\delta}^{\gamma} 2\mu_{\alpha\beta}\,\delta_{\alpha\delta}\,tr\{[\mathbf{D}_\beta - \frac{3+\sqrt{6}}{3}(tr\mathbf{D}_\beta)\mathbf{I}][\mathbf{D}_\delta - \frac{3+\sqrt{6}}{3}(tr\mathbf{D}_\delta)\mathbf{I}]\} \\ &= \sum_{\beta,\delta}^{\gamma} C^\alpha_{\beta\delta}\,tr\{[\mathbf{D}_\beta - \frac{3+\sqrt{6}}{3}(tr\mathbf{D}_\beta)\mathbf{I}][\mathbf{D}_\delta - \frac{3+\sqrt{6}}{3}(tr\mathbf{D}_\delta)\mathbf{I}]\} \geq 0 \end{aligned} \tag{7.6.23}$$

yields the following results *for a two-phase mixture*

$$C^{\alpha}_{\delta\delta} = 2\mu_{\alpha\alpha} \geq 0; \quad \mu_{11} \geq 0, \quad \mu_{22} \geq 0 \tag{7.6.24}$$

and

$$C^{\alpha}_{11}C^{\alpha}_{22} - \frac{1}{4}(C^{\alpha}_{12} + C^{\alpha}_{21})^2 = 2\mu_{\alpha 1}\delta_{\alpha 1}2\mu_{\alpha 2}\delta_{\alpha 2} - \frac{1}{4}(2\mu_{\alpha 1}\delta_{\alpha 2} + 2\mu_{\alpha 2}\delta_{\alpha 1})^2 \geq 0$$

giving for $\alpha = 1$, and $\alpha = 2$, respectively

$$\mu_{12} = 0, \quad \mu_{21} = 0 \tag{7.6.25}$$

Combining (7.6.21), (7.6.22), (7.6.24), and (7.6.25) gives the following restrictions on the two-phase "viscosity" coefficients

$$\begin{gathered} \mu_{11} \geq 0, \quad \mu_{22} \geq 0, \quad \lambda_{11} \geq -\frac{2}{3}\mu_{11}, \quad \lambda_{22} \geq -\frac{2}{3}\mu_{22} \\ \mu_{12} = \mu_{21} = \lambda_{12} = \lambda_{21} = 0 \end{gathered} \tag{7.6.26}$$

The condition

$$\sum_{\beta,\delta}^{\gamma} 2\hat{e}_{\alpha\beta}\,\delta_{\alpha\delta}\, tr(\mathbf{W}_{\beta}^T\mathbf{W}_{\delta}) = \sum_{\beta,\delta}^{\gamma} D^{\alpha}_{\beta\delta}\, tr(\mathbf{W}_{\beta}^T\mathbf{W}_{\delta}) \geq 0$$

requires that

$$D^{\alpha}_{\delta\delta} = 2\hat{e}_{\alpha\alpha} \geq 0; \quad e_{11} \leq 0 \tag{7.6.27}$$

where use was made of $(7.6.7)_1$. Furthermore,

$$D^{\alpha}_{11}D^{\alpha}_{22} - \frac{1}{4}(D^{\alpha}_{12} + D^{\alpha}_{21})^2 = 2\hat{e}_{\alpha 1}\delta_{\alpha 1}2\hat{e}_{\alpha 2}\delta_{\alpha 2} - \frac{1}{4}(2\hat{e}_{\alpha 2}\delta_{\alpha 1} + 2\hat{e}_{\alpha 1}\delta_{\alpha 2})^2 \geq 0$$

gives

$$\hat{e}_{12} = 0; \quad \hat{e}_{21} = 0 \tag{7.6.28}$$

Using now (7.6.7) it is thus established that

$$e_{11} = 0, \quad e_{21} = 0, \quad e_{12} = arbitrary \tag{7.6.29}$$

The components of the vector $\mathbf{V}$ are

$$\begin{gathered} \mathbf{V} = (\nabla\bar{\bar{\theta}}_1, \ldots, \nabla\bar{\bar{\theta}}_{\gamma}, \nabla\bar{\bar{\rho}}_1, \ldots, \nabla\bar{\bar{\rho}}_{\gamma}, \tilde{\mathbf{v}}_1, \ldots, \tilde{\mathbf{v}}_{\gamma}, \\ \nabla\phi_1, \ldots, \nabla\phi_{\gamma}, \grave{\phi}_1, \ldots, \grave{\phi}_{\gamma}, \tilde{i}_1, \ldots, \tilde{i}_{\gamma}) \end{gathered}$$

and upon using (2.3.12), equation (7.6.4) may be transformed into the following bilinear form:

$$\begin{aligned}
\mathbf{VAV}^T = \sum_{\beta,\delta}^{\gamma}\{&\frac{\nu_{\alpha\beta}}{\bar{\bar{\theta}}_\alpha}\delta_{\alpha\delta}\nabla\bar{\bar{\theta}}_\delta\cdot\nabla\bar{\bar{\rho}}_\beta + (\frac{\zeta_{\alpha\beta}}{\bar{\bar{\theta}}_\alpha}\delta_{\alpha\delta} + (\delta_{\alpha\beta} - \frac{\bar{\rho}_\beta}{\rho})\gamma_{\alpha\delta})\nabla\bar{\bar{\theta}}_\delta\cdot\tilde{\mathbf{v}}_\beta \\
&+\frac{\kappa_{\alpha\beta}}{\bar{\bar{\theta}}_\alpha}\delta_{\alpha\delta}\nabla\bar{\bar{\theta}}_\delta\cdot\nabla\bar{\bar{\theta}}_\beta + \frac{\Gamma_{\alpha\beta}}{\bar{\bar{\theta}}_\alpha}\delta_{\alpha\delta}\nabla\bar{\bar{\theta}}_\delta\cdot\nabla\phi_\beta + \frac{K_{\alpha\beta}}{\phi_\alpha}\delta_{\alpha\delta}\dot{\phi}_\delta\tilde{i}_\beta \\
&+\frac{H_{\alpha\beta}}{\phi_\alpha}\delta_{\alpha\delta}\dot{\phi}_\delta\dot{\phi}_\beta + [(\delta_{\alpha\delta} - \frac{\bar{\rho}_\delta}{\rho})\Delta_{\alpha\beta} - \bar{\rho}_\alpha\phi_\alpha\frac{\partial\tilde{\psi}_\alpha}{\partial\bar{\bar{\rho}}_\beta}(\delta_{\alpha\delta} - \delta_{\beta\delta})]\nabla\bar{\bar{\rho}}_\beta\cdot\tilde{\mathbf{v}}_\delta \\
&+(\delta_{\alpha\delta} - \frac{\bar{\rho}_\delta}{\rho})(\xi_{\alpha\beta} - \delta_{\gamma\beta}\sum_\sigma^\gamma \xi_{\alpha\sigma})\tilde{\mathbf{v}}_\delta\cdot\tilde{\mathbf{v}}_\beta \\
&+[(\delta_{\alpha\delta} - \frac{\bar{\rho}_\delta}{\rho})M_{\alpha\beta} - \bar{\rho}_\alpha\phi_\alpha\frac{\partial\tilde{\psi}_\alpha}{\partial\phi_\beta}(\delta_{\alpha\delta} - \delta_{\beta\delta})]\nabla\phi_\beta\cdot\tilde{\mathbf{v}}_\delta\} + O(\varepsilon^3)
\end{aligned} \tag{7.6.30}$$

where, for $\beta \neq \alpha$, use was made of the result

$$\dot{\phi}_\beta(\frac{\partial\tilde{\psi}_\alpha}{\partial\bar{\bar{\rho}}_\beta}\frac{\bar{\rho}_\beta}{\phi_\beta} - \frac{\partial\tilde{\psi}_\alpha}{\partial\phi_\beta}) = O(\varepsilon^3) \tag{7.6.31}$$

which follows by utilizing (7.5.26) and (7.6.19). The *necessary conditions* for $\mathbf{V}A\mathbf{V}^T \geq 0$ thus are

$$\frac{\kappa_{\alpha\alpha}}{\bar{\bar{\theta}}_\alpha} \geq 0, \quad \frac{H_{\alpha\alpha}}{\phi_\alpha} \geq 0 \tag{7.6.32}$$

$$(\delta_{\alpha\beta} - \frac{\bar{\rho}_\beta}{\rho})(\xi_{\alpha\beta} - \delta_{\gamma\beta}\sum_\sigma^\gamma \xi_{\alpha\sigma}) \geq 0; \quad \alpha,\ \beta,\ \sigma = 1,\ldots,\gamma \tag{7.6.33}$$

from where the following restrictions on the material coefficients $\xi_{\alpha\beta}$ are obtained

$$\begin{aligned}
\alpha = \beta \neq \gamma: &\quad \xi_{\alpha\alpha} \geq 0 \\
\alpha = \beta = \gamma: &\quad \xi_{\gamma\gamma} \geq \textstyle\sum_\sigma^\gamma \xi_{\gamma\sigma};\ \sum_\sigma^{\gamma-1}\xi_{\gamma\sigma} \leq 0 \\
\alpha \neq \beta: &\quad \xi_{\alpha\beta} \leq \delta_{\gamma\beta}\textstyle\sum_\sigma^\gamma \xi_{\alpha\sigma} \\
\alpha \neq \beta \neq \gamma: &\quad \xi_{\alpha\beta} \leq 0
\end{aligned} \tag{7.6.34}$$

## 7.7 Two-Phase Structured Models Without Phase Change

After summarizing the linear constitutive equations developed in the previous section, considerations will be given in this section to two-phase compressible, incompressible, dilute, and concentrated mixtures without the phase change. This consideration is necessary in order to determine the utility of the theory before developing more complex models.

### 7.7.1 Summary of the Linearized Constitutive Equations

The restrictions of the material coefficients (7.6.13), (7.6.14), (7.6.26), and (7.6.29) produce a simple form of the stress tensor (7.5.16), *i.e.*

$$\bar{T}_{\alpha ij} = -\phi_\alpha(p_\alpha + \bar{\bar{\pi}}_\alpha)\delta_{ij} - 2\bar{\bar{\rho}}_\alpha \phi_\alpha I_\alpha \phi_{\alpha,i}\phi_{\alpha,j} + [\lambda_{\alpha\alpha}(tr\mathbf{D}_\alpha)\delta_{ij} + 2\mu_{\alpha\alpha}D_{\alpha ij}] + O_{\alpha\alpha}\dot{\phi}_\alpha\delta_{ij} + O(\varepsilon^2) \tag{7.7.1}$$

where use was also made of (7.5.26). Because this stress tensor is symmetric, the angular momentum condition (7.5.17) is reduced to

$$\hat{M}_{\alpha ij} = O(\varepsilon^2) \tag{7.7.2}$$

The constitutive equation for the surface traction moment $\bar{\mathbf{S}}_\alpha$ is given by (7.5.19), and by making use of (7.6.18) and (7.6.12) it is reduced to the following form

$$\bar{S}_{\alpha ii} = -\phi_\alpha P_\alpha(\{S'_{\beta 0}\}) - O_{\alpha\alpha}\phi_\alpha(tr\mathbf{D}_\alpha) + \sum_\beta^2 K_{\alpha\beta}\,\bar{i}_\beta + \sum_\beta^2 H_{\alpha\beta}\,\dot{\phi}_\beta + O(\varepsilon^2) \tag{7.7.3}$$

Since from (7.4.39)

$$\bar{\lambda}_{\alpha kkm,m}(\{S_{\beta 0}\}) = -\phi_\alpha\beta_\alpha(\{S_{\beta 0}\}) + \phi_\alpha\bar{\bar{\lambda}}^R_{\alpha kkm,m}(\{S_{\beta 0}\}) + \phi_\alpha(p_\alpha + \bar{\bar{\pi}}_\alpha(\{S_{\beta 0}\}) \tag{7.7.4}$$

it follows that by combining this result with (7.7.3), gives

$$\bar{S}_{\alpha kk} + \bar{\lambda}_{\alpha kkm,m} = -\phi_\alpha P_\alpha(\{S'_{\beta 0}\}) + \phi_\alpha(p_\alpha + \bar{\bar{\pi}}_\alpha(\{S_{\beta 0}\})) - \phi_\alpha\beta_\alpha(\{S_{\beta 0}\}) + \phi_\alpha\bar{\bar{\lambda}}^R_{\alpha kkm,m}(\{S_{\beta 0}\}) - O_{\alpha\alpha}\phi_\alpha(tr\mathbf{D}_\alpha) + \sum_\beta^2(K_{\alpha\beta}\,\bar{i}_\beta + H_{\alpha\beta}\,\dot{\phi}_\beta) + O(\varepsilon^2) \tag{7.7.5}$$

where the second order effect of $\bar{\bar{\lambda}}^R_{\alpha km,m}$ has been retained.

The interphase interaction force $\bar{\mathbf{t}}_\alpha$ is given by (7.5.7) and it can be written as

$$\bar{t}_{\alpha i} = (M_{\alpha 2} - M_{\alpha 1})\phi_{1,i} - M_{\alpha 2}(\phi_1 + \phi_2)_{,i} - \xi_{\alpha 1}(\tilde{v}_{1i} - \tilde{v}_{2i}) - \sum_\beta^2 \gamma_{\alpha\beta}\,\bar{\bar{\theta}}_{\beta,i} - \sum_\beta^2 \Delta_{\alpha\beta}\,\bar{\bar{\rho}}_{\beta,i} + O(\varepsilon^2) \tag{7.7.6}$$

with restrictions imposed on $\xi_{\alpha 1}$ as given by (7.6.34). For a saturated mixture $\phi_1 + \phi_2 = 1$, and specializing the result to incompressible phases in a state of equilibrium $\{S_{\beta 0}\}$, equations (7.7.6), (7.7.1), (7.3.4), (3.4.1), and the momentum equation (3.4.2) yield

$$-\phi_{\alpha,i}(p_\alpha + \bar{\bar{\pi}}_\alpha) - \phi_\alpha(p_\alpha + \bar{\bar{\pi}}_\alpha)_{,i} + \bar{\bar{\rho}}_\alpha \phi_\alpha \tilde{b}_{\alpha i} + (M_{\alpha 2} - M_{\alpha 1})\phi_{1,i}$$
$$-(\bar{\bar{\rho}}_\alpha \phi_\alpha I_\alpha \phi_{\alpha,i} \phi_{\alpha,j})_{,j} = 0 \qquad (7.7.7)$$

In an equilibrium single phase flow, the absolute pressure level is not responsible for flow and the pressure gradient is balanced by the body forces. As seen from (7.7.7), however, this result should be modified for two-phase mixtures where the pressure gradient can also be balanced by the volumetric fraction gradient expressing the structural property of the mixture. Assuming that this is the case and noting that $M_\alpha$ are the equilibrium state properties, implies that

$$-\phi_{\alpha,i}(p_\alpha + \bar{\bar{\pi}}_\alpha) + (M_{\alpha 2} - M_{\alpha 1})\phi_{1,i} = 0$$

or

$$M_{12} - M_{11} = p_1 + \bar{\bar{\pi}}_1, \qquad M_{21} - M_{22} = p_2 + \bar{\bar{\pi}}_2 \qquad (7.7.b)$$

reducing (7.7.6) to the following form:

$$\bar{t}_{\alpha i} = \phi_{\alpha,i}(p_\alpha + \bar{\bar{\pi}}_\alpha) - \xi_{\alpha i}(\tilde{v}_{1i} - \tilde{v}_{2i}) - \sum_\beta^2 \gamma_{\alpha\beta}\,\bar{\bar{\theta}}_{\beta,i} - \sum_\beta^2 \Delta_{\alpha\beta}\,\bar{\bar{\rho}}_{\beta,i} + O(\varepsilon^2) \qquad (7.7.8)$$

As in the single phase multicomponent mixture theories (BOWEN,1976) and dispersed multiphase mixtures without structure (DOBRAN,1984b), the interphase interaction force (7.7.6) accounts for the viscous drag and diffusion of phases, $\xi_{\alpha 1}(\tilde{\mathbf{v}}_1 - \tilde{\mathbf{v}}_2)$, and for the temperature gradient or Soret effect. The dependence of $\bar{\mathbf{t}}_\alpha$ on the volumetric fraction and density gradients reflects the multiphase mixtures' structural characteristics.

The Helmholtz potential of phase $\alpha$ is obtained from (7.5.26), (7.6.19), and (7.6.20), *i.e.*

$$\tilde{\psi}_\alpha = \tilde{\bar{\bar{\psi}}}_\alpha(\bar{\bar{\rho}}_\alpha, \bar{\bar{\theta}}_\alpha, \phi_\alpha) + I_\alpha(\bar{\bar{\rho}}_\alpha, \bar{\bar{\theta}}_\alpha, \phi_\alpha)\,\phi_{\alpha,i}\phi_{\alpha,i} + O(\varepsilon^4)$$
$$= \tilde{\bar{\bar{\psi}}}_\alpha(\bar{\bar{\rho}}_\alpha, \bar{\bar{\theta}}_\alpha, \phi_\alpha) + O(\varepsilon^2) \qquad (7.7.9)$$

This result shows that to a first order approximation the Helmholtz potential is determined from the equilibrium state properties $\{S'_{\beta 0}\}$ of the mixture and that it satisfies the principle of phase separation (see section 6.13).

The heat flux vector of phase $\alpha$ expressed by (7.5.6) contains terms which are proportional to the temperature gradient (the Fourier effect) and to the difference in velocities (the Duffour effect), as in single phase multicomponent and dispersed multiphase mixtures (BOWEN,1976, and DOBRAN,1984b). The dependence of $\bar{\mathrm{q}}_\alpha$ on the volumetric fraction and density gradients reflects the multiphase mixtures' structural properties. Notice that the thermal conductivity coefficient $\kappa_{\alpha\alpha}$ in (7.5.6) is predicted by the theory to be positive (*cit.* (7.6.32)$_1$), as physically required for the heat to flow along the decreasing temperature gradient. *For a saturated two-phase mixture*, equation (7.5.6) can be written as

$$\bar{q}_{\alpha i} = -\sum_{\beta}^{2} \kappa_{\alpha\beta}\,\bar{\bar{\theta}}_{\beta,i} - (\Gamma_{\alpha 1} - \Gamma_{\alpha 2})\phi_{1,i} - \sum_{\beta}^{2} \nu_{\alpha\beta}\,\bar{\bar{\rho}}_{\beta,i} - \zeta_{\alpha 1}(\tilde{v}_{1i} - \tilde{v}_{2i}) + O(\varepsilon^2) \tag{7.7.10}$$

A constitutive equation for the reduced intrinsic stress vector $\bar{\bar{\lambda}}^{R}_{\alpha kkm}$ is obtained from (7.4.23), (7.5.26), and (7.6.19)$_2$. Thus,

$$\bar{\bar{\lambda}}^{R}_{\alpha kkm} = 2\bar{\bar{\rho}}_\alpha \phi_\alpha \phi_{\alpha,m} I_\alpha(\bar{\bar{\rho}}_\alpha, \bar{\bar{\theta}}_\alpha, \phi_\alpha) + O(\varepsilon^4) \tag{7.7.11}$$

which can be used in (7.4.25) to establish a relation between $\bar{\lambda}_{\alpha kkm}$ and $\tilde{\ell\epsilon}_{\alpha m}$, and the restriction on the material coefficients in the constitutive relations (7.5.8) and (7.5.22), *i.e.*

$$\begin{aligned} \bar{\lambda}_{\alpha kkm} &= \phi_\alpha \bar{\bar{\lambda}}^{R}_{\alpha kkm} + \bar{\bar{\rho}}_\alpha \phi_\alpha \tilde{\ell\epsilon}_{\alpha m} = 2\bar{\bar{\rho}}_\alpha \phi_\alpha^2 I_\alpha \phi_{\alpha,m} \\ &+ \sum_{\beta}^{2} \epsilon^{\alpha}_{0,\beta}\,\bar{\bar{\rho}}_{\beta,m} + \epsilon^{\alpha}_{1,1}(\tilde{v}_{1m} - \tilde{v}_{2m}) + \sum_{\beta}^{2} \epsilon^{\alpha}_{2,\beta}\,\bar{\bar{\rho}}_{\beta,m} + \sum_{\beta}^{2} \epsilon^{\alpha}_{3,\beta}\,\phi_{\beta,m} + O(\varepsilon^2) \\ &= \sum_{\beta}^{2} a^{\alpha}_{1,\beta}\,\bar{\bar{\rho}}_{\beta,m} + b^{\alpha}_{1,1}(\tilde{v}_{1m} - \tilde{v}_{2m}) + \sum_{\beta}^{2} c^{\alpha}_{1,\beta}\,\bar{\bar{\theta}}_{\beta,m} + \sum_{\beta}^{2} d^{\alpha}_{1,\beta}\,\phi_{\beta,m} + O(\varepsilon^2) \end{aligned} \tag{7.7.12}$$

Thus, *for a saturated two-phase mixture* the following relations may be established:

$$\begin{aligned} &\epsilon^{\alpha}_{0,\beta} = c^{\alpha}_{1,\beta}, \quad \epsilon^{\alpha}_{2,\beta} = a^{\alpha}_{1,\beta}, \quad \epsilon^{\alpha}_{1,1} = b^{\alpha}_{1,1}; \quad \alpha, \beta = 1, 2 \\ &2\bar{\bar{\rho}}_1 \phi_1^2 I_1 + \epsilon^{1}_{3,1} - \epsilon^{1}_{3,2} = d^{1}_{1,1} - d^{1}_{1,2}, \\ &-2\bar{\bar{\rho}}_2(1-\phi_1)^2 I_2 + \epsilon^{2}_{3,1} - \epsilon^{2}_{3,2} = d^{2}_{1,1} - d^{2}_{1,2} \end{aligned} \tag{7.7.13}$$

Equations (7.3.1) and (7.3.2), when combined with equations (7.3.8), (7.6.1), (7.5.18), (7.5.19), and (7.5.22) where it is assumed that $\bar{\lambda}_{\alpha ijk,k} = O(\varepsilon^2)$, yield restrictions on the coefficients $\Phi_{\alpha\beta}$ and $T_{\alpha\beta}$. Thus,

$$\phi_{\alpha 1} D_{1jk} + \Phi_{\alpha 2} D_{2jk} + O(\varepsilon^2) = 0, \quad T_{\alpha 1}(W_{1jk} - W_{2jk}) + O(\varepsilon^2) = 0; \quad j \neq k$$

(7.7.14)

from where

$$\Phi_{11} = \Phi_{12} = \Phi_{21} = \Phi_{22} = 0, \quad T_{11} = T_{21} = 0 \tag{7.7.15}$$

such that upon combining these results with (7.6.12) and (7.6.18), gives

$$\Pi_{11} + \phi_1 O_{11} = 0, \;\; \Pi_{12} = 0, \;\; \Pi_{22} + \phi_2 O_{22} = 0, \;\; \Pi_{21} = 0 \tag{7.7.16}$$

The following relations can also be established from section 7.3 and the above constitutive equations for use in the balance equations (7.2.3), (7.3.1), and (7.3.3).

$$k_{\alpha jji,i} = -(\frac{\grave{\phi}_\alpha}{\phi_\alpha} U_{\alpha jji})_{,i} = O(\varepsilon^2) \tag{7.7.17}$$

$$\check{k}_{\alpha jj} = 0 \tag{7.7.18}$$

$$\breve{g}_{\alpha kk} = \bar{\rho}_\alpha (\frac{\grave{\phi}_\alpha}{\phi_\alpha})^2 \bar{\bar{i}}_\alpha + \bar{v}_{\alpha j}(\bar{\rho}_\alpha \bar{\bar{i}}_\alpha \frac{\grave{\phi}_\alpha}{\phi_\alpha})_{,j} - [(\frac{\grave{\phi}_\alpha}{\phi_\alpha})^2 U_{\alpha jjm}]_{,m} = O(\varepsilon^3) \tag{7.7.19}$$

$$\hat{\epsilon}_\alpha = (\frac{\grave{\phi}_\alpha}{\phi_\alpha})_{,i} \phi_\alpha \overset{=R}{\lambda}_{\alpha jji} - \frac{\grave{\phi}_\alpha}{\phi_\alpha} \epsilon^\alpha_{1,1}(tr\mathbf{D}_1 - tr\mathbf{D}_2) + O(\varepsilon^3) \tag{7.7.20}$$

$$\bar{h}_{\alpha kkm,m} = \bar{\lambda}_{\alpha kkm,m} - (\bar{\rho}_\alpha \frac{\grave{\phi}_\alpha}{\phi_\alpha} \bar{\bar{i}}_\alpha \bar{v}_{\alpha m})_{,m} = \bar{\lambda}_{\alpha kkm,m} + O(\varepsilon^3) \tag{7.7.21}$$

$$U_{\alpha ijk,k} = O(\varepsilon^2) \;\; (assumed\; to\; follow\; from\; (7.5.23)) \tag{7.7.22}$$

Notice that in deriving (7.7.20) use was made of (7.3.9), (7.7.12), (7.6.1), (7.6.2), and (7.5.10).

### 7.7.2 Two-Phase Modeling Equations

The constitutive equations of the last section can be combined with the conservation and balance equations of section 7.3 to produce compressible and incompressible models without the phase change. Given below is a summary of these equations for the purpose of collecting them in one place.

$$\grave{\bar{\rho}}_\alpha + \bar{\rho}_\alpha \bar{v}_{\alpha i,i} = 0, \quad \bar{\rho}_\alpha = \phi_\alpha \bar{\bar{\rho}}_\alpha \tag{7.7.23}$$

$$\bar{\rho}_\alpha \grave{\tilde{v}}_{\alpha i} = \bar{T}_{\alpha ij,j} + \bar{\rho}_\alpha \tilde{b}_{\alpha i} + \bar{t}_{\alpha i} - \left(\frac{\grave{\phi}_\alpha}{\phi_\alpha}\right)_{,i} \bar{\rho}_\alpha \frac{\grave{\phi}_\alpha}{\phi_\alpha} \tilde{i}_\alpha \tag{7.7.24}$$

$$\bar{\rho}_\alpha \grave{\tilde{i}}_\alpha = 2\bar{\rho}_\alpha \frac{\grave{\phi}_\alpha}{\phi_\alpha} \tilde{i}_\alpha + k_{\alpha jji,i} \tag{7.7.25}$$

$$\bar{\rho}_\alpha \overline{\left(\tilde{i}_\alpha \frac{\grave{\phi}_\alpha}{\phi_\alpha}\right)}^{\grave{}} = \bar{S}_{\alpha kk} + \bar{\lambda}_{\alpha kkm,m} - \left(\bar{\rho}_\alpha \frac{\grave{\phi}_\alpha}{\phi_\alpha} \tilde{i}_\alpha \tilde{v}_{\alpha m}\right)_{,m} + \breve{g}_{\alpha kk} \tag{7.7.26}$$

$$\bar{\rho}_\alpha \grave{\tilde{\epsilon}}_\alpha = \bar{T}_{\alpha ij} \tilde{v}_{\alpha i,j} - \bar{q}_{\alpha k,k} + \bar{\rho}_\alpha \tilde{r}_\alpha + \hat{\epsilon}_\alpha \tag{7.7.27}$$

$$\begin{aligned} \bar{T}_{\alpha ij} = {} & -\phi_\alpha (p_\alpha + \bar{\bar{\pi}}_\alpha)\delta_{ij} - 2\bar{\bar{\rho}}_\alpha \phi_\alpha I_\alpha \phi_{\alpha,i}\phi_{\alpha,j} + [\lambda_{\alpha\alpha}(tr\mathbf{D}_\alpha)\delta_{ij} + 2\mu_{\alpha\alpha} D_{\alpha ij}] \\ & + O_{\alpha\alpha}\grave{\phi}_\alpha \delta_{ij} + O(\varepsilon^2) \end{aligned} \tag{7.7.1}$$

$$\bar{t}_{\alpha i} = \phi_{\alpha,i}(p_\alpha + \bar{\bar{\pi}}_\alpha) - \xi_{\alpha 1}(\tilde{v}_{1i} - \tilde{v}_{2i}) - \sum_\beta^2 \gamma_{\alpha\beta} \bar{\bar{\theta}}_{\beta,i} - \sum_\beta^2 \Delta_{\alpha\beta} \bar{\bar{\rho}}_{\beta,i} + O(\varepsilon^2) \tag{7.7.8}$$

$$k_{\alpha jji,i} = -\left(\frac{\grave{\phi}_\alpha}{\phi_\alpha}\right)_{,i} U_{\alpha jji} = O(\varepsilon^3) \tag{7.7.17}$$

$$\bar{S}_{\alpha kk} = -\phi_\alpha P_\alpha(\{S'_{\beta 0}\}) - O_{\alpha\alpha}\phi_\alpha(tr\mathbf{D}_\alpha) + \sum_\beta^2 (K_{\alpha\beta}\tilde{i}_\beta + H_{\alpha\beta}\grave{\phi}_\beta) + O(\varepsilon^2) \tag{7.7.3}$$

$$\bar{\rho}_\alpha \tilde{\ell}_{\alpha kk} = 0 \tag{7.2.10}$$

$$\begin{aligned} \bar{\lambda}_{\alpha kkm,m}(\{S_{\beta 0}\}) = {} & -\phi_\alpha \beta_\alpha(\{S_{\beta 0}\}) + \phi_\alpha \bar{\bar{\lambda}}^R_{\alpha kkm,m}(\{S_{\beta 0}\}) \\ & + \phi_\alpha(p_\alpha + \bar{\bar{\pi}}_\alpha(\{S_{\beta 0}\})) \end{aligned} \tag{7.7.4}$$

$$\bar{\bar{\lambda}}^R_{\alpha kkm}(\{S_{\beta 0}\}) = 2\bar{\bar{\rho}}_\alpha \phi_\alpha I_\alpha \phi_{\alpha,m} + O(\varepsilon^4) \tag{7.7.11}$$

$$\breve{g}_{\alpha kk} = \bar{\rho}_\alpha \left(\frac{\grave{\phi}_\alpha}{\phi_\alpha}\right)^2 \tilde{i}_\alpha + \tilde{v}_{\alpha j}\left(\bar{\rho}_\alpha \tilde{i}_\alpha \frac{\grave{\phi}_\alpha}{\phi_\alpha}\right)_{,j} - \left(\left(\frac{\grave{\phi}_\alpha}{\phi_\alpha}\right)^2 U_{\alpha jjm}\right)_{,m} \tag{7.7.19}$$

$$\bar{q}_{\alpha i} = -\sum_{\beta}^{2} \kappa_{\alpha\beta}\bar{\bar{\theta}}_{\beta,i} - (\Gamma_{\alpha 1} - \Gamma_{\alpha 2})\phi_{1,i} - \sum_{\beta}^{2} \nu_{\alpha\beta}\bar{\bar{\rho}}_{\beta,i}$$
$$-\zeta_{\alpha 1}(\tilde{v}_{1i} - \tilde{v}_{2i}) + O(\varepsilon^2) \tag{7.7.10}$$

$$\hat{\epsilon}_{\alpha} = (\frac{\grave{\phi}_\alpha}{\phi_\alpha})_{,i}\,\phi_\alpha \bar{\bar{\lambda}}^{R}_{\alpha jji} - \frac{\grave{\phi}_\alpha}{\phi_\alpha}\epsilon^{\alpha}_{1,1}(tr\mathbf{D}_1 - tr\mathbf{D}_2) + O(\varepsilon^3) \tag{7.7.20}$$

Combining (7.7.25), (7.7.26), (7.7.4), (7.7.17), and (7.7.18) gives an alternate form of the equilibrium moments equation (7.7.26), *i.e.*

$$\bar{\bar{\rho}}_\alpha \bar{\bar{i}}_\alpha \grave{\grave{\phi}}_\alpha = \bar{S}_{\alpha kk} - \phi_\alpha\beta_\alpha(\{S_{\beta 0}\}) + \phi_\alpha(p_\alpha + \bar{\bar{\pi}}_\alpha(\{S_{\beta 0}\}))$$
$$+\phi_\alpha\bar{\bar{\lambda}}^{R}_{\alpha kkm,m}(\{S_{\beta 0}\}) - \bar{\bar{\rho}}_\alpha\bar{\bar{i}}_\alpha\grave{\phi}_\alpha(tr\mathbf{D}_\alpha) + O(\epsilon^4) \tag{7.7.28}$$

and upon substituting for $\bar{S}_{\alpha kk}$ from (7.7.3) it follows that

$$\bar{\bar{\rho}}_\alpha \bar{\bar{i}}_\alpha \grave{\grave{\phi}}_\alpha = -\phi_\alpha P_\alpha(\{S'_{\beta 0}\}) + \phi_\alpha(p_\alpha + \bar{\bar{\pi}}_\alpha(\{S_{\beta 0}\})) - \phi_\alpha\beta_\alpha(\{S_{\beta 0}\})$$
$$+\phi_\alpha\bar{\bar{\lambda}}^{R}_{\alpha kkm,m}(\{S_{\beta 0}\}) - O_{\alpha\alpha}\phi_\alpha(tr\mathbf{D}_\alpha) + \sum_{\beta}^{2}(K_{\alpha\beta}\,\bar{i}_\beta + H_{\alpha\beta}\,\grave{\phi}_\beta)$$
$$-\bar{\bar{\rho}}_\alpha\bar{\bar{i}}_\alpha\grave{\phi}_\alpha(tr\mathbf{D}_\alpha) + O(\varepsilon^2) \tag{7.7.29}$$

Notice in the above equation that the second order effect of $\bar{\bar{\lambda}}^{R}_{\alpha kkm,m}$ is retained. If this effect is ignored, then according to (7.4.39)

$$\beta_\alpha(\{S'_{\beta 0}\}) - (p_\alpha + \bar{\bar{\pi}}_\alpha(\{S'_{\beta 0}\})) + P_\alpha(\{S'_{\beta 0}) = 0$$

From this result it is easy to see that the difference in equilibrium pressures of two phases is related to the difference in the configuration and interfacial pressures. Since the difference in interfacial pressures may be ignored in equilibrium, it follows that in this state the difference of configuration pressures can be related to the surface tension and curvature of the interface.

### 7.7.3 Examples of Two-Phase Models

#### 7.7.3.1 SATURATED AND CONCENTRATED MIXTURES WITH INCOMPRESSIBLE PHASES AND NEGLIGIBLE INERTIAL EFFECTS

Many two-phase flow mixtures can be assumed to be saturated and phases incompressible. For incompressible phases, $\bar{\bar{\rho}}_1$ and $\bar{\bar{\rho}}_2$ are constant and the continuity equation $(7.7.23)_1$ is reduced to

$$\grave{\phi}_\alpha + \phi_\alpha\,(tr\mathbf{D}_\alpha) = 0, \quad \alpha = 1,2 \tag{7.7.30}$$

With the additional assumptions of saturation and negligible inertia, *i.e.*

$$\phi_1 + \phi_2 = 1; \quad \bar{i}_\alpha = 0; \quad \alpha = 1,2 \tag{7.7.31}$$

the balance equation (7.7.29) is reduced to the following form:

$$P_\alpha = p_\alpha - \beta_\alpha + O_{\alpha\alpha}\grave{\phi}_\alpha \frac{1}{\phi_\alpha} + \bar{\bar{\lambda}}^R_{\alpha kkm,m} + \frac{1}{\phi_\alpha}(H_{\alpha 1}\grave{\phi}_1 + H_{\alpha 2}\grave{\phi}_2) \tag{7.7.32}$$

where for an incompressible phase we set $\bar{\bar{\pi}}_\alpha = 0$. The saturation constraint $(7.7.31)_1$ implies that

$$\grave{\phi}_2 = -\grave{\phi}_1 + (\bar{\mathbf{v}}_1 - \bar{\mathbf{v}}_2)\cdot\nabla\phi_1 \tag{7.7.33}$$

$$\begin{aligned} \grave{\grave{\phi}}_2 = &-\grave{\grave{\phi}}_1 + [\grave{\bar{\mathbf{v}}}_1 - \grave{\bar{\mathbf{v}}}_2 + (\nabla\bar{\mathbf{v}}_1)(\bar{\mathbf{v}}_2 - \bar{\mathbf{v}}_1)]\cdot\nabla\phi_1 \\ &+ [(\grave{\nabla\phi_1}) + \nabla(\grave{\phi}_1) + \nabla(\nabla\phi_1)(\bar{\mathbf{v}}_2 - \bar{\mathbf{v}}_1)](\bar{\mathbf{v}}_1 - \bar{\mathbf{v}}_2) \end{aligned} \tag{7.7.34}$$

Using (7.7.30), (7.7.33), defining in (7.7.11)

$$A_\alpha = 2\bar{\bar{\rho}}_\alpha \phi_\alpha I_\alpha \tag{7.7.35}$$

and assuming that

$$P_1 = P_2 \tag{7.7.36}$$

it follows from (7.7.32) that the pressure difference between the phases is

$$\begin{aligned} p_1 - p_2 = &\beta_1 - \beta_2 - ((A_1 + A_2)\phi_{1,m})_{,m} + (O_{11} + H_{11} - H_{12})\,tr\mathbf{D}_1 \\ &-(O_{22} - H_{21} + H_{22})\,tr\mathbf{D}_2 - \left(\frac{H_{12}}{\phi_1} - \frac{H_{21}}{\phi_2}\right)(\bar{\mathbf{v}}_1 - \bar{\mathbf{v}}_2)\cdot\nabla\phi_1 \end{aligned} \tag{7.7.37}$$

This result is similar to a result obtained by PASSMAN *et al.* (1984) in their theory of mixtures (equation 5C.7.11[1]) for saturated and concentrated incompressible mixtures. It shows that the difference in phasic pressures results from the intergranular forces represented by $\beta_1 - \beta_2$, nonhomogeneous particle distribution represented by the volumetric fraction gradient $\nabla\phi_1$, intergranular friction represented by the viscosity coefficients $B_1$ and $B_2$, *i.e.*

$$B_1 = -(O_{11} + H_{11} - H_{12}) \tag{7.7.38}$$

$$B_2 = -(O_{22} + H_{22} - H_{21}) \tag{7.7.39}$$

[1]Notice, however, that this equation contains sign errors in the third, fourth, and fifth terms, if one assumes the validity of (5C.7.10).

and diffusion of phases represented by $(\tilde{\mathbf{v}}_1 - \tilde{\mathbf{v}}_2)$.

Solving for the pressure in (7.7.32), using (7.7.33), and substituting into (7.7.1) yields the phasic stress tensors

$$\begin{aligned}\bar{\mathbf{T}}_1 = -\{&\phi_1(P_1+\beta_1) + [2\phi_1 O_{11} + \phi_1(H_{11} - H_{12}) - \lambda_{11}]tr\mathbf{D}_1 \\ -\nabla\cdot(\phi_1 A_1 \nabla\phi_1)\}\mathbf{I} &+ [H_{12}(\tilde{\mathbf{v}}_1 - \tilde{\mathbf{v}}_2)\cdot\nabla\phi_1]\mathbf{I} + 2\mu_{11}\mathbf{D}_1 - A_1(\nabla\phi_1\cdot\nabla\phi_1)\mathbf{I} \\ &-\bar{\bar{\rho}}_1\phi_1 I_1 \nabla\phi_1\nabla\phi_1 + O(\varepsilon^2)\end{aligned} \tag{7.7.40}$$

$$\begin{aligned}\bar{\mathbf{T}}_2 = -\{&\phi_2(P_2+\beta_2) + [2\phi_2 O_{22} + \phi_2(H_{22} - H_{21}) - \lambda_{22}]tr\mathbf{D}_2 \\ -\nabla\cdot(\phi_2 A_2 \nabla\phi_2)\}\mathbf{I} &+ [H_{21}(\tilde{\mathbf{v}}_1 - \tilde{\mathbf{v}}_2)\cdot\nabla\phi_1]\mathbf{I} + 2\mu_{22}\mathbf{D}_2 - A_2(\nabla\phi_2\cdot\nabla\phi_2)\mathbf{I} \\ &-\bar{\bar{\rho}}_2\phi_2 I_2 \nabla\phi_2\nabla\phi_2 + O(\varepsilon^2)\end{aligned} \tag{7.7.41}$$

These results resemble those obtained by PASSMAN *et al.* (1984) (equation 5C.7.12). They contain the second order effects of diffusion and concentration gradients, and do not necessarily satisfy the principle of phase separation.

### 7.7.3.2 SATURATED AND DILUTE MIXTURES WITH INCOMPRESSIBLE PHASES AND NEGLIGIBLE INERTIAL EFFECTS

Saturated and dilute mixtures with incompressible phases and negligible inertial effects may be modeled using equations of section 7.7.3.1, with $\phi_\alpha << 1$ for the dilute phase. A saturated and dilute mixture may have significant inertial effects if the grains or particles are large, and produce significant particle concentration regions where the diluteness assumption may break down. The *a priori* estimation of the importance of inertial effects in the field equations requires more detailed knowledge of the problem, such as the geometry and flow characteristics (laminar and turbulent flow).

### 7.7.3.3 SATURATED AND DILUTE MIXTURES WITH COMPRESSIBLE PHASES AND THE RAYLEIGH-TYPE BUBBLE EQUATION

For dilute mixtures, we may set in (7.7.29)

$$\bar{\bar{\lambda}}^{R}_{\alpha kkm} = A_\alpha \phi_{\alpha,m} = 0, \qquad A_\alpha = 2\bar{\bar{\rho}}_\alpha \phi_\alpha I_\alpha = 0 \tag{7.7.42}$$

Assuming the equality of interfacial pressures ($P_1 = P_2$), using (7.7.33), (7.7.34), (7.7.42), and ignoring the diffusion of phases ($\tilde{\mathbf{v}}_1 = \tilde{\mathbf{v}}_2$), the following result is obtained from (7.7.29)

$$\bar{\bar{\pi}}_1 - \bar{\bar{\pi}}_2 = \beta_1 - \beta_2 + \dot{\phi}_1\left(\frac{\bar{\bar{\rho}}_1 \bar{i}_1}{\phi_1} + \frac{\bar{\bar{\rho}}_2 \bar{i}_2}{\phi_2}\right) + O_{11}(tr\mathbf{D}_1) - O_{22}(tr\mathbf{D}_2)$$

$$-\bar{i}_1(\frac{K_{11}}{\phi_1} - \frac{K_{21}}{\phi_2}) + \bar{i}_2(\frac{K_{22}}{\phi_2} - \frac{K_{12}}{\phi_1}) - \frac{\grave{\phi}_1}{\phi_1}[H_{11} - H_{12} + \frac{\phi_1}{\phi_2}(H_{22} - H_{21})]$$
$$+\frac{1}{\phi_1}\bar{\bar{\rho}}_1\bar{i}_1\grave{\phi}_1(tr\mathbf{D}_1) + \frac{1}{\phi_2}\bar{\bar{\rho}}_2\bar{i}_2\grave{\phi}_1(tr\mathbf{D}_2) \tag{7.7.43}$$

This equation can be reduced into the Rayleigh-type bubble equation (*cit.* VAN WIJNGAARDEN,1972) by assuming: (1) spherical bubbles (phase 1) of radius $R$, (2) a constant number density of bubbles $Z$, and (3) that the liquid phase inertia can be computed from the flow field around a single bubble, *i.e.*

$$\frac{1}{3}4\pi R^3\bar{\bar{\rho}}_1 = constant \tag{7.7.44}$$

$$\frac{3\phi_1}{4\pi R^3} = Z = constant \tag{7.7.45}$$

$$\bar{i}_2 = \frac{1}{4\pi RZ} \tag{7.7.46}$$

From (7.7.44) it follows that

$$\frac{\grave{\bar{\bar{\rho}}}_1}{\bar{\bar{\rho}}_1} = -3\frac{\dot{R}}{R} \tag{7.7.47}$$

whereas from (7.7.45) we have

$$\frac{\grave{\phi}_1}{\phi_1} = 3\frac{\dot{R}}{R} \tag{7.7.48}$$

The conservation of mass equation $(7.7.23)_1$ yields

$$tr\mathbf{D}_\alpha = -\frac{\grave{\bar{\bar{\rho}}}_\alpha}{\bar{\bar{\rho}}_\alpha} - \frac{\grave{\phi}_\alpha}{\phi_\alpha} \tag{7.7.49}$$

such that when combined with (7.7.33), (7.7.47), (7.7.48), and with $\tilde{\mathbf{v}}_1 = \tilde{\mathbf{v}}_2$, gives

$$tr\mathbf{D}_1 = 0, \quad tr\mathbf{D}_2 = -\frac{\grave{\bar{\bar{\rho}}}_2}{\bar{\bar{\rho}}_2} + \frac{\grave{\phi}_1}{\phi_2} \tag{7.7.50}$$

Substituting the above relations into (7.7.43) results in the following equation

$$\bar{\bar{\pi}}_1 - \bar{\bar{\pi}}_2 = \beta_1 - \beta_2 + \bar{\bar{\rho}}_2 R\ddot{R}(\frac{1}{\phi_2} + \frac{\bar{\bar{\rho}}_1}{\bar{\bar{\rho}}_2}\bar{i}_1\frac{3}{R^2})$$

$$
\begin{aligned}
&-3\frac{\dot{R}}{R}[\frac{\phi_1}{\phi_2}(O_{22}+H_{22}-H_{21})+H_{11}-H_{12}] \\
&-\bar{\bar{i}}_1(\frac{K_{11}}{\phi_1}-\frac{K_{21}}{\phi_2})+\bar{\bar{i}}_2(\frac{K_{22}}{\phi_2}-\frac{K_{12}}{\phi_1})-\frac{\grave{\bar{\bar{\rho}}}_2}{\bar{\bar{\rho}}_2}(\frac{\bar{\bar{\rho}}_2}{\phi_2}R\dot{R}-O_{22}) \\
&+2\bar{\bar{\rho}}_2\dot{R}\dot{R}(\frac{1}{\phi_2}+\frac{3\phi_1}{2\phi_2^2}+3\frac{\bar{\bar{\rho}}_1\bar{\bar{i}}_1}{\bar{\bar{\rho}}_2R^2})
\end{aligned}
\tag{7.7.51}
$$

Since the mixture is assumed to be dilute and bubbles sufficiently small,

$$
\phi_1 = 1-\phi_2 << 1 \tag{7.7.52}
$$

$$
\frac{\bar{\bar{\rho}}_1}{\bar{\bar{\rho}}_2}\frac{3\bar{\bar{i}}_1}{R^2} << 1 \tag{7.7.53}
$$

it follows from (7.7.51) that

$$
\bar{\bar{\pi}}_1-\bar{\bar{\pi}}_2=\beta_1-\beta_2+\bar{\bar{\rho}}_2R\ddot{R}+2\bar{\bar{\rho}}_2\dot{R}\dot{R}+\frac{4\mu_2^*}{R}\dot{R}+i^*-\frac{\grave{\bar{\bar{\rho}}}_2}{\bar{\bar{\rho}}_2}(\bar{\bar{\rho}}_2R\dot{R}-O_{22}) \tag{7.7.54}
$$

where

$$
\mu_2^* = -\frac{3}{4}[\phi_1(O_{22}+H_{22}-H_{21})+H_{11}-H_{12}] \tag{7.7.55}
$$

is an *effective viscosity coefficient*, and

$$
i^* = -\bar{\bar{i}}_1(\frac{K_{11}}{\phi_1}-\frac{K_{21}}{\phi_2})+\bar{\bar{i}}_2(\frac{K_{22}}{\phi_2}-\frac{K_{12}}{\phi_1}) \tag{7.7.56}
$$

represents an *effective inertia coefficient* for liquid and bubbles.

Equation (7.7.54) is similar in form to the Rayleigh's bubble equation with dissipation (VAN WIJNGAARDEN,1972) *if*: (1) $i^*=0$; (2) $\bar{\bar{\rho}}_2=$ *constant*; (3) $\beta_1-\beta_2=2\sigma/R$, with $\sigma$ being the surface tension; (4) $\mu_2^*=\mu_2$ being the liquid viscosity; and (5) the factor 2 multiplying the inertia term $\dot{R}\dot{R}$ being replaced by 3/2. This difference of factors multiplying the inertia terms may be associated with the averaging procedure used to establish equation (7.7.54). The viscous dissipation in liquid and gas is accounted in (7.7.54) by the viscosity coefficient $\mu_2^*$. The positivity of this coefficient may be established from (7.7.40) and (7.7.41) by setting[2]

$$
\lambda_{\alpha\alpha}-\phi_\alpha(2O_{\alpha\alpha}+H_{\alpha\alpha}-H_{\alpha\beta})+\frac{2}{3}\mu_{\alpha\alpha}\geq 0; \quad \beta\neq\alpha \tag{7.7.57}
$$

[2]This condition is necessary but not sufficient for (7.7.40) and (7.7.41) to retain a formal similarity with the single phase incompressible stress tensor form when the structural effects of the mixture are negligible, since in that case (7.7.32) shows that $P_\alpha+\beta_\alpha\approx p_\alpha$.

According to (7.6.21) we may select $\lambda_{\alpha\alpha} + 2\mu_{\alpha\alpha}/3 = 0$ in order for these equations to retain a formal similarity with the corresponding stress tensor equation of single phase flow when the structural effects of the mixture are "small". This necessary condition thus gives from (7.7.57)

$$-O_{\alpha\alpha} \geq O_{\alpha\alpha} + H_{\alpha\alpha} - H_{\alpha\beta}; \quad \beta \neq \alpha \tag{7.7.58}$$

and using $(7.6.32)_2$ it is established that

$$O_{\alpha\alpha} \leq 0 \tag{7.7.59}$$

*if* the crosscoupling terms $H_{12}$ and $H_{21}$ are ignored. From these results it may be, therefore, permissible to state that

$$\mu_2^* \geq 0 \tag{7.7.60}$$

The explicit effect of inertia $i^*$ appearing in (7.7.54) has not been discussed in the literature in the Rayleigh-type bubble equations. The compressibility of the liquid phase in this equation is represented by the term containing $\grave{\bar{\rho}}_2$ and it has been discussed previously by DRUMHELLER, KIPP & BEDFORD (1982). This term represents the effect caused by the propagation of pressure waves in the liquid and it does not appear to have been quantified by experiments due, possibly, to the inability of measuring instruments to resolve the liquid phase pressure fluctuations.

The relative motion between the phases can have a significant effect on the wave propagation or bubble motion, and it should be studied using the general forms of (7.7.29), (7.7.33), and (7.7.34). The "bubble equation" (7.7.54) contains the difference of configuration pressures $(\beta_1 - \beta_2)$ which may be associated with the surface tension effect in the equilibrium state $\{S'_{\beta o}\}$ as discussed above. This association is, however, somewhat artificial, since the surface tension effect should enter into the theory more explicitly through a nonlocal theory of constitutive equations which will not be discussed in this book.

PASSMAN *et al.* (1984) claim to recover the Reyleigh-type bubble equation with dissipation in their theory of mixtures as discussed in chapter 5. A close examination of the derivation of their bubble equation reveals an error in their second and third equations following (5C.7.18), where 8 should be 4. Consequently, their bubble equation involves the coefficients 2 multiplying $\bar{\bar{\rho}}_2 R\ddot{R}$ and 3 multiplying $\bar{\bar{\rho}}_2 \dot{R}\dot{R}$, instead of 1 and 3/2, respectvely, as they claim. Moreover, their derivation of the bubble equation rests upon their results (5C.7.14) and (5.C.7.15), *i.e.*

$$\bar{\bar{\rho}}_1 \phi_1 i_1 = \bar{\bar{\rho}}_2 \phi_2 i_2$$

$$\bar{\bar{\rho}}_1 \phi_1 \grave{i}_1 = \bar{\bar{\rho}}_2 \phi_2 \grave{i}_2 + \bar{\bar{\rho}}_1 \phi_1 i_1 \nabla \cdot (\tilde{\mathbf{v}}_1 - \tilde{\mathbf{v}}_2)$$

which are claimed to follow from (3.4.21) when $\phi = \phi_1 + \phi_2$. These results are not clear, but by setting in (3.4.21) $\nu_{mn}^{(\alpha)} = \dot{\phi}_\alpha \delta_{mn}$, $\nu_{mn} = \dot{\phi}$ with $\dot{\phi} = 0$, it may be possible to claim these results, *if* indeed $\nu_{mn} = \dot{\phi}$!

## 7.8 Discussion of Results

The linearized constitutive equations developed in this chapter have practical limitations. In particular, the drag coefficient $\xi_{\alpha 1}$ in (7.7.8) is also known to depend on $|\tilde{\mathbf{v}}_1 - \tilde{\mathbf{v}}_2|$ which can be predicted by a theory of nonlinear constitutive equations where the coefficients multiplying the independent variables in constitutive equations depend on invariants of these variables. From (7.6.34), $\xi_{11} \geq 0$ and $\xi_{21} \leq 0$, as it is physically required in (7.7.6) for the drag of phase $\beta$ to oppose the motion of phase $\alpha$. Moreover, the expressions for the viscosity and thermal conductivity coefficients (equations (7.6.26) and $(7.6.32)_1$) have a formal similarity with the single phase flow, and depend, in general, on temperature, density and volumetric fractions of *all* phases in the mixture. In the first approximation and involving only the dilatation, the theory predicts that the angular momentum tensor $\hat{\mathbf{M}}_\alpha$ is symmetric (*cit.* (7.7.2)). This result would be, of course, expected if we were to volume average $(T_{ij}^{(\alpha\delta)} - T_{ji}^{(\alpha\delta)})$, with $T_{ij}^{(\alpha\delta)}$ symmetric, and argue that the averaged stress tensor $\bar{\mathbf{T}}_\alpha$ can be completely reconstructed from the information of its local macroscopic counterpart. As discussed by DOBRAN (1984a), it is, however, approprite to expect that the global (averaged) theory of mixtures contains information inherent of its global structure which cannot be completely realized by a contruction from its local macroscopic information.

The manner in which the interfacial pressure is introduced into the postulatory theory of mixtures is criticized in section 6.12.2, since it is unphysical to define, in general, a *single* interface pressure for more than two phases in a mixture. In the present theory of mixtures, the interphase pressure is identified with the surface traction pressure $P_\alpha$ defined by (7.4.40). This pressure is defined for *each phase* in the mixture and does not produce the unphysical result noted above.

According to (7.4.42), the difference of configuration pressures $(\beta_\alpha - \beta_\beta)$ can be related in the equilibrium state $\{S'_{\beta 0}\}$ to the difference of phasic pressures $(p_\alpha - p_\beta)$, which in turn is related to the surface tension between the phases and curvature of the interface (see also the discussion at the end of section 7.4). In this sense it is, therefore, appropriate to relate $(\beta_\alpha - \beta_\beta)$ to the average surface tension force and curvature, in general.

The recovery of the Rayleigh-type bubble equation (7.7.54) is an important result in this chapter for it gives credibility to the simple theory without the rotational effects. The assumptions leading to this equation starting from (7.7.29) are physically reasonable. The Rayleigh bubble equation has the coefficient 3/2 multiplying the inertia term, whereas in (7.7.54) this coefficient is 2. This difference in coefficients is not large, but it is sufficiently important to be noted, although arguments can be presented (as stated in the previous section) for possible reasons of this discrepancy. The last term on the right of equation (7.7.54) represents the effect of the liquid phase compressibility which is shown to be affected by the bubble motion and liquid viscosity. The former effect has an opposite sign to the corresponding term in the bubble equation of DRUMHELLER, KIPP & BEDFORD (1982) who showed that liquid pressure variations can be significant at early times during the initiation of pressure pulses in bubbly liquids, and that at later times the liquid pressure rapidly oscillates about the gas pressure. When comparing the gas pressure oscillation results with the transient data of air, carbon monoxide, and helium bubbles in water, DRUMHELLER and co-workers obtained good comparisons in some situations and associated the disrepancies of this comparison with the breakup of bubbles and lack of including the relative motion between the liquid and bubbles in their bubble modeling equation. According to the present result (7.7.54), the pressure propagation mechanisms in bubbly liquids appear to be accounted in a more complete manner than in previous models, as attested by the additional effects of bubble and liquid inertias, and by the liquid viscosity/compressibility coupling. For small amplitude disturbances and phase velocities removed from the natural frequency associated with the oscillation of bubbles, the effects of inertia, compressibility, and heat transfer are probably not too important, and a simple bubble model may suffice. Large amplitude pressure disturbances may, however, require the use of (7.7.29) instead of (7.7.54), and the solution of this equation together with the inertia equation (7.7.25), continuity equation $(7.7.23)_1$, and the momentum equation (7.7.24). Since the transfer of heat between the gas and liquid can result in an important dissipation effect, the propagation of large amplitude waves in bubbly liquids may also require the use of energy equation (7.7.27). For the reasons discussed in chapter 1, the time-averaged multiphase model is not appropriate to model rapidly changing phenomena and it does not appear to recover the Rayleigh-type bubble equation. Consequently, this model appears to be inferior to a theory of multiphase mixtures based on the volume averaging approach.

The theory of granular materials of GOODMAN & COWIN (1972) predicts a failure criterion in accordance with the Mohr-Coulomb theory of limiting equilibrium (SOKOLOVSKI,1965). This criterion states that yielding or granular "flow" will occur at a point on a plane element where the shearing stress $S$ exceeds a certain value of the normal stress $N$, with the coefficient of proportionality depending on the cohesion $c$ and the critical *static* angle of internal friction $\Theta_i$ of the bulk solid, *i.e.*

$$S = c - N \tan\Theta_i \tag{7.8.1}$$

The yield criterion (7.8.1) follows from a stress constitutive equation, and for the case of incompressible grains it was shown by COWIN & NUNZIATO (1981) that the GOODMAN and COWIN's theory does not predict the yield criterion in the form of (7.8.1). SAVAGE (1979) has also questioned the general validity of (7.8.1) and notes the difficulties of establishing the parameters $c$ and $\Theta_i$.

The theory of mixtures discussed in previous sections may also be examined for the existence of a yield criterion similar to (7.8.1). Towards this end it is necessary to invoke the equilibrium state of the mixture $\{S_{\beta 0}\}$ and examine the constitutive equations (7.7.1), (7.7.4), (7.7.11), (7.7.12), (7.4.22), (7.5.26), and $(7.6.19)_1$. By combining (7.7.4), (7.7.12) and (7.7.11), and ignoring the thermal effect due to $\tilde{\ell}\epsilon_\alpha$, it follows that the pressures can be expressed as

$$\phi_\alpha(p_\alpha + \bar{\bar{\pi}}_\alpha) = \phi_\alpha\beta_\alpha + 2\bar{\bar{\rho}}_\alpha\phi_\alpha I_\alpha \nabla\phi_\alpha\cdot\nabla\phi_\alpha \tag{7.8.2}$$

which upon substitution into the stress constitutive equation (7.7.1), gives

$$\bar{T}_{\alpha ij} = -(\phi_\alpha\beta_\alpha + 2\bar{\bar{\rho}}_\alpha\phi_\alpha I_\alpha\phi_{\alpha,k}\phi_{\alpha,k})\delta_{ij} - \bar{\bar{\rho}}_\alpha\phi_\alpha I_\alpha\phi_{\alpha,i}\phi_{\alpha,j} \tag{7.8.3}$$

On a plane with normal $\mathbf{n}$, the stress force is given by $\bar{\mathbf{T}}_\alpha\mathbf{n}$, *i.e.*

$$\bar{\mathbf{T}}_\alpha\mathbf{n} = -(\phi_\alpha\beta_\alpha + 2\bar{\bar{\rho}}_\alpha\phi_\alpha I_\alpha \nabla\phi_\alpha\cdot\nabla\phi_\alpha)\mathbf{n} - \bar{\bar{\rho}}_\alpha\phi_\alpha I_\alpha \nabla\phi_\alpha(\nabla\phi_\alpha\cdot\mathbf{n}) \tag{7.8.4}$$

whereas the normal stress $N_\alpha$ is found from

$$N_\alpha = (\bar{\mathbf{T}}_\alpha\mathbf{n})\cdot\mathbf{n} = -(\phi_\alpha\beta_\alpha + 2\bar{\bar{\rho}}_\alpha\phi_\alpha I_\alpha \nabla\phi_\alpha\cdot\nabla\phi_\alpha) - (\bar{\bar{\rho}}_\alpha\phi_\alpha I_\alpha)(\nabla\phi_\alpha\cdot\mathbf{n})^2 \tag{7.8.5}$$

Since the square of the shearing stress $S_\alpha$ and the normal stress $N_\alpha$ is equal to the square of the stress force, we obtain, using (7.8.4),

$$\begin{aligned} S_\alpha^2 + N_\alpha^2 = (\bar{\mathbf{T}}_\alpha\mathbf{n})\cdot(\bar{\mathbf{T}}_\alpha\mathbf{n}) &= (\phi_\alpha\beta_\alpha + 2\bar{\bar{\rho}}_\alpha\phi_\alpha I_\alpha \nabla\phi_\alpha\cdot\nabla\phi_\alpha)^2 \\ &+2(\phi_\alpha\beta_\alpha + 2\bar{\bar{\rho}}_\alpha\phi_\alpha I_\alpha \nabla\phi_\alpha\cdot\nabla\phi_\alpha)\bar{\bar{\rho}}_\alpha\phi_\alpha I_\alpha(\nabla\phi_\alpha\cdot\mathbf{n})^2 \\ &+(\bar{\bar{\rho}}_\alpha\phi_\alpha I_\alpha)^2(\nabla\phi_\alpha\cdot\nabla\phi_\alpha)(\nabla\phi_\alpha\cdot\mathbf{n})^2 \end{aligned} \tag{7.8.6}$$

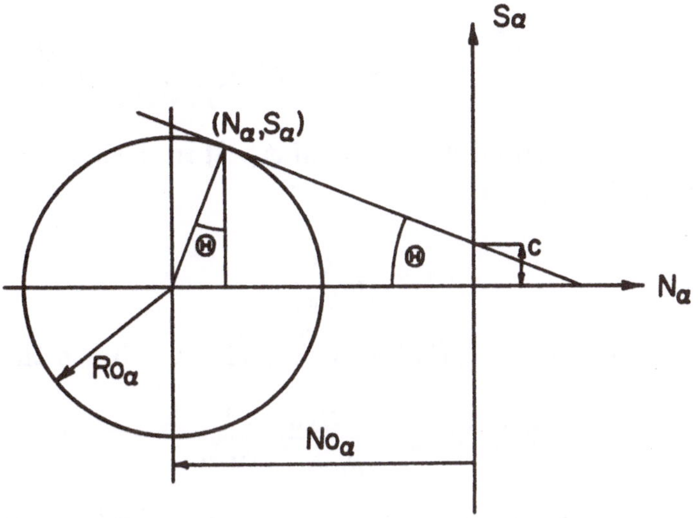


**Figure 7.1: Illustration of Mohr-Coulomb criterion for limiting equilibrium on a Mohr circle diagram**

Completing the square in the above equation, using (7.8.5), and defining

$$No_\alpha = -(\phi_\alpha\beta_\alpha + 2\bar{\bar{\rho}}_\alpha\phi_\alpha I_\alpha \nabla\phi_\alpha \cdot \nabla\phi_\alpha) \tag{7.8.7}$$

$$Ro_\alpha^2 = (\bar{\bar{\rho}}_\alpha\phi_\alpha I_\alpha)^2(\nabla\phi_\alpha \cdot \nabla\phi_\alpha)(\nabla\phi_\alpha \cdot \mathbf{n})^2 \tag{7.8.8}$$

gives

$$S_\alpha^2 + (N_\alpha - No_\alpha)^2 = Ro_\alpha^2 \tag{7.8.9}$$

a shear/normal stress relation suitable for representation on a Mohr circle as shown in Figure 7.1. This circle consists of radius $Ro_\alpha$ and it is located on a horizontal (normal stress) axis at a position $No_\alpha$ units displaced from the origin. To prove the existence of the traditional Mohr-Coulomb relation (7.8.1) it is necessary to eliminate $\beta_\alpha$ in (7.8.7) using (7.4.22), (7.5.26), and $(7.6.19)_1$, *i.e.*

$$\beta_\alpha = \bar{\bar{\rho}}_\alpha\phi_\alpha\left(\frac{\partial\tilde{\bar{\psi}}_\alpha}{\partial\phi_\alpha} + \frac{\partial I_\alpha}{\partial\phi_\alpha}\nabla\phi_\alpha \cdot \nabla\phi_\alpha\right) \tag{7.8.10}$$

Substituting this result into (7.8.7) yields

$$No_\alpha = -\bar{\bar{\rho}}_\alpha\phi_\alpha\left[\phi_\alpha\frac{\partial\tilde{\bar{\psi}}_\alpha}{\partial\phi_\alpha} + \nabla\phi_\alpha \cdot \nabla\phi_\alpha\left(\phi_\alpha\frac{\partial I_\alpha}{\partial\phi_\alpha} + 2I_\alpha\right)\right] \tag{7.8.11}$$

Eliminating $\nabla\phi_\alpha \cdot \nabla\phi_\alpha$ using (7.8.8) gives

$$No_\alpha = -\bar{\bar{\rho}}_\alpha \phi_\alpha [\phi_\alpha \frac{\partial \tilde{\bar{\psi}}_\alpha}{\partial \phi_\alpha} + (\phi_\alpha \frac{\partial I_\alpha}{\partial \phi_\alpha} + 2I_\alpha)(\frac{Ro_\alpha}{\bar{\bar{\rho}}_\alpha \phi_\alpha I_\alpha})^2 / (\frac{\partial \phi_\alpha}{\partial n})^2] \quad (7.8.12)$$

where $\partial\phi_\alpha/\partial n$ is the normal derivative of $\phi_\alpha$. If $\mathbf{m}$ designates a unit vector in the direction of $\nabla\phi_\alpha$, *i.e.*

$$\mathbf{m} = \frac{\nabla\phi_\alpha}{|\nabla\phi_\alpha|} \quad (7.8.13)$$

then using this equation and (7.8.8), it can also be shown that

$$\nabla\phi_\alpha \cdot \nabla\phi_\alpha = \frac{Ro_\alpha}{\bar{\bar{\rho}}_\alpha \phi_\alpha I_\alpha} \frac{1}{\mathbf{m}\cdot\mathbf{n}} \quad (7.8.14)$$

such that (7.8.11) has an equivalent expression of the following form:

$$No_\alpha = -\bar{\bar{\rho}}_\alpha \phi_\alpha [\phi_\alpha \frac{\partial \tilde{\bar{\psi}}_\alpha}{\partial \phi_\alpha} + (\phi_\alpha \frac{\partial I_\alpha}{\partial \phi_\alpha} + 2I_\alpha) \frac{Ro_\alpha}{\bar{\bar{\rho}}_\alpha \phi_\alpha I_\alpha} \frac{1}{\mathbf{m}\cdot\mathbf{n}}] \quad (7.8.15)$$

or

$$Ro_\alpha = -No_\alpha \, sin\Theta + c \, cos\Theta \quad (7.8.16)$$

where

$$sin\Theta = \frac{I_\alpha \, \mathbf{m}\cdot\mathbf{n}}{\phi_\alpha \frac{\partial I_\alpha}{\partial \phi_\alpha} + 2I_\alpha}, \qquad c \, cos\Theta = -\bar{\bar{\rho}}_\alpha \phi_\alpha^2 \frac{\tilde{\bar{\psi}}_\alpha}{\partial \phi_\alpha} sin\Theta \quad (7.8.17)$$

If $\mathbf{m}\cdot\mathbf{n}$ is independent of $Ro_\alpha$, (7.8.16) can be represented in terms of $S_\alpha$ and $N_\alpha$ by the use of formulas

$$Ro_\alpha = \frac{S_\alpha}{cos\Theta}, \qquad tan\Theta = -\frac{No_\alpha - N_\alpha}{S_\alpha} \quad (7.8.18)$$

as shown in Figure 7.1 and which satisfy (7.8.9). Thus, it follows that

$$S_\alpha = c - N_\alpha \, tan\Theta \quad (7.8.19)$$

which is similar to the traditional Mohr-Coulomb yield criterion (7.8.1). The condition that $\mathbf{m}\cdot\mathbf{n}$ be independent of $Ro_\alpha$ may occur when the volumetric fraction gradient is normal ($\mathbf{m}\cdot\mathbf{n} = 1$), irrespective of whether or not the phase is compressible or incompressible. The present theory of mixtures contains, therefore, the Mohr-Coulomb yield-type criterion only under special circumstances of *normal-volumetric-fraction-gradient.*

The yield criterion (7.8.19) should not be interpreted as an equation of static equilibrium of the mixture (for it is the product of a rate-dependent theory), but rather as a limiting form of the stress tensor (7.7.1) when the medium has just come to rest ($\mathbf{D} \to \mathbf{0}$). That is, the condition (7.8.19) is too special to represent the static equilibrium of mixtures in general, since this state should also be represented by the elastic properties of the medium (see chapter 6). As SAVAGE (1979) has also pointed out, the constitutive equations suitable for flowing materials need not necessarily be appropriate to describe the states of static equilibrium, as in the case of most theories of liquids which are inappropriate for modeling the solids. The constitutive equations can be, of course, constructed in terms of rate-dependent and rate-independent constitutive variables as discussed in chapter 6, but only at the expense of greater complexity of the theory. For these reasons, it is only appropriate to consider the validity of the rate-dependent theory of this chapter for nonzero flow regimes where the dissipation of mechanical energy is by particle-particle collision rather than by particle-particle rubbing, with the angles $\Theta_i$ and $\Theta$ in equations (7.8.1) and (7.8.19) being different. HANES & INMAN (1985a,b) associate $\Theta$ with the critical *dynamic* angle of internal friction which, as shown by BAGNOLD (1956), is less than the static angle $\Theta_i$ by a few degrees. The granular experiments of HANES and INMAN involving glass beads and sand in air and water not only demonstrate the dependence of the dynamic angle of internal friction on the mean particle diameter, density, and interstitial fluid, but also on the *particle concentration at the yield plane* of the flow. In these rheological experiments, the particle concentration varied linearly across the granular shear region and approached a constant value below the yield plane, where no flow of granules occurred. However, no particle concentration gradients at the yield plane were measured to justify the prediction of (7.8.19).

The experiments of BAGNOLD (1956), SAVAGE and coworkers (1979, 1983,1984), and HANES & INMAN (1985a,b) show that at high shear rates, or grain inertia regime, the stresses in granular flow are primarily transmitted by the intergranular collisions and depend quadratically on the mean shear rate. At low shear rates or in a macroviscous flow regime, however, the stresses are primarily transmitted by the interstitial fluid and depend linearly on the mean shear rate or velocity gradient and properties of the fluid. In an attempt to account for these exprimental observations, SAVAGE & COWIN (SAVAGE,1979) extended the theory of GOODMAN & COWIN (1971,1972) to incorporate the nonlinear shear-rate effects through the dependence of the coefficients of independent variables in the constitutive equation for shear stress on the invariants of these variables. More recent models of granular flow are, however, constructed using the kinetic theory model, whereby the field and constitutive equations are derived for idealized granular materials consisting of uniform, rough, and inelastic spherical particles with binary collision modeling (SAVAGE & JEFFREY,1981; JENKINS & SAVAGE,1983; LUN *et al.*,1984; LUN & SAVAGE,1987; JENKINS & MANCINI,1987). In these theories, the kinetic energy of random motion of grains plays the role of internal energy with which is associated a "grain temperature", whereas the effect of incorporating the rotary particle inertia into the theory results in producing a non-symmetric stress tensor (LUN & SAVAGE,1987). Although the kinetic theory results may adequately represent short time duration collisions between particles, they do not account for long time frictional contacts which can occur even in rapid granular flows in the vicinity of a solid boundary. JOHNSON & JACKSON (1987) allowed, therefore, for the superposition of these short and long time effects and proposed constitutive equations for shear stress, total energy, and the total heat flux to be used in the momentum and energy field equations.

The kinetic theory or, more generally, the statistical mechanics has a potential of producing the macroscopic field equations and clearly identify the transport properties and phenomenological coefficients in the constitutive equations. For dense multiphase suspensions, the binary collision modeling is not adequate and should involve multiple collisions between grains and account for the particle nonuniformity and nonrigidity. At the present, the kinetic theory is useful in providing detailed information for very *idealized* granular flows which exclude the interstitial fluid from playing a significant role.

A fully linear theory of constitutive equations is, in general, not adequate for modeling multiphase flows and can be extended to account for nonlinear phenomena by using the tensor representation theorems of section 4 of the appendix. The simplest extension of the linear theory is to allow the phenomenological coefficients in the constitutive equations to depend on the invariants of independent variables. Thus, in the equation for stress tensor (7.7.1), the set of phenomenological coefficients

$$\{S_\alpha\} = \{I_\alpha, \lambda_{\alpha\alpha}, \mu_{\alpha\alpha}, O_{\alpha\alpha}\} \tag{7.8.20}$$

can be allowed to depend on the following invariants of independent constitutive variables:

$$\{S_\alpha\} = \{\phi_\beta, \bar{\bar{\theta}}_\beta, \grave{\phi}_\alpha, |\nabla\phi_\alpha|, tr\mathbf{D}_\alpha, tr\mathbf{D}_\alpha^2, tr[(\nabla\phi_\alpha)^2\mathbf{D}_\alpha], tr[(\nabla\phi_\alpha)^2\mathbf{D}_\alpha^2]\} \tag{7.8.21}$$

producing in this manner a nonlinear dependence of the stress tensor on the strain rate. Further nonlinear effects into the stress tensor can be introduced by such terms as $(\nabla\phi_\alpha)^2\mathbf{D}_\alpha$ and $\mathbf{D}_\alpha^2$ from the full tensor representation form of (7.7.1), using equation (A.42), or more generally by using equations (A.42) and (A.44) of the appendix.

The solutions of conservation and balance equations depend on the prescribed initial and boundary conditions. By considering a simple granular flow it is apparent from the particles' point of view that they may stick or slip relative to a surface. The continuum theory of mixtures does not deal with the details of granules close to the surface and may need a reconciliation with the particles' point of view in order to supply physically reasonable solutions. HUTTER *et al.* (1986a,b), using the modeling equations of JENKINS & SAVAGE (1983), studied numerically chute flows with different types of assumed boundary conditions and showed that some of these conditions may not even produce solutions, let alone produce physically meaningful results.

Another characteristic of a rate-dependent theory is that it cannot model fast waves associated with the elasticity of the granular material, but only slower waves associated with the dilatancy (changes in the volumetric fraction) of the medium. As noted by COWIN & NUNZIATO (1981), dilatant waves have been observed by a number of investigators, such as in the gravity flow of dry powders and bulk solids, fluidized beds, *etc.*

## 7.9 Application of the Theory to Shearing Flow of a Fluid-Saturated Granular Material

Fluid-saturated granular materials occur naturally in many situations where the suspension of particles may be dilute or concentrated. In this section we will examine a shearing flow whereby the motion of a suspension is achieved by placing it between two parallel plates, one of which is held stationary and the other of which is moved at a constant speed (Figure 7.2). In a single phase Newtonian flow situation, the velocity profile between the plates is linear, whereas for the suspension flows this is not true, even if considerable simplifications are made in the field and constitutive equations as will be shown below.

For an isothermal, saturated, incompressible solid and fluid phases, and inertionless ($\tilde{i}_\alpha = 0,\ \alpha = s, f$) granular material it follows from (7.7.30) and (7.7.31) that

$$\frac{\partial\phi_s}{\partial t} + \nabla\cdot(\phi_s\tilde{\mathbf{v}}_s) = 0 \tag{7.9.1}$$

$$\frac{\partial\phi_f}{\partial t} + \nabla\cdot(\phi_f\tilde{\mathbf{v}}_f) = 0 \tag{7.9.2}$$

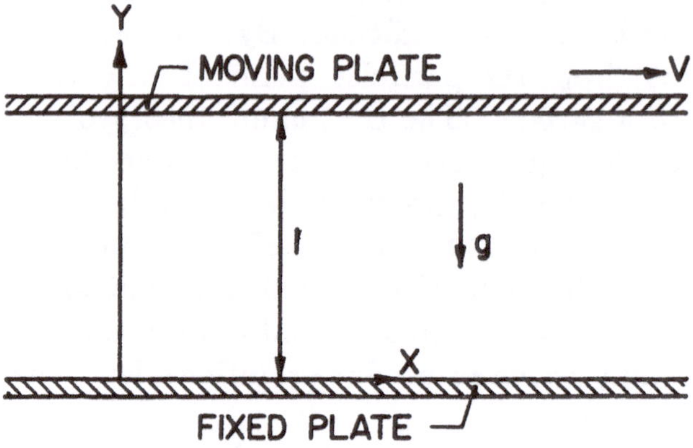


Figure 7.2: Geometry of a two-phase shear flow between parallel plates

$$\phi_s + \phi_f = 1 \tag{7.9.3}$$

whereas from (7.7.37) and (7.7.32) it follows, respectively, that

$$P_c = \beta_s - \beta_f = p_s - p_f + (2(\alpha_s + \alpha_f)\phi_{s,m})_{,m} \tag{7.9.4}$$

$$\pi = -P_f = -p_f + \beta_f - (2\alpha_f\phi_{f,m})_{,m} \tag{7.9.5}$$

where it is assumed that

$$H_{\alpha\beta} = 0, \quad O_{\alpha\alpha} = 0; \quad \alpha, \beta = s, f \tag{7.9.6}$$

which is consistent with $(7.6.32)_2$ and (7.7.59). In (7.9.4) and (7.9.5) the coefficients $\alpha_s$ and $\alpha_f$ are defined as follows

$$2\alpha_s = A_s, \qquad 2\alpha_f = A_f \tag{7.9.7}$$

and $P_c = \beta_s - \beta_f$ represents the *capillary pressure*. The momentum equation (7.7.24) is reduced to

$$\bar{\rho}_s \grave{\bar{v}}_{si} = \bar{T}_{sij,j} + \bar{\rho}_s \bar{b}_{si} + \bar{t}_{si} \tag{7.9.8}$$

$$\bar{\rho}_f \grave{\bar{v}}_{fi} = \bar{T}_{fij,j} + \bar{\rho}_f \bar{b}_{fi} + \bar{t}_{fi} \tag{7.9.9}$$

where the constitutive equations for stresses are obtained from (7.7.1), and interphase momentum sources from (7.7.6), *i.e.*

$$\bar{T}_{sij} = -\phi_s p_s \delta_{ij} - 2\alpha_s \phi_{s,i}\phi_{s,j} + 2\phi_s \mu_s D_{sij} \tag{7.9.10}$$

$$\bar{T}_{fij} = -\phi_f p_f \delta_{ij} - 2\alpha_f \phi_{f,i}\phi_{f,j} + 2\phi_f \mu_f D_{fij} \tag{7.9.11}$$

$$\bar{t}_{si} = -\pi\, \phi_{s,i} - D(\tilde{v}_{si} - \tilde{v}_{fi}) \tag{7.9.12}$$

$$\bar{t}_{fi} = -\pi\, \phi_{f,i} - D(\tilde{v}_{fi} - \tilde{v}_{si}) \tag{7.9.13}$$

The above equations contain the assumptions of

$$\lambda_{\alpha\alpha} = 0, \qquad \mu_{\alpha\alpha} = \phi_\alpha \mu_\alpha; \qquad \alpha = s, f \tag{7.9.14}$$

$$M_{sf} - M_{ss} = -\pi \tag{7.9.15}$$

$$M_{ff} - M_{fs} = \pi \tag{7.9.16}$$

whereas the drag coefficient $D$ is positive as required by $(7.6.34)_{1,4}$, *i.e.*

$$D = \xi_{ss} = -\xi_{fs} \geq 0 \tag{7.9.17}$$

Combining the above field and constitutive equations produces the following modeling equations:

$$\frac{\partial \phi_s}{\partial t} + \nabla\cdot(\phi_s \tilde{\mathbf{v}}_s) = 0 \tag{7.9.1}$$

$$\frac{\partial \phi_f}{\partial t} + \nabla\cdot(\phi_f \tilde{\mathbf{v}}_f) = 0 \tag{7.9.2}$$

$$\phi_s + \phi_f = 1 \tag{7.9.3}$$

$$P_c = 2\nabla\cdot[(\alpha_s + \alpha_f)\nabla\phi_s] + p_s - p_f \tag{7.9.4}$$

$$\pi = -p_f - 2\nabla\cdot(\alpha_f \nabla\phi_f) + \beta_f \tag{7.9.5}$$

$$\begin{aligned} \bar{\rho}_s \dot{\tilde{\mathbf{v}}}_s = {} & -\nabla(\phi_s p_s) - 2\nabla\cdot(\alpha_s \nabla\phi_s \nabla\phi_s) + \nabla\cdot[\phi_s \mu_s(\nabla\tilde{\mathbf{v}}_s + \nabla\tilde{\mathbf{v}}_s^T)] \\ & + \bar{\rho}_s \tilde{\mathbf{b}}_s - \pi\nabla\phi_s - D(\tilde{\mathbf{v}}_s - \tilde{\mathbf{v}}_f) \end{aligned} \tag{7.9.18}$$

$$\begin{aligned} \bar{\rho}_f \dot{\tilde{\mathbf{v}}}_f = {} & -\nabla(\phi_f p_f) - 2\nabla\cdot(\alpha_f \nabla\phi_f \nabla\phi_f) + \nabla\cdot[\phi_f \mu_f(\nabla\tilde{\mathbf{v}}_f + \nabla\tilde{\mathbf{v}}_f^T)] \\ & + \bar{\rho}_f \tilde{\mathbf{b}}_f - \pi\nabla\phi_f - D(\tilde{\mathbf{v}}_f - \tilde{\mathbf{v}}_s) \end{aligned} \tag{7.9.19}$$

$$\bar{\rho}_s = \phi_s \bar{\bar{\rho}}_s, \quad \bar{\rho}_f = \phi_f \bar{\bar{\rho}}_f, \quad \bar{\bar{\rho}}_s = const., \quad \bar{\bar{\rho}}_f = const. \tag{7.9.20}$$

Equations (7.9.1)-(7.9.5) and (7.9.18)-(7.9.20) are formally identical to the equations proposed by PASSMAN & NUNZIATO (1981) to study flows of fluid-saturated granular materials. They also assumed that $\mu_s(\phi_s)$, $\mu_f(\phi_f)$, $P_c(\phi_f)$, and $\beta_f(\phi_f)$, which may only be true in special circumstances in the flow of granular materials. The viscosity of the solid is assumed to be governed by the following expression:

$$\mu_s = \frac{\bar{\mu}_s}{(\phi_m - \phi_s)^2} \tag{7.9.21}$$

where $\bar{\mu}_s$ =constant and $\mu_m$ denotes the maximum value of $\phi_s$, namely $\phi_m = 0.74$ for a hexagonal close-packed configuration of solids. The constitutive equation (7.9.21) has the property that $\mu_s \to \infty$ as $\phi_s \to \phi_m$, which can be physically visualized as representing a "locking" phenomenon. As the particles come close together they become more closely interlocked, so that it becomes more difficult to move one particle relative to the other. SAVAGE (1979) also proposed a constitutive equation for $\mu_s$ which is similar to (7.9.21).

Assuming that $\mu_f$, $\alpha_s$, $\alpha_f$, and $D$ are constant and that the capillary effects are negligible ($\beta_f = P_c = 0$), PASSMAN & NUNZIATO (1981) used the above modeling equations to analyze a shearing flow as illustrated in Figure 7.2, with solid and fluid velocities given by

$$\tilde{\mathbf{v}}_s = (v_s(y), 0, 0), \quad \tilde{\mathbf{v}}_f = (v_f(y), 0, 0) \tag{7.9.22}$$

and

$$\phi_s,\ \phi_f,\ p_s,\ p_f,\ \textit{and}\ \pi\ \textit{as functions of}\ y\ \textit{only} \tag{7.9.23}$$

The distance between the plates is $\ell$ and the flow is in the $x - y$ plane. The equations governing this flow situation are, therefore, as follows:

$$\phi_s + \phi_f = 1 \tag{7.9.24}$$

$$2(\alpha_s + \alpha_f)\frac{d^2\phi_s}{dy^2} + p_s - p_f = 0 \tag{7.9.25}$$

$$\pi = -p_f - 2\alpha_f \frac{d^2\phi_f}{dy^2} \tag{7.9.26}$$

$$\frac{d}{dy}[2\alpha_s(\frac{d\phi_s}{dy})^2 + \phi_s p_s] + \pi\frac{d\phi_s}{dy} + g\phi_s\bar{\bar{\rho}}_s = 0 \tag{7.9.27}$$

$$\frac{d}{dy}[2\alpha_f(\frac{d\phi_f}{dy})^2 + \phi_f p_f] + \pi\frac{d\phi_f}{dy} + g\phi_f\bar{\bar{\rho}}_f = 0 \tag{7.9.28}$$

$$\frac{d}{dy}(\phi_s\mu_s\frac{dv_s}{dy}) - D(v_s - v_f) = 0 \tag{7.9.29}$$

$$\frac{d}{dy}(\phi_f\mu_f\frac{dv_f}{dy}) - D(v_f - v_s) = 0 \tag{7.9.30}$$

with $\mu_s$ given by (7.9.21)

The seven modeling equations (7.9.24)-(7.9.30) contain seven unknowns $\phi_s$, $\phi_f$, $\pi$, $p_s$, $p_f$, $v_s$, and $v_f$, and require boundary conditions for a solution. PASSMAN and NUNZIATO chose to specify different values of solid fractions

$$\phi_s(0), \quad \phi_s(\ell) \tag{7.9.31}$$

and for the velocities assumed the no-slip condition, *i.e.*

$$v_s(0) = 0, \quad v_s(\ell) = 1, \quad v_f(0) = 0, \quad v_f(\ell) = 1 \tag{7.9.32}$$

The no-slip boundary condition may not be appropriate, and by imposing a slip condition is equivalent to adding a rigid motion to the solution. The boundary conditions for pressures $p_s$ and $p_f$ may be determined from (7.9.10) and (7.9.11),

$$\bar{T}_{syy} = -\phi_s p_s - 2\alpha_s(\frac{d\phi_s}{dy})^2 \tag{7.9.33}$$

$$\bar{T}_{fyy} = -\phi_f p_f - 2\alpha_f(\frac{d\phi_f}{dy})^2 \tag{7.9.34}$$

by taking

$$\bar{T}_{syy}(\ell) = \bar{T}_{fyy}(\ell) = 0 \tag{7.9.35}$$

Using the material parameters for a sand-water mixture as given in Table 7.1, PASSMAN & NUNZIATO (1981) numerically solved (7.9.24)-(7.9.30) subject to (7.9.21), boundary conditions (7.9.31)-(7.9.32) and (7.9.35), obtaining the results as illustrted in Figures 7.3-7.6.

Figures 7.3 and 7.4 show distributions of the solid fraction $\phi_s$ and velocity $v_s$ for the case of $\phi_s(0) = \phi_s(\ell) = 0.4$ and $\ell = 1cm$. With these boundary conditions the particles are shown to be perfectly entrained in the flow ($v_s = v_f$) and there are high shear rates close to the walls, whereas

| | $\bar{\bar{\rho}}$ $\frac{g}{cm^3}$ | $\alpha$ $\frac{g\ cm}{s^2}$ | $\bar{\mu}$ $\frac{g}{cm\ s}$ | $\phi_m$ | $D$ $\frac{g}{cm^3 s}$ | $g$ $\frac{cm}{s^2}$ |
|---|---|---|---|---|---|---|
| Sand (s) | 2.2 | 4.00 | 7230 | 0.74 | | |
| Water (f) | 1.0 | 2.92 | 0.01 | - | | |
| Interaction | | | | | 112 | 980 |

**Table 7.1: Material parameters for a sand-water mixture**

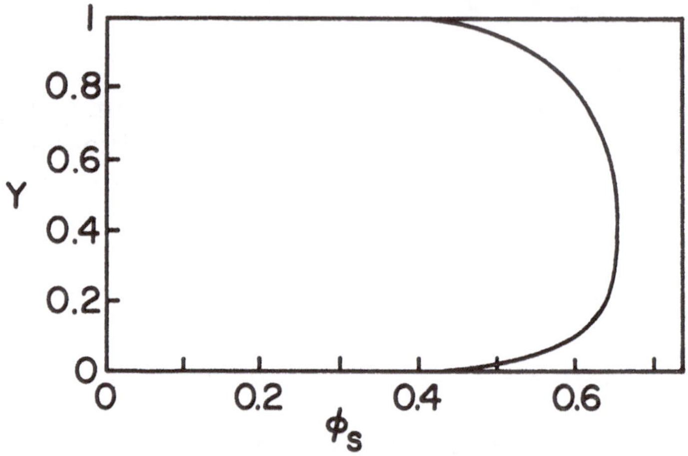


**Figure 7.3: Distribution of the solid fraction $\phi_s$ for the case $\phi_s(0) = 0.4$, $\phi_s(\ell) = 0.4$, $\ell = 1cm$.**

the material near the middle of the channel is relatively undisturbed. The solid volume fraction exhibits the "dilatancy" effect, since it exhibits high gradients and low values in the regions of high shear, whereas it is essentially constant in the low shear region. The skewness of the graphs about the centerline comes about due to the effect of the gravity force. The solutions for $\phi_s$ and $v_s$ with $\phi_s(\ell) = 0.2$, $\phi_s(0) = \phi_m = 0.74$, and $\ell = 0.5cm$ are illustrated in Figures 7.5 and 7.6. The solid fraction distribution becomes almost linear, with high concentration values or "locking" phenomenon at the bottom and producing almost a rigid body motion. The results presented in Figures 7.3-7.6 also appear to agree qualitatively with experiments (PASSMAN & NUNZIATO,1981).

## 7.10 Concluding Remarks

The linear theory of constitutive equations involving the deformation postulate (6.12.36) without the rotational effect and using the field equations of chapters 2 and 3, reproduces many physical results and allows for many

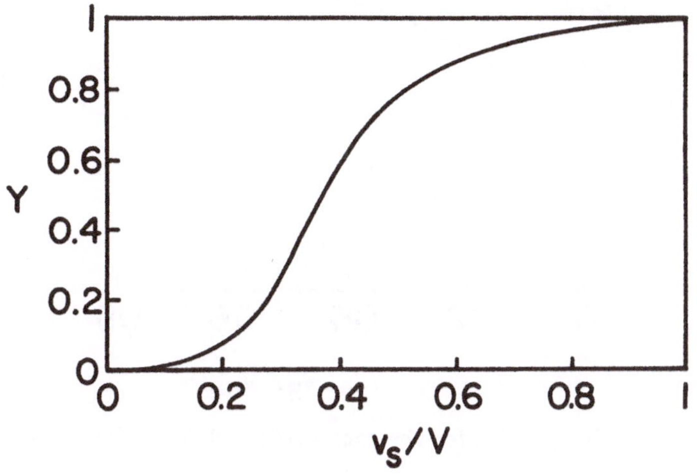


Figure 7.4: Velocity profiles for the case $\phi_s(0) = \phi_s(\ell) = 0.4$, $\ell = 1cm$.

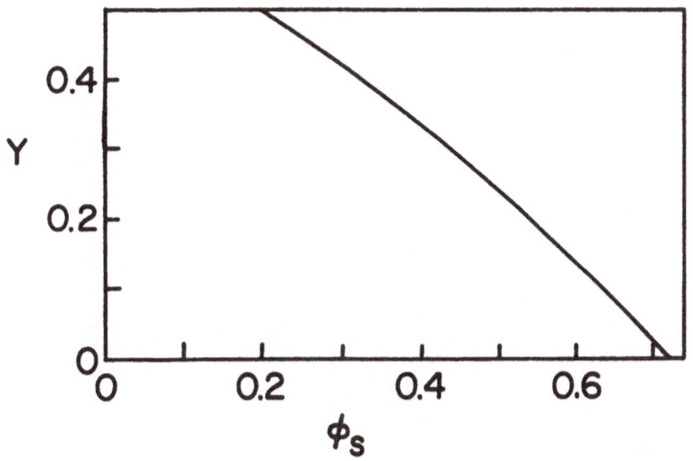


Figure 7.5: Distribution of the solid fraction $\phi_s$ for the case $\phi_s(0) = 0.74$, $\phi_s(\ell) = 0.2$, $\ell = 0.5cm$.

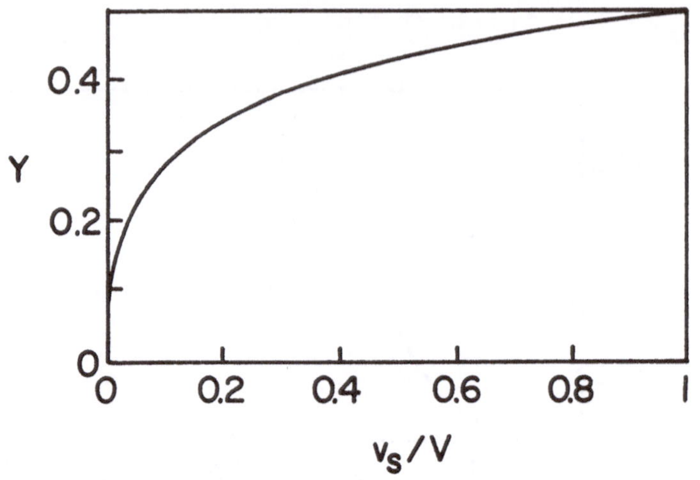


Figure 7.6: Velocity profiles for the case $\phi_s(0) = 0.74$, $\phi_s(\ell) = 0.2$, $\ell = 0.5cm$.

structural effects in the mixture. The simplest extension of the linear constitutive theory is to allow for the coefficients multiplying independent variables to depend on the invariants of these variables and selectively introduce nonlinear terms in the constitutive equations as it becomes appropriate. In the following chapter, considerations are given to more complete modeling of the structured multiphase mixtures. This modeling involves the construction of constitutive equations with dilatation and rotation.

CHAPTER 8

# MIXTURES WITH ROTATION AND DILATATION

## 8.1 Purpose of the Chapter

This chapter provides a continuation of chapter 7 dealing with the development of constitutive equations of multiphase mixtures with structure. Considerations are given to mixtures with rotational and dilatational effects relative to the center of mass of each phase by utilizing the isotropy assumption of the inertia tensor (5.1.9) and deformation assumption (6.12.36). As such, the chapter forms a starting point for the development of more complete models of fluidlike and solidlike mixtures with linear and nonlinear constitutive equations, studies involving wave propagation, and applications of modeling equations to different problems.

## 8.2 Constitutive Equations for Mixtures of Isotropic Fluids

Equations (2.4.14), (3.4.2), (7.2.3), (7.2.8), (7.2.12), and (7.2.16) provide a set of $13\gamma$ algebraic equations for $13\gamma$ unknowns

$$\bar{\rho}_\alpha,\ \tilde{v}_{\alpha i},\ \tilde{b}_{\alpha i}\ (or\ \tilde{\ell}_{\alpha[jk]}),\ \phi_\alpha,\ \mu_i^{(\alpha)},\ \bar{\bar{\theta}}_\alpha,\ \tilde{\imath}_\alpha$$

With the variables $\tilde{r}_\alpha$ and $\tilde{\ell}_{\alpha[jk]}$ (or $\tilde{b}_{\alpha i}$) assumed to be given, it is sufficient to determine the remaining variables

$$\Upsilon_\alpha = (\hat{c}_\alpha,\ \bar{\mathbf{T}}_\alpha,\ \hat{\mathbf{M}}_\alpha,\ \bar{\mathbf{q}}_\alpha,\ \bar{s}_\alpha,\ \tilde{\psi}_\alpha,\ \hat{s}_\alpha,\ \hat{\mathbf{i}}_\alpha,\ \mathbf{U}_\alpha,\ \bar{\boldsymbol{\lambda}}_\alpha,\ \bar{\mathbf{S}}_\alpha,\ \bar{\mathbf{t}}_\alpha,\ \tilde{\boldsymbol{\ell}}\epsilon_\alpha,\ \hat{\hat{\epsilon}}_\alpha,\ \bar{q}_{s\alpha})$$
$$\alpha = 1, \ldots, \gamma \tag{8.2.1}$$

by the constitutive equations which are restricted by the second law of thermodynamics. As in section (7.4) and based on the developments in chapter 6, it will be assumed that the mixture properties satisfy the principles of local action and smooth and local memory, and that the independent variables in the constitutive equations are

$$\Upsilon_\alpha = G_{\alpha\boldsymbol{\kappa}}[\bar{\bar{\rho}}_\beta,\ \nabla\bar{\bar{\rho}}_\beta,\ \mathbf{D}_\beta,\ \mathbf{W}_\beta - \mathbf{W}_\gamma,\ \bar{\bar{\theta}}_\beta,\ \nabla\bar{\bar{\theta}}_\beta,\ \tilde{\mathbf{v}}_\beta - \tilde{\mathbf{v}}_\gamma,\ \tilde{\imath}_\beta,\ \phi_\beta,$$
$$\nabla\phi_\beta,\ \dot{\phi}_\beta,\ \mathbf{b}^{(\beta)},\ \nabla\hat{\boldsymbol{\nu}}^{(\beta)}] = G_{\alpha\boldsymbol{\kappa}}[\{S_\beta\}] \tag{8.2.2}$$

where for each $\alpha$, $\alpha=1,\ldots,\gamma$, $\beta=1,\ldots,\gamma$. The constitutive assumption (8.2.2) is, therefore, an extension of (7.4.3) that includes the additional effects of rotational motion relative to the center of mass of each phase as specified by $\mathbf{b}^{(\beta)}$ and $\nabla\hat{\boldsymbol{\nu}}^{(\beta)}$.

The energy equation (7.2.16) can be further reduced by adding to the right side

$$\frac{1}{2}(\hat{\nu}_{im}^{(\alpha)}\hat{\nu}_{im}^{(\alpha)}+(\frac{\grave{\phi}_\alpha}{\phi_\alpha})^2)$$

multiplied by equation (7.2.3) and

$$(\hat{\nu}_{kj}^{(\alpha)}+\frac{\grave{\phi}_\alpha}{\phi_\alpha}\delta_{kj})$$

multiplied by equation (3.2.13), and using the inertia tensor isotropy (5.1.9) and deformation assumption (6.12.36). The result of this algebraic manipulation is

$$\begin{aligned}
\bar{\rho}_\alpha\grave{\bar{\epsilon}}_\alpha &= -\bar{q}_{\alpha k,k}+\bar{\rho}_\alpha\tilde{r}_\alpha-\hat{c}_\alpha(\tilde{\epsilon}_\alpha-\hat{\tilde{\epsilon}}_\alpha)-\bar{q}_{s\alpha}+\hat{\nu}_{mj,i}^{(\alpha)}\bar{\lambda}_{\alpha jmi}+(\frac{\grave{\phi}_\alpha}{\phi_\alpha})_{,i}\bar{\lambda}_{\alpha jji}\\
&-\hat{\nu}_{im}^{(\alpha)}(\bar{\rho}_\alpha\tilde{\ell}\epsilon_{am})_{,i}-\hat{\nu}_{im,i}^{(\alpha)}\bar{\rho}_\alpha\tilde{\ell}\epsilon_{am}-(\frac{\grave{\phi}_\alpha}{\phi_\alpha}\bar{\rho}_\alpha\tilde{\ell}\epsilon_{ai})_{,i}+(\frac{\grave{\phi}_\alpha}{\phi_\alpha})^2U_{\alpha mnj}\hat{\nu}_{mn,j}^{(\alpha)}\\
&+\bar{T}_{\alpha ij}\bar{v}_{\alpha ij}+\frac{1}{2}\hat{c}_\alpha\hat{\nu}_{im}^{(\alpha)}\hat{\nu}_{iq}^{(\alpha)}(\hat{i}_{\alpha mq})_{q\neq m}-\frac{1}{2}(\hat{\nu}_{im}^{(\alpha)}\hat{\nu}_{in}^{(\alpha)}\hat{\nu}_{\ell q}^{(\alpha)})_{,\ell}U_{\alpha mnq}\\
&-\frac{1}{2}(\frac{\grave{\phi}_\alpha}{\phi_\alpha}\hat{\nu}_{im}^{(\alpha)}\hat{\nu}_{in}^{(\alpha)}U_{\alpha mn\ell})_{,\ell}+\frac{1}{2}\hat{\nu}_{im}^{(\alpha)}\hat{\nu}_{im}^{(\alpha)}\hat{\nu}_{\ell n,\ell}^{(\alpha)}U_{\alpha jjn}+\frac{1}{2}\hat{\nu}_{im}^{(\alpha)}\hat{\nu}_{im}^{(\alpha)}(\frac{\grave{\phi}_\alpha}{\phi_\alpha}U_{\alpha jj\ell})_{,\ell}\\
&+\hat{\nu}_{im}^{(\alpha)}\hat{\nu}_{in,j}^{(\alpha)}\hat{\nu}_{j\ell}^{(\alpha)}U_{\alpha mn\ell}+\hat{\nu}_{im}^{(\alpha)}(\frac{\grave{\phi}_\alpha}{\phi_\alpha})_{,j}\hat{\nu}_{j\ell}^{(\alpha)}U_{\alpha mi\ell}+\hat{\nu}_{im}^{(\alpha)}\hat{\nu}_{in,j}^{(\alpha)}\frac{\grave{\phi}_\alpha}{\phi_\alpha}U_{\alpha mnj}\\
&+\hat{\nu}_{im}^{(\alpha)}(\frac{\grave{\phi}_\alpha}{\phi_\alpha})_{,j}\frac{\grave{\phi}_\alpha}{\phi_\alpha}U_{\alpha mij}+\frac{\grave{\phi}_\alpha}{\phi_\alpha}\hat{\nu}_{ab,i}^{(\alpha)}\hat{\nu}_{im}^{(\alpha)}U_{\alpha abm}
\end{aligned}\tag{8.2.3}$$

The condition $(5.2.9)_2$ and frame-indifference requirement (4.3.49)

$$\hat{\nu}_{im}^{(\alpha)}U_{\alpha k\ell i,m}=0\quad\rightarrow\quad U_{\alpha k\ell i,m}=U_{\alpha k\ell m,i}\tag{8.2.4}$$

require that

$$\hat{c}_\alpha\hat{i}_{\alpha k\ell}-[\hat{\nu}_{mi,m}^{(\alpha)}+(\frac{\grave{\phi}_\alpha}{\phi_\alpha})_{,m}\delta_{mi}]U_{\alpha k\ell i}-\frac{\grave{\phi}_\alpha}{\phi_\alpha}U_{\alpha k\ell i,i}=0,\quad k\neq\ell\tag{8.2.5}$$

from where follow the sufficient conditions

$$\hat{i}_{\alpha k\ell} = \hat{i}_{\alpha ii}\delta_{k\ell}, \qquad U_{\alpha k\ell i} = U_{\alpha jji}\delta_{k\ell} \tag{8.2.6}$$

such that the energy equation (8.2.3) is reduced to

$$\bar{\rho}_\alpha \grave{\tilde{\epsilon}}_\alpha = -\bar{q}_{\alpha k,k} + \bar{\rho}_\alpha \bar{r}_\alpha - \hat{c}_\alpha(\tilde{\epsilon}_\alpha - \hat{\tilde{\epsilon}}_\alpha) - \bar{q}_{s\alpha} + \hat{\nu}^{(\alpha)}_{mj,i}\bar{\lambda}_{\alpha jmi} - \hat{\nu}^{(\alpha)}_{im,i}\bar{\rho}_\alpha \tilde{\ell}\epsilon_{\alpha m}$$
$$+(\frac{\grave{\phi}_\alpha}{\phi_\alpha})_{,i}\bar{\lambda}_{\alpha jji} - (\frac{\grave{\phi}_\alpha}{\phi_\alpha}\bar{\rho}_\alpha\tilde{\ell}\epsilon_{\alpha i})_{,i} + \bar{T}_{\alpha ij}\tilde{v}_{\alpha i,j} - \hat{\nu}^{(\alpha)}_{im}(\bar{\rho}_\alpha\tilde{\ell}\epsilon_{\alpha m})_{,i} \tag{8.2.7}$$

Using the Helmholtz potential (6.12.18) and energy equation (8.2.7) in the entropy inequality (2.4.45) to eliminate $\bar{\rho}_\alpha\bar{r}_\alpha$, yields

$$-\bar{\rho}_\alpha(\grave{\tilde{\psi}}_\alpha + \tilde{s}_\alpha\grave{\bar{\bar{\theta}}}_\alpha) - \frac{1}{\bar{\bar{\theta}}_\alpha}\bar{\mathbf{q}}_\alpha\cdot\nabla\bar{\bar{\theta}}_\alpha + \hat{s}_\alpha\bar{\bar{\theta}}_\alpha + \bar{T}_{\alpha ij}\tilde{v}_{\alpha ij} + \hat{\nu}^{(\alpha)}_{ij}(\bar{\rho}_\alpha\tilde{\ell}\epsilon_{\alpha i})_{,j}$$
$$+(\frac{\grave{\phi}_\alpha}{\phi_\alpha})_{,m}\bar{\lambda}_{\alpha jjm} + \hat{\nu}^{(\alpha)}_{mj,i}\bar{\lambda}_{\alpha jmi} + \hat{c}_\alpha(\hat{\tilde{\epsilon}}_\alpha - \tilde{\psi}_\alpha) - \bar{q}_{s\alpha} - (\frac{\grave{\phi}_\alpha}{\phi_\alpha}\bar{\rho}_\alpha\tilde{\ell}\epsilon_{\alpha i})_{,i}$$
$$-\hat{\nu}^{(\alpha)}_{im,i}\bar{\rho}_\alpha\tilde{\ell}\epsilon_{\alpha m} + p_\alpha(\grave{\phi}_\alpha + \phi_\alpha\tilde{v}_{\alpha i,i} - \frac{\hat{c}_\alpha}{\bar{\bar{\rho}}_\alpha}) \geq 0 \tag{8.2.8}$$

where use is made of the incompressibility constraint (6.12.40). Comparing (8.2.8) with (7.3.10) shows that the former equation reduces to the latter equation in the absence of the rotational effect ($\hat{\boldsymbol{\nu}}^{(\alpha)} = \mathbf{0}$).

Repeating the procedure of section (7.4) by computing $\grave{\tilde{\psi}}_\alpha$ with independent variables given by (8.2.2), substituting the result into (8.2.8), and requiring that the entropy inequality holds for arbitrary variations of

$$\{\grave{\bar{\bar{\rho}}}_{\beta,i}\},\ \bar{\bar{\rho}}_{\beta,ij},\ \grave{D}_{\beta ij},\ D_{\beta ij,k},\ \grave{W}_{\beta ij} - \grave{W}_{\gamma ij},\ W_{\beta ij,k} - W_{\gamma ij,k},\ \grave{\bar{\bar{\theta}}}_\beta,\ \{\grave{\bar{\bar{\theta}}}_{\beta,i}\},$$
$$\bar{\bar{\theta}}_{\beta,ij},\ \grave{\tilde{v}}_{\beta i} - \grave{\tilde{v}}_{\gamma i},\ \tilde{i}_{\beta,i},\ \grave{\tilde{i}}_\beta,\ \grave{\phi}_\beta,\ \{\grave{\phi}_\beta\}_{,i},\ \phi_{\beta,ij},\ \grave{b}^{(\beta)}_{ij},\ b^{(\beta)}_{ij,k},\ \{\grave{\hat{\nu}}^{(\beta)}_{ij,k}\},\ \hat{\nu}^{(\beta)}_{ij,k\ell}$$

produces results (7.4.8)-(7.4.18) *and*

$$\frac{\partial\tilde{\psi}_\alpha}{\partial b^{(\beta)}_{ij}} = \frac{\partial\tilde{\psi}_\alpha}{\partial\hat{\nu}^{(\beta)}_{ij,k}} = 0, \qquad \beta = 1,\ldots,\gamma \tag{8.2.9}$$

Utilizing the definitions (7.4.20)-(7.4.24) reduces the entropy inequality (8.2.8) to

$$-\frac{1}{\bar{\bar{\theta}}_\alpha}\bar{\mathbf{q}}_\alpha\cdot\nabla\bar{\bar{\theta}}_\alpha - \bar{\rho}_\alpha\sum_{\beta\neq\alpha}^{\gamma}[\frac{\partial\tilde{\psi}_\alpha}{\partial\bar{\bar{\rho}}_\beta}\frac{\hat{c}_\beta}{\phi_\beta} - (\frac{\partial\tilde{\psi}_\alpha}{\partial\bar{\bar{\rho}}_\beta}\frac{\bar{\bar{\rho}}_\beta}{\phi_\beta} - \frac{\partial\tilde{\psi}_\alpha}{\partial\phi_\beta})\grave{\phi}_\beta$$
$$+(\frac{\partial\tilde{\psi}_\alpha}{\partial\bar{\bar{\rho}}_\beta}\bar{\bar{\rho}}_{\beta,i} + \frac{\partial\tilde{\psi}_\alpha}{\partial\phi_\beta}\phi_{\beta,i})(\tilde{v}_{\alpha i} - \tilde{v}_{\beta i}) - \frac{\partial\tilde{\psi}_\alpha}{\partial\bar{\bar{\rho}}_\beta}\bar{\bar{\rho}}_\beta tr\mathbf{D}_\beta)] + \hat{f}_\alpha - u_{\alpha i}\bar{t}_{\alpha i}$$

$$+\frac{\grave{\phi}_\alpha}{\phi_\alpha}[\phi_\alpha(p_\alpha + \bar{\bar{\pi}}_\alpha) - \bar{\bar{\lambda}}^R_{\alpha kkm}\phi_{\alpha,m} - \phi_\alpha\beta_\alpha - (\bar{\rho}_\alpha\tilde{\ell}\epsilon_{\alpha i})_{,i}]$$

$$+\hat{c}_\alpha[-\tilde{\psi}_\alpha - \frac{p_\alpha}{\bar{\bar{\rho}}_\alpha} - \frac{\bar{\bar{\pi}}_\alpha}{\bar{\bar{\rho}}_\alpha} + \hat{\hat{\epsilon}}_\alpha] + \bar{T}^e_{\alpha ij}\tilde{v}_{\alpha i,j} + \hat{\nu}^{(\alpha)}_{ij}(\bar{\rho}_\alpha\tilde{\ell}\epsilon_{\alpha i})_{,j}$$

$$+\hat{\nu}^{(\alpha)}_{mj,i}\bar{\lambda}_{\alpha jmi} - \hat{\nu}^{(\alpha)}_{im,i}\bar{\rho}_\alpha\tilde{\ell}\epsilon_{\alpha m} = R_\alpha(\{S_\beta\}) \geq 0 \qquad (8.2.10)$$

Moreover, the rotational effect $\hat{\nu}^{(\beta)}$, does not affect the functional form of $\tilde{\psi}_\alpha$, (*cit.* (7.4.26)), *i.e.*

$$\tilde{\psi}_\alpha = \tilde{\psi}_\alpha(\bar{\bar{\rho}}_\beta, \bar{\bar{\theta}}_\alpha, \phi_\beta, \nabla\phi_\alpha) \qquad (8.2.11)$$

and the result (7.4.25) remains valid.

By defining an *equilibrium thermokinetic process* $\{S_{\beta 0}\}$ as in section 7.4,

$$R_\alpha(\bar{\bar{\rho}}_\beta, \nabla\bar{\bar{\rho}}_\beta, \mathbf{D}_\beta = 0, \mathbf{W}_\beta - \mathbf{W}_\gamma = 0, \bar{\bar{\theta}}_\beta, \nabla\bar{\bar{\theta}}_\beta = 0, \tilde{\mathbf{v}}_\beta - \tilde{\mathbf{v}}_\gamma = 0,$$
$$\tilde{i}_\beta, \phi_\beta, \nabla\phi_\beta, \grave{\phi}_\beta = 0, \mathbf{b}^{(\beta)} = 0, \nabla\hat{\nu}^{(\beta)} = 0) = R_\alpha(\{S_{\beta 0}\})$$
$$\beta = 1, \ldots, \gamma \qquad (8.2.12)$$

and reproducing the steps as in (7.4.28) and using (7.4.29) produces results (7.4.30)-(7.4.36), *and*

$$\bar{\bar{\rho}}_\alpha\phi_\alpha\tilde{\ell}\epsilon_{\alpha i}(\{S_{\beta 0}\}) = 0; \quad (\bar{\bar{\rho}}_\alpha\phi_\alpha\tilde{\ell}\epsilon_{\alpha i}(\{S_{\beta 0}\}))_{,j} = 0, \quad j \neq i \qquad (8.2.13)$$

$$\bar{\lambda}_{\alpha jmi}(\{S_{\beta 0}\}) = 0; \quad j \neq m \qquad (8.2.14)$$

Moreover, (7.2.7)-(7.2-10) yield

$$\bar{S}_{\alpha(jk)}(\{S_{\beta 0}\}) + \bar{\lambda}_{\alpha(jk)m,m}(\{S_{\beta 0}\}) = 0 \qquad (8.2.15)$$

$$\bar{S}_{\alpha[jk]}(\{S_{\beta 0}\}) + \bar{\lambda}_{\alpha[jk]m,m}(\{S_{\beta 0}\}) + \bar{\rho}_\alpha\tilde{\ell}_{\alpha[jk]}(\{S_{\beta 0}\}) = 0 \qquad (8.2.16)$$

$$\tilde{\ell}_{\alpha(jk)}(\{S_{\beta 0}\}) = 0 \qquad (8.2.17)$$

## 8.3 Linearized Constitutive Equations with Rotation and Dilatation

The linearized constitutive equations considerably simplify the theories of multiphase mixtures. Following the development of section 7.5, the *reduced equilibrium state* $\{S'_{\beta 0}\}$ is defined as follows

$$\{S'_{\beta 0}\} = \{\bar{\bar{\rho}}_\beta, \nabla\bar{\bar{\rho}}_\beta = 0, \mathbf{D}_\beta = 0, \mathbf{W}_\beta - \mathbf{W}_\gamma = 0, \bar{\bar{\theta}}_\beta, \nabla\bar{\bar{\theta}}_\beta = 0,$$
$$\tilde{\mathbf{v}}_\beta - \tilde{\mathbf{v}}_\gamma = 0, \tilde{\mathbf{i}}_\beta = 0, \phi_\beta, \nabla\phi_\beta = 0, \grave{\phi}_\beta = 0, \mathbf{b}^{(\beta)} = 0, \nabla\hat{\nu}^{(\beta)} = 0\}$$
$$(8.3.1)$$

and is useful as a state about which the linearization of constitutive equations can be performed. In developing the linear constitutive equations it is more useful, however, to replace $\mathbf{b}^{(\beta)}$ by $(\mathbf{b}^{(\beta)} + \mathbf{D}_\beta)$, since the latter quantity is generally smaller in magnitude, as an independent variable in (8.2.2) and define a measure $\varepsilon^2$ equivalent to (7.5.3) as follows

$$\varepsilon^2 = \sum_\beta^\gamma C_{1\beta} \nabla \bar{\bar{\rho}}_\beta \cdot \nabla \bar{\bar{\rho}}_\beta + \sum_\beta^\gamma C_{2\beta}\, tr(\mathbf{D}_\beta \mathbf{D}_\beta)$$
$$+ \sum_\beta^{\gamma-1} C_{3\beta}\, tr[(\mathbf{W}_\beta - \mathbf{W}_\gamma)(\mathbf{W}_\beta - \mathbf{W}_\gamma)] + \sum_\beta^\gamma C_{4\beta} \nabla \bar{\bar{\theta}}_\beta \cdot \nabla \bar{\bar{\theta}}_\beta$$
$$+ \sum_\beta^\gamma C_{5\beta} (\tilde{\mathbf{v}}_\beta - \tilde{\mathbf{v}}_\gamma) \cdot (\tilde{\mathbf{v}}_\beta - \tilde{\mathbf{v}}_\gamma) + \sum_\beta^\gamma C_{6\beta}\, tr(\bar{\mathbf{i}}_\beta \bar{\mathbf{i}}_\beta) + \sum_\beta^\gamma C_{7\beta} \nabla \phi_\beta \cdot \nabla \phi_\beta$$
$$+ \sum_\beta^\gamma C_{8\beta}\, \dot{\phi}_\beta \dot{\phi}_\beta + \sum_\beta^\gamma C_{9\beta}\, tr[(\mathbf{b}^{(\beta)} + \mathbf{D}_\beta)(\mathbf{b}^{(\beta)} + \mathbf{D}_\beta)] + \sum_\beta^\gamma C_{10\beta}\, \hat{\nu}^{(\beta)}_{ij,k} \hat{\nu}^{(\beta)}_{ij,k} \tag{8.3.2}$$

Employing $(7.4.31)_1$, $(7.4.31)_3$, $(8.2.13)_1$, and representation theorems of isotropic tensors as in section 7.5 it can be shown that the rotational effect changes (7.5.6)-(7.5.8) to

$$\bar{q}_{\alpha i}(\{S_\beta\}) = -\sum_\beta^\gamma \nu_{\alpha\beta}\, \bar{\bar{\rho}}_{\beta,i} - \sum_\beta^{\gamma-1} \zeta_{\alpha\beta} (\tilde{v}_{\beta i} - \tilde{v}_{\gamma i}) - \sum_\beta^\gamma \kappa_{\alpha\beta}\, \bar{\bar{\theta}}_{\beta,i}$$
$$- \sum_\beta^\gamma \Gamma_{\alpha\beta}\, \phi_{\beta,i} - \sum_\beta^\gamma A_{\alpha\beta}\, \hat{\nu}^{(\beta)}_{il,l} + O(\varepsilon^2) \tag{8.3.3}$$

$$\bar{t}_{\alpha i}(\{S_\beta\}) = -\sum_\beta^\gamma \gamma_{\alpha\beta}\, \bar{\bar{\theta}}_{\beta,i} - \sum_\beta^{\gamma-1} \xi_{\alpha\beta} (\tilde{v}_{\beta i} - \tilde{v}_{\gamma i}) - \sum_\beta^\gamma \Delta_{\alpha\beta}\, \bar{\bar{\rho}}_{\beta,i}$$
$$- \sum_\beta^\gamma M_{\alpha\beta}\, \phi_{\beta,i} - \sum_\beta^\gamma N_{\alpha\beta}\, \hat{\nu}^{(\beta)}_{il,l} + O(\varepsilon^2) \tag{8.3.4}$$

$$\bar{\bar{\rho}}_\alpha \phi_\alpha \bar{l} \epsilon_{\alpha i}(\{S_\beta\}) = \sum_\beta^\gamma \epsilon^\alpha_{0,\beta}\, \bar{\bar{\theta}}_{\beta,i} + \sum_\beta^{\gamma-1} \epsilon^\alpha_{1,\beta} (\tilde{v}_{\beta i} - \tilde{v}_{\gamma i}) + \sum_\beta^\gamma \epsilon^\alpha_{2,\beta}\, \bar{\bar{\rho}}_{\beta,i}$$
$$+ \sum_\beta^\gamma \epsilon^\alpha_{3,\beta}\, \phi_{\beta,i} + \sum_\beta^\gamma \epsilon^\alpha_{4,\beta}\, \hat{\nu}^{(\beta)}_{il,l} + O(\varepsilon^2) \tag{8.3.5}$$

Hence, from equation (8.3.5) we obtain

$$(\bar{\bar{\rho}}_\alpha \phi_\alpha \tilde{\ell} \epsilon_{\alpha i})_{,j} = \sum_{\beta}^{\gamma-1} \epsilon^{\alpha}_{1,\beta}(D_{\beta ij} - D_{\gamma ij} + W_{\beta ij} - W_{\gamma ij}) + O(\varepsilon^2) \qquad (8.3.6)$$

Linearizing $\bar{T}^e_{\alpha ij}$ and using $(7.4.31)_2$ and (7.4.20) gives

$$\begin{aligned}\bar{T}_{\alpha ij}(\{S_\beta\}) = &-\phi_\alpha(p_\alpha + \bar{\bar{\pi}}_\alpha)\delta_{ij} - \bar{\bar{\rho}}_\alpha \phi_\alpha \frac{\partial \tilde{\psi}_\alpha}{\partial \phi_{\alpha,j}} \phi_{\alpha,i} + 2\sum_{\beta}^{\gamma-1} e_{\alpha\beta}(W_{\beta ij} - W_{\gamma ij}) \\ &+ \sum_{\beta}^{\gamma}[\lambda_{\alpha\beta}(tr\mathbf{D}_\beta)\delta_{ij} + 2\mu_{\alpha\beta}D_{\beta ij}] + \sum_{\beta}^{\gamma} \iota_{\alpha\beta}\,\bar{i}_\beta \delta_{ij} + \sum_{\beta}^{\gamma} O_{\alpha\beta}\,\grave{\phi}_\beta \delta_{ij} \\ &+ 2\sum_{\beta}^{\gamma} \mu^0_{\alpha\beta}(b^{(\beta)}_{ij} + D_{\beta ij}) + O(\varepsilon^2)\end{aligned} \qquad (8.3.7)$$

whereas by using (7.4.40) it follows that

$$\begin{aligned}\bar{S}_{\alpha ij}(\{S_\beta\}) = &-\phi_\alpha P_\alpha(\{S'_{\beta 0}\})\delta_{ij} + \sum_{\beta}^{\gamma}[\Pi_{\alpha\beta}(tr\mathbf{D}_\beta)\delta_{ij} + \Phi_{\alpha\beta}D_{\beta ij}] \\ &+ \sum_{\beta}^{\gamma-1} T_{\alpha\beta}(W_{\beta ij} - W_{\gamma ij}) + \sum_{\beta}^{\gamma} K_{\alpha\beta}\,\bar{i}_\beta \delta_{ij} + \sum_{\beta}^{\gamma} H_{\alpha\beta}\,\grave{\phi}_\beta \delta_{ij} \\ &+ \sum_{\beta}^{\gamma} \Phi^0_{\alpha\beta}(b^{(\beta)}_{ij} + D_{\beta ij}) + O(\varepsilon^2)\end{aligned} \qquad (8.3.8)$$

Equations $(7.4.33)_1$ and $(8.2.6)_1$ can be used to show that

$$\hat{c}_\alpha \hat{i}_{\alpha ij}(\{S_\beta\}) = \sum_{\beta}^{\gamma} \Psi_{\alpha\beta}(tr\mathbf{D}_\beta)\delta_{ij} + \sum_{\beta}^{\gamma} E_{\alpha\beta}\,\bar{i}_\beta \delta_{ij} + \sum_{\beta}^{\gamma} Z_{\alpha\beta}\grave{\phi}_\beta \delta_{ij} + O(\varepsilon^2) \qquad (8.3.9)$$

whereas (7.5.21) and $(8.2.6)_2$ give

$$\begin{aligned}\bar{\lambda}_{\alpha ijk}(\{S_\beta\}) = &\sum_{\beta}^{\gamma} \delta_{ij}[a^{\alpha}_{1,\beta}\,\bar{\bar{\rho}}_{\beta,k} + b^{\alpha}_{1,\beta}(\tilde{v}_{\beta k} - \tilde{v}_{\gamma k}) + c^{\alpha}_{1,\beta}\,\bar{\bar{\theta}}_{\beta,k} + d^{\alpha}_{1,\beta}\,\phi_{\beta,k}] \\ &+ \sum_{\beta}^{\gamma} \delta_{ik}[a^{\alpha}_{2,\beta}\,\bar{\bar{\rho}}_{\beta,j} + b^{\alpha}_{2,\beta}(\tilde{v}_{\beta j} - \tilde{v}_{\gamma j}) + c^{\alpha}_{2,\beta}\,\bar{\bar{\theta}}_{\beta,j} + d^{\alpha}_{2,\beta}\,\phi_{\beta,j}] \\ &+ \sum_{\beta}^{\gamma} \delta_{jk}[a^{\alpha}_{3,\beta}\,\bar{\bar{\rho}}_{\beta,i} + b^{\alpha}_{3,\beta}(\tilde{v}_{\beta i} - \tilde{v}_{\gamma i}) + c^{\alpha}_{3,\beta}\,\bar{\bar{\theta}}_{\beta,i} + d^{\alpha}_{3,\beta}\,\phi_{\beta,i}] \\ &+ \sum_{\beta}^{\gamma}[\delta_{ij}A^{\alpha}_{1,\beta}\,\hat{\nu}^{(\beta)}_{k\ell,\ell} + \delta_{ik}A^{\alpha}_{2,\beta}\,\hat{\nu}^{(\beta)}_{\ell j,\ell} + \delta_{jk}A^{\alpha}_{3,\beta}\,\hat{\nu}^{(\beta)}_{\ell i,\ell} \\ &+ A^{\alpha}_{4,\beta}\,\hat{\nu}^{(\beta)}_{ij,k} + A^{\alpha}_{5,\beta}\,\hat{\nu}^{(\beta)}_{ik,j} + A^{\alpha}_{6,\beta}\,\hat{\nu}^{(\beta)}_{jk,i}] + O(\varepsilon^2)\end{aligned} \qquad (8.3.10)$$

$$U_{\alpha iik}(\{S_\beta\}) = \sum_\beta^\gamma [a_{4,\beta}^\alpha \bar{\bar{\rho}}_{\beta,k} + b_{4,\beta}^\alpha (\tilde{v}_{\beta k} - \tilde{v}_{\gamma k}) + c_{4,\beta}^\alpha \bar{\bar{\theta}}_{\beta,k} + d_{4,\beta}^\alpha \phi_{\beta,k}$$

$$+ B_{1,\beta}^\alpha \hat{\nu}_{k\ell,\ell}^{(\beta)}] + O(\varepsilon^2) \quad (8.3.11)$$

Equations (7.5.26)-(7.5.30) do not change form with the inclusion of the rotational effect $\hat{\nu}^{(\beta)}$, and for completeness only $\tilde{\psi}_\alpha$ will be listed, *i.e.*

$$\tilde{\psi}_\alpha(\{S_\beta\}) = \bar{\bar{\tilde{\psi}}}_\alpha(\bar{\bar{\rho}}_\beta, \bar{\bar{\theta}}_\alpha, \phi_\beta) + I_\alpha(\bar{\bar{\rho}}_\beta, \bar{\bar{\theta}}_\alpha, \phi_\beta)\,\phi_{\alpha,i}\phi_{\alpha,i} + O(\varepsilon^4) \quad (8.3.12)$$

For mixtures without the phase change $\hat{c}_\alpha = 0$, and assuming that $\hat{f}_\alpha = O(\varepsilon^3)$ as in section 7.6, further restrictions of material coefficients in the linearized constitutive equations may be obtained from the entropy inequality. Each of these coefficients may depend, in general, on all reduced equilibrium state properties of the mixture $\{S'_{\beta 0}\}$. With simplifications (7.6.1) and (7.6.2), the entropy inequality (8.2.10), combined with the above constitutive equations, becomes

$$\{\sum_{\beta,\delta}^\gamma [(\delta_{\alpha\delta}\frac{A_{\alpha\beta}}{\bar{\bar{\theta}}_\alpha} + (\epsilon_{0,\delta}^\alpha + c_{2,\delta}^\alpha - c_{3,\delta}^\alpha)\delta_{\alpha\beta})\bar{\bar{\theta}}_{\delta,i} + (\epsilon_{2,\delta}^\alpha + a_{2,\delta}^\alpha - a_{3,\delta}^\alpha)\delta_{\alpha\beta}\bar{\bar{\rho}}_{\delta,i}$$

$$+(\epsilon_{3,\delta}^\alpha + d_{2,\delta}^\alpha - d_{3,\delta}^\alpha)\delta_{\alpha\beta}\phi_{\delta,i} + (\epsilon_{4,\beta}^\alpha - A_{2,\beta}^\alpha + A_{3,\beta}^\alpha)\delta_{\alpha\delta}\hat{\nu}_{i\ell,\ell}^{(\delta)}\hat{\nu}_{im,m}^{(\beta)}$$

$$+[(\delta_{\alpha\delta} - \frac{\bar{\rho}_\delta}{\rho})N_{\alpha\beta} + \delta_{\alpha\beta}(\epsilon_{1,\delta}^\alpha + b_{2,\delta}^\alpha - b_{3,\delta}^\alpha)(1-\delta_{\delta\gamma})]\tilde{v}_{\delta i}]\hat{\nu}_{im,m}^{(\beta)} + \mathbf{VAV}^T\}$$

$$\{1\}$$

$$+\{\sum_{\beta\neq\alpha}^\gamma [\bar{\rho}_\alpha \frac{\partial\tilde{\psi}_\alpha}{\partial\bar{\bar{\rho}}_\beta}\bar{\bar{\rho}}_\beta + \frac{\dot{\phi}_\alpha}{\phi_\alpha}(\Pi_{\alpha\beta} + \Phi_{\alpha\beta})]tr\mathbf{D}_\beta\}$$

$$\{2\}$$

$$+\{[\frac{\dot{\phi}_\alpha}{\phi_\alpha}(\Pi_{\alpha\alpha} + \Phi_{\alpha\alpha}) + \sum_\beta^\gamma(\iota_{\alpha\beta}\,\bar{i}_\beta + O_{\alpha\beta}\,\dot{\phi}_\beta)]tr\mathbf{D}_\alpha\}$$

$$\{3\}$$

$$+\{\sum_{\beta,\delta}^\gamma \delta_{\alpha\delta}(\lambda_{\alpha\beta} + \frac{2}{3}\mu_{\alpha\beta})(tr\mathbf{D}_\beta)(tr\mathbf{D}_\delta)\} + \{\sum_{\beta,\delta}^\gamma (2\hat{e}_{\alpha\beta} + \hat{\epsilon}_{1,\beta}^\alpha)\delta_{\alpha\delta}\, tr(\mathbf{W}_\beta^T\mathbf{W}_\delta)\}$$

$$\{4\} \qquad\qquad \{5\}$$

$$+\{\sum_{\beta,\delta}^\gamma 2\mu_{\alpha\beta}\delta_{\alpha\delta}\, tr[(\mathbf{D}_\beta - \frac{3+\sqrt{6}}{3}(tr\mathbf{D}_\beta)\mathbf{I})(\mathbf{D}_\delta - \frac{3+\sqrt{6}}{3}(tr\mathbf{D}_\delta)\mathbf{I})]\}$$

$$\{6\}$$

$$+\{\sum_{\beta,\delta}^\gamma (\hat{\epsilon}_{1,\beta}^\alpha\delta_{\alpha\delta} + 2\mu_{\beta\delta}^0\delta_{\alpha\beta})\, tr[\mathbf{W}_\beta^T(\mathbf{b}^{(\delta)} + \mathbf{D}_\delta)]\}$$

$$\{7\}$$

$$+\{\sum_{\beta,\delta}^{\gamma}(A^{\alpha}_{4,\beta}\hat{\nu}^{(\beta)}_{ji,k}+A^{\alpha}_{5,\beta}\hat{\nu}^{(\beta)}_{ki,j}+A^{\alpha}_{6,\beta}\hat{\nu}^{(\beta)}_{kj,i})\delta_{\alpha\delta}\hat{\nu}^{(\delta)}_{ij,k}\}+O(\varepsilon^3)=R_\alpha\geq 0$$

$$\{8\}$$

$$(8.3.13)$$

where $\mathbf{V}A\mathbf{V}^T$ is given by (7.6.30) and use is made of (7.6.31). $\delta_{\alpha\delta}$ is the Kronecker delta defined by (7.6.9), $\hat{e}_{\alpha\beta}$ is defined by (7.6.7), *and*

$$\hat{\epsilon}^{\alpha}_{1,\beta}=\epsilon^{\alpha}_{1,\beta},\quad \beta=1,\ldots,\gamma-1;\qquad \hat{\epsilon}^{\alpha}_{1,\gamma}=-\sum_{\beta=1}^{\gamma-1}\epsilon^{\alpha}_{1,\beta} \tag{8.3.14}$$

As in section 7.6, the necessary conditions for the function $R_\alpha$ in (8.3.13) to be positive semidefinite are that each of the eight terms which are enclosed by the brackets $\{\ldots\}$ be positive semidefinite. Thus, for a two-phase mixture, setting in (8.3.13) the term $\{3\}$ equal to zero recovers conditions (7.6.12)-(7.6.14), setting the term $\{2\}$ equal to zero recovers (7.6.18) and (7.6.19), and setting the terms $\{4\}$ and $\{6\}$ equal to zero recovers (7.6.26). When the terms $\{5\}$ and $\{7\}$ are set equal to zero and use made of (7.6.7) and (8.3.14) we obtain

$$\epsilon^{1}_{1,1}=\epsilon^{2}_{1,1}=e_{11}=e_{21}=0 \tag{8.3.15}$$

$$\overset{0}{\mu}_{11}=\overset{0}{\mu}_{12}=\overset{0}{\mu}_{21}=\overset{0}{\mu}_{22}=0 \tag{8.3.16}$$

The sufficient and necessary conditions for positive semidefinite form of the term $\{8\}$ are

$$A^{\alpha}_{4,\alpha}\leq 0;\quad A^{\alpha}_{4,\beta}=0,\quad \alpha\neq\beta;\quad A^{\alpha}_{5,\beta}=A^{\alpha}_{6,\beta}=0 \tag{8.3.17}$$

By defining a new vector $\tilde{\mathbf{V}}$ as

$$\tilde{\mathbf{V}}=(\nabla\bar{\bar{\theta}}_1,\ldots,\nabla\bar{\bar{\theta}}_\gamma,\nabla\bar{\rho}_1,\ldots,\nabla\bar{\rho}_\gamma,\tilde{\mathbf{v}}_1,\ldots,\tilde{\mathbf{v}}_\gamma,\nabla\phi_1,\ldots,\nabla\phi_\gamma,$$
$$\hat{\nu}^{(1)}_{im,m},\ldots,\hat{\nu}^{(\gamma)}_{im,m}) \tag{8.3.18}$$

such that

$$\tilde{\mathbf{V}}A\tilde{\mathbf{V}}^T=\sum_{\beta,\delta}^{\gamma}[\frac{\nu_{\alpha\beta}}{\bar{\bar{\theta}}_\alpha}\delta_{\alpha\delta}\bar{\bar{\theta}}_{\delta,i}\bar{\bar{\rho}}_{\beta,i}+(\frac{\zeta_{\alpha\beta}}{\bar{\bar{\theta}}_\alpha}\delta_{\alpha\delta}+(\delta_{\alpha\beta}-\frac{\bar{\rho}_\beta}{\rho})\gamma_{\alpha\delta})\bar{\bar{\theta}}_{\delta,i}\tilde{v}_{\beta i}$$

$$+\frac{\kappa_{\alpha\beta}}{\bar{\bar{\theta}}_\alpha}\delta_{\alpha\delta}\bar{\bar{\theta}}_{\delta,i}\bar{\bar{\theta}}_{\beta,i}+\frac{\Gamma_{\alpha\beta}}{\bar{\bar{\theta}}_\alpha}\delta_{\alpha\delta}\bar{\bar{\theta}}_{\delta,i}\phi_{\beta,i}+\frac{K_{\alpha\beta}}{\bar{\bar{\theta}}_\alpha}\delta_{\alpha\delta}\grave{\phi}_\delta\grave{\bar{\bar{i}}}_\beta+\frac{H_{\alpha\beta}}{\bar{\bar{\theta}}_\alpha}\delta_{\alpha\delta}\grave{\phi}_\delta\grave{\phi}_\beta$$

$$+[(\delta_{\alpha\delta} - \frac{\bar{\rho}_\delta}{\rho})\Delta_{\alpha\beta} - \bar{\bar{\rho}}_\alpha \phi_\alpha \frac{\partial \tilde{\psi}_\alpha}{\partial \bar{\bar{\rho}}_\beta}(\delta_{\alpha\delta} - \delta_{\beta\delta})]\bar{\bar{\rho}}_{\beta,i}\tilde{v}_{\delta i}$$
$$+(\delta_{\alpha\delta} - \frac{\bar{\rho}_\delta}{\rho})(\xi_{\alpha\beta} - \delta_{\gamma\beta}\sum_\sigma^\gamma \xi_{\alpha\sigma})\tilde{v}_{\delta i}\tilde{v}_{\beta i} + [(\delta_{\alpha\delta} - \frac{\bar{\rho}_\delta}{\rho})M_{\alpha\beta}$$
$$-\bar{\bar{\rho}}_\alpha \phi_\alpha \frac{\partial \tilde{\psi}_\alpha}{\partial \phi_\beta}(\delta_{\alpha\delta} - \delta_{\beta\delta})]\phi_{\beta,i}\tilde{v}_{\delta i} + \{[\delta_{\alpha\delta}\frac{A_{\alpha\beta}}{\bar{\bar{\theta}}_\alpha} + (\epsilon^\alpha_{0,\delta} + c^\alpha_{2,\delta} - c^\alpha_{3,\delta})\delta_{\alpha\beta}]\bar{\bar{\theta}}_{\delta,i}$$
$$+(\epsilon^\alpha_{2,\delta} + a^\alpha_{2,\delta} - a^\alpha_{3,\delta})\delta_{\alpha\beta}\bar{\bar{\rho}}_{\delta,i} + (\epsilon^\alpha_{3,\delta} + d^\alpha_{2,\delta} - d^\alpha_{3,\delta})\delta_{\alpha\beta}\phi_{\delta,i} + [(\delta_{\alpha\delta} - \frac{\bar{\rho}_\delta}{\rho})N_{\alpha\beta}$$
$$+(\epsilon^\alpha_{1,\delta} + b^\alpha_{2,\delta} - b^\alpha_{3,\delta})\delta_{\alpha\beta}(1 - \delta_{\gamma\delta})]\tilde{v}_{\delta i} + (\epsilon^\alpha_{4,\beta} - A^\alpha_{2,\beta} + A^\alpha_{3,\beta})\delta_{\alpha\delta}\hat{v}^{(\delta)}_{i\ell,\ell}\}\hat{v}^{(\beta)}_{im,m}]$$
$$(8.3.19)$$

the entropy inequality (8.3.13) becomes

$$\tilde{\mathbf{V}}A\tilde{\mathbf{V}}^T + O(\epsilon^3) \geq 0 \qquad (8.3.20)$$

The *necessary conditions* for (8.3.20) and for two-phase mixtures are identical to (7.6.32) and (7.6.33), producing restrictions on the coefficients $\xi_{\alpha\beta}$ as given by (7.6.34). Moreover

$$\epsilon^\alpha_{4,\alpha} - A^\alpha_{2,\alpha} + A^\alpha_{3,\alpha} \geq 0, \quad \alpha = 1,\ldots,\gamma \qquad (8.3.21)$$

The condition (7.4.25) and relations (7.4.23), (8.3.12), (8.3.10), (8.3.5), and (8.3.15) yield, *for two-phase mixtures*:

$$b^\alpha_{1,\beta} + b^\alpha_{2,\beta} + b^\alpha_{3,\beta} = 0, \quad \alpha,\beta = 1,\ldots,\gamma \qquad (8.3.22)$$

$$a^\alpha_{1,\beta} + a^\alpha_{2,\beta} + a^\alpha_{3,\beta} - \epsilon^\alpha_{2,\beta} = 0, \quad \alpha,\beta = 1,\ldots,\gamma \qquad (8.3.23)$$

$$c^\alpha_{1,\beta} + c^\alpha_{2,\beta} + c^\alpha_{3,\beta} - \epsilon^\alpha_{0,\beta} = 0, \quad \alpha,\beta = 1,\ldots,\gamma \qquad (8.3.24)$$

$$d^\alpha_{1,\beta} + d^\alpha_{2,\beta} + d^\alpha_{3,\beta} - \epsilon^\alpha_{3,\beta} = 0, \quad \alpha \neq \beta, \quad \beta = 1,\ldots,\gamma$$
$$d^\alpha_{1,\alpha} + d^\alpha_{2,\alpha} + d^\alpha_{3,\alpha} - \epsilon^\alpha_{3,\alpha} = 2\bar{\bar{\rho}}_\alpha \phi^2_\alpha I_\alpha, \quad \alpha = \beta \qquad (8.3.25)$$

$$A^\alpha_{1,\beta} - A^\alpha_{2,\beta} - A^\alpha_{3,\beta} - A^\alpha_{5,\beta} - A^\alpha_{6,\beta} - \epsilon^\alpha_{4,\beta} = 0; \quad \alpha,\beta = 1,\ldots,\gamma \quad (8.3.26)$$

where (8.3.26) can be further reduced by using (8.3.17) and (8.3.21).

*For saturated two-phase mixtures* $\phi_{2,m} = -\phi_{1,m}$, and it follows that

$$(d^\alpha_{1,1} - d^\alpha_{1,2}) + (d^\alpha_{2,1} - d^\alpha_{2,2}) + (d^\alpha_{3,1} - d^\alpha_{3,2}) - (\epsilon^\alpha_{0,1} - \epsilon^\alpha_{0,2})$$
$$= 2\bar{\bar{\rho}}_\alpha \phi^2_\alpha I_\alpha(\delta_{\alpha 1} - \delta_{\alpha 2})$$
$$(8.3.27)$$

As sufficient conditions we may take

$$a^\alpha_{1,\beta} = a^\alpha_{2,\beta} = a^\alpha_{3,\beta} = b^\alpha_{1,\beta} = b^\alpha_{2,\beta} = b^\alpha_{3,\beta} = c^\alpha_{1,\beta} = c^\alpha_{2,\beta} = c^\alpha_{3,\beta} = 0$$
$$\epsilon^\alpha_{0,\beta} = \epsilon^\alpha_{2,\beta} = \epsilon^\alpha_{3,\beta} = \epsilon^\alpha_{4,\beta} = A^\alpha_{1,\beta} = A^\alpha_{2,\beta} = A^\alpha_{3,\beta} = A^\alpha_{5,\beta} = A^\alpha_{6,\beta} = 0 \tag{8.3.28}$$

and

$$d^\alpha_{1,1} - d^\alpha_{1,2} = d^\alpha_{2,1} - d^\alpha_{2,2} = d^\alpha_{3,1} - d^\alpha_{3,2} = \frac{1}{3} 2\bar{\bar{\rho}}_\alpha \phi^2_\alpha I_\alpha (\delta_{\alpha 1} - \delta_{\alpha 2}) \tag{8.3.29}$$

such that with (8.3.17) they reduce (8.3.10) to

$$\bar{\lambda}_{\alpha ijk} = A^\alpha_{4,\alpha}\, \hat{\nu}^{(\alpha)}_{ij,k} + \frac{2}{3}\bar{\bar{\rho}}_\alpha \phi^2_\alpha I_\alpha (\delta_{\alpha 1} - \delta_{\alpha 2})(\delta_{ij}\phi_{1,k} + \delta_{ik}\phi_{1,j} + \delta_{jk}\phi_{1,i}) + O(\varepsilon^2) \tag{8.3.30}$$

whereas (8.3.5) and (8.3.6) become

$$\bar{\bar{\rho}}_\alpha \phi_\alpha \tilde{l} \epsilon_{\alpha i} = O(\varepsilon^2), \quad (\bar{\bar{\rho}}_\alpha \phi_\alpha \tilde{l} \epsilon_{\alpha i})_{,j} = O(\varepsilon^2) \tag{8.3.31}$$

since from (8.3.15) $\epsilon^\alpha_{1,1} = 0$.

## 8.4 Two-Phase Modeling Equations with Rotational and Dilatational Effects

The restrictions on the material coefficients expressed by (7.6.13), (7.6.14), (7.6.26), (8.3.15), and (8.3.16) reduce the stress tensor (8.3.7) into a form independent of the rotational effects in the linear approximation, *i.e.*

$$\bar{T}_{\alpha ij} = -\phi_\alpha (p_\alpha + \bar{\bar{\pi}}_\alpha)\delta_{ij} - 2\bar{\bar{\rho}}_\alpha \phi_\alpha I_\alpha \phi_{\alpha,i}\phi_{\alpha,j} + [\lambda_{\alpha\alpha}(tr\mathbf{D}_\alpha)\delta_{ij} + 2\mu_{\alpha\alpha} D_{\alpha ij}]$$
$$+ O_{\alpha\alpha}\dot{\phi}_\alpha \delta_{ij} + O(\varepsilon^2) \tag{8.4.1}$$

and hence

$$\hat{M}_{\alpha ij} = \bar{T}_{\alpha ij} - \bar{T}_{\alpha ji} = O(\varepsilon^2) \tag{8.4.2}$$

The surface traction moment $\bar{\mathbf{S}}_\alpha$ is given by (8.3.8), and using (7.6.12) and (7.6.18) it is reduced into the following form

$$\bar{S}_{\alpha ij} = -\phi_\alpha P_\alpha \delta_{ij} + \Pi_{\alpha\alpha}[(tr\mathbf{D}_\alpha)\delta_{ij} - D_{\alpha ij}] - O_{\alpha\alpha}\phi_\alpha D_{\alpha ij}$$
$$+ \Pi_{\alpha\beta}(1 - \delta_{\alpha\beta})[(tr\mathbf{D}_\beta)\delta_{ij} - D_{\beta ij}] + \sum_\beta^2 [K_{\alpha\beta}\, \dot{i}_\beta + H_{\alpha\beta}\, \dot{\phi}_\beta]\delta_{ij}$$
$$+ T_{\alpha 1}(W_{1ij} - W_{2ij}) + \sum_\beta^2 \Phi^0_{\alpha\beta}(b^{(\beta)}_{ij} + D_{\beta ij}) + O(\varepsilon^2) \tag{8.4.3}$$

But from (7.2.12), $(8.2.6)_2$, and (6.12.36) it follows that

$$\bar{S}_{\alpha(jk)} + \bar{\lambda}_{\alpha(jk)m,m} - \bar{\rho}_\alpha \bar{i}_\alpha \frac{\grave{\phi}_\alpha}{\phi_\alpha} D_{\alpha jk} = 0, \quad j \neq k \tag{8.4.4}$$

which when combined with (8.3.30)

$$\bar{\lambda}_{\alpha(jk)m} = \frac{2}{3}\bar{\bar{\rho}}_\alpha \phi_\alpha^2 I_\alpha (\delta_{\alpha 1} - \delta_{\alpha 2})(\delta_{km}\phi_{1,j} + \delta_{jm}\phi_{1,k}) + O(\varepsilon^2), \quad j \neq k \tag{8.4.5}$$

gives

$$\Phi_{\alpha\alpha} D_{\alpha jk} + \Phi_{\alpha\beta}(1 - \delta_{\alpha\beta}) D_{\beta jk} + O(\varepsilon^2) = 0, \quad j \neq k \tag{8.4.6}$$

producing the following *necessary conditions*

$$\Phi_{\alpha\alpha} = \Phi_{\alpha\beta} = 0, \quad \alpha, \beta = 1, 2 \quad \text{or} \quad \Pi_{\alpha\alpha} + \phi_\alpha O_{\alpha\alpha} = 0; \quad \Pi_{\alpha\beta} = 0, \quad \alpha \neq \beta \tag{8.4.7}$$

Equation (8.4.3) is, therefore, reduced to

$$\begin{aligned} \bar{S}_{\alpha ij} = &-\phi_\alpha P_\alpha \delta_{ij} - \phi_\alpha O_{\alpha\alpha}(tr\mathbf{D}_\alpha)\delta_{ij} + \sum_\beta^2 (K_{\alpha\beta}\dot{\bar{i}}_\beta + H_{\alpha\beta}\grave{\phi}_\beta)\delta_{ij} \\ &+ T_{\alpha 1}(W_{1ij} - W_{2ij}) + \sum_\beta^2 \Phi^0_{\alpha\beta}(b^{(\beta)}_{ij} + D_{\beta ij}) + O(\varepsilon^2) \end{aligned} \tag{8.4.8}$$

and reduces to (7.7.3) in the absence of the rotational effect (this limit is obtained by setting $\mathbf{b}^{(\beta)} = -\mathbf{D}_\beta$). The arguments leading to the development of (7.7.8) can also be employed to reduce the interphase interaction force $\bar{\mathbf{t}}_\alpha$ given by (8.3.4). Thus,

$$\begin{aligned} \bar{t}_{\alpha i} = &\ \phi_{\alpha,i}(p_\alpha + \bar{\bar{\pi}}_\alpha) - \xi_{\alpha 1}(\tilde{v}_{1i} - \tilde{v}_{2i}) - \sum_\beta^2 \gamma_{\alpha\beta}\bar{\bar{\theta}}_{\beta,i} - \sum_\beta^2 \Delta_{\alpha\beta}\bar{\bar{\rho}}_{\beta,i} \\ &- \sum_\beta^2 N_{\alpha\beta}\hat{\nu}^{(\beta)}_{i\ell,\ell} + O(\varepsilon^2) \end{aligned} \tag{8.4.9}$$

For a *saturated mixture*, the heat flux vector (8.3.3) becomes

$$\begin{aligned} \bar{q}_{\alpha i} = &-\sum_\beta^2 \kappa_{\alpha\beta}\bar{\bar{\theta}}_{\beta,i} - (\Gamma_{\alpha 1} - \Gamma_{\alpha 2})\phi_{1,i} - \sum_\beta^2 \nu_{\alpha\beta}\bar{\bar{\rho}}_{\beta,i} - \zeta_{\alpha 1}(\tilde{v}_{1i} - \tilde{v}_{2i}) \\ &- \sum_\beta^2 A_{\alpha\beta}\hat{\nu}^{(\beta)}_{i\ell,\ell} + O(\varepsilon^2) \end{aligned} \tag{8.4.10}$$

and if it is assumed that $\mathbf{U}_\alpha$ is only affected by the mixture's structural characteristics $\nabla\phi_\beta$ and $\hat{\boldsymbol{\nu}}^{(\beta)}$, then from (8.3.11) use may be made of

$$U_{\alpha iik} = (d^\alpha_{4,1} - d^\alpha_{4,2})\phi_{1,k} + \sum_\beta^2 B^\alpha_{1,\beta}\hat{\nu}^{(\beta)}_{k\ell,\ell} + O(\varepsilon^2) \tag{8.4.11}$$

The energy source $\hat{\epsilon}_\alpha$ which is expressed by (7.2.14) can be more conveniently found from (8.2.7) and (2.4.35). Hence

$$\hat{\epsilon}_\alpha = -\hat{c}_\alpha(\bar{\epsilon}_\alpha - \hat{\bar{\epsilon}}_\alpha) - \bar{q}_{s\alpha} + \hat{\nu}^{(\alpha)}_{mj,i}\bar{\lambda}_{\alpha jmi} - \hat{\nu}^{(\alpha)}_{im,i}\bar{\bar{\rho}}_\alpha\phi_\alpha\bar{\ell}\epsilon_{\alpha m}$$
$$+(\frac{\grave{\phi}_\alpha}{\phi_\alpha})_{,i}\bar{\lambda}_{\alpha jji} - (\frac{\grave{\phi}_\alpha}{\phi_\alpha}\bar{\bar{\rho}}_\alpha\phi_\alpha\bar{\ell}\epsilon_{\alpha i})_{,i} + O(\varepsilon^3) \tag{8.4.12}$$

where use is made of (8.3.6) and (8.3.15). This result can be further simplified by utilizing $(8.3.28)_2$ in (8.3.5), giving

$$\bar{\bar{\rho}}_\alpha\phi_\alpha\bar{\ell}\epsilon_{\alpha i} = O(\varepsilon^2)$$

Employing (6.12.36), $(8.2.6)_2$, and skew-symmetry of $\hat{\boldsymbol{\nu}}^{(\alpha)}$, equation (7.2.10) is reduced to

$$\bar{\rho}_\alpha\bar{\ell}_{\alpha(jk)} = \bar{\rho}_\alpha\bar{i}_\alpha(\mu^{(\alpha)}_j\mu^{(\alpha)}_k - \mu^{(\alpha)}_i\mu^{(\alpha)}_i\delta_{jk}) - \mu^{(\alpha)}_i\epsilon_{m\ell i}(\frac{\grave{\phi}_\alpha}{\phi_\alpha})_{,m}U_{\alpha pp\ell}\delta_{jk}$$
$$-\frac{1}{2}\mu^{(\alpha)}_i\bar{\rho}_\alpha\bar{i}_\alpha(\bar{v}_{\alpha k,m}\epsilon_{mji} + \bar{v}_{\alpha j,m}\epsilon_{mki}) \tag{8.4.13}$$

Without neglecting the phase change effect, given below is a summary of the field equations of chapter 2 and section 7.2 with the rotational and dilatational effects included.

$$\grave{\bar{\rho}}_\alpha + \bar{\rho}_\alpha\bar{v}_{\alpha i,i} = \hat{c}_\alpha, \quad \bar{\rho}_\alpha = \bar{\bar{\rho}}_\alpha\phi_\alpha \tag{2.4.14}$$

$$\bar{\rho}_\alpha\grave{\bar{v}}_{\alpha i} = \bar{T}_{\alpha ij,j} + \bar{\rho}_\alpha\bar{b}_{\alpha i} + \hat{p}_{\alpha i} \tag{2.4.19}$$

$$\bar{\rho}_\alpha\grave{\bar{i}}_\alpha = -\hat{c}_\alpha\bar{i}_\alpha + K_{\alpha jj} + k_{\alpha jji,i} + \check{k}_{\alpha jj} \tag{7.2.3}$$

$$\bar{\rho}_\alpha\grave{\bar{\epsilon}}_\alpha = \bar{T}_{\alpha ij}\bar{v}_{\alpha i,j} - \bar{q}_{\alpha k,k} + \bar{\rho}_\alpha\bar{r}_\alpha + \hat{\epsilon}_\alpha \tag{2.4.35}$$

$$\bar{\rho}_\alpha(\overline{\bar{i}_\alpha\frac{\grave{\phi}_\alpha}{\phi_\alpha}}^{\grave{}} - \bar{i}_\alpha(\frac{\grave{\phi}_\alpha}{\phi_\alpha})^2)\delta_{jk} = \frac{\grave{\phi}_\alpha}{\phi_\alpha}[-\hat{c}_\alpha\bar{i}_\alpha + \hat{k}_{\alpha ii} - (\frac{\grave{\phi}_\alpha}{\phi_\alpha})_{,i}U_{\alpha mmi}]\delta_{jk}$$
$$+\bar{S}_{\alpha(jk)} + \bar{\lambda}_{\alpha(jk)m,m} - \bar{\rho}_\alpha\bar{i}_\alpha\frac{\grave{\phi}_\alpha}{\phi_\alpha}D_{\alpha jk} \tag{7.2.12}$$

$$\bar{\rho}_\alpha \bar{i}_\alpha [\grave{\mu}_\ell^{(\alpha)} + 2\frac{\grave{\phi}_\alpha}{\phi_\alpha}\mu_\ell^{(\alpha)}] = \frac{1}{2}\epsilon_{\ell jk}[\bar{S}_{\alpha[jk]} + \bar{\rho}_\alpha \bar{\ell}_{\alpha[jk]} + \bar{\lambda}_{\alpha[jk]m,m}$$
$$+\bar{\rho}_\alpha \bar{i}_\alpha \frac{\grave{\phi}_\alpha}{\phi_\alpha} W_{\alpha jk} - \frac{\grave{\phi}_\alpha}{\phi_\alpha}\hat{\nu}_{kj,i}^{(\alpha)} U_{\alpha ppi} + \epsilon_{m\ell i}\mu_i^{(\alpha)} U_{\alpha pp\ell}\hat{\nu}_{kj,m}^{(\alpha)}$$
$$+\frac{1}{2}\bar{\rho}_\alpha \bar{i}_\alpha \mu_i^{(\alpha)}(\bar{v}_{\alpha k,m}\epsilon_{mji} - \bar{v}_{\alpha j,m}\epsilon_{mki})] \tag{7.2.8}$$

where use is made of $(8.2.6)_2$, (6.12.36), (5.1.3), and (5.1.4). Moreover,

$$\hat{p}_{\alpha i} = \bar{t}_{\alpha i} - (\hat{\nu}_{im,j}^{(\alpha)} + (\frac{\grave{\phi}_\alpha}{\phi_\alpha})_{,j}\delta_{im})\bar{\rho}_\alpha \bar{i}_\alpha (\hat{\nu}_{jm}^{(\alpha)} + \frac{\grave{\phi}_\alpha}{\phi_\alpha}\delta_{jm}) \tag{7.2.17}$$

$$K_{\alpha jj} = 2\bar{\rho}_\alpha \frac{\grave{\phi}_\alpha}{\phi_\alpha}\bar{i}_\alpha \tag{7.2.6}$$

$$k_{\alpha jji,i} = -\hat{\nu}_{in,i}^{(\alpha)} U_{\alpha jjn} - (\frac{\grave{\phi}_\alpha}{\phi_\alpha} U_{\alpha jji})_{,i} \tag{7.2.5}$$

$$\check{k}_{\alpha jj} = \hat{c}_\alpha \hat{i}_{\alpha jj} \tag{5.2.8}$$

$$\hat{k}_{\alpha jj} = \hat{c}_\alpha \hat{i}_{\alpha jj} - \hat{\nu}_{mn,m}^{(\alpha)} U_{\alpha jjn} - (\frac{\grave{\phi}_\alpha}{\phi_\alpha} U_{\alpha jjm})_{,m} \tag{7.2.4}$$

These field equations together with the constitutive equations developed above include the effects of rotation and dilatation and should be useful to model complex two-phase mixtures, including the phase change processes when supplied by the constitutive equations for $\hat{c}_\alpha$, $\hat{\hat{\epsilon}}_\alpha$, $\bar{q}_{s\alpha}$, and $\hat{i}_{\alpha jj}$. As in section 7.5, these constitutive equations can be easily developed by extending (7.5.20) and (7.5.24) to include the rotational effect. The coefficients in the constitutive equations can then be restricted by the entropy inequality (8.2.10).

## 8.5 Nonlinear Constitutive Equations

At high shear rates, the granular media exhibit a nonlinear behavior of the shear stress with the shear rate, as discussed in section 7.8. Nonlinear constitutive equations considerably complicate the theory of multiphase mixtures and can render it unpractical if all nonlinear effects are retained.

In this section, some central ideas employed in the construction of nonlinear constitutive equations are discussed which may be utilized for the construction of very complete models of multiphase mixtures.

For a multiphase mixture of isotropic fluids, use can be made of (6.10.1) with independent constitutive variables given by

$$\Upsilon_\alpha = G_{\alpha\boldsymbol{\kappa}}[\bar{\bar{\rho}}_\beta, \nabla\bar{\bar{\rho}}_\beta, \mathbf{D}_\beta, \mathbf{W}_\beta - \mathbf{W}_\gamma, \bar{\bar{\theta}}_\beta, \nabla\bar{\bar{\theta}}_\beta, \tilde{\mathbf{v}}_\beta - \tilde{\mathbf{v}}_\gamma, \bar{\mathbf{i}}_\beta, \phi_\beta, \mathbf{b}^{(\beta)} + \mathbf{D}_\beta, \nabla\boldsymbol{\nu}^{(\beta)}] \tag{8.5.1}$$

where use is made of $\bar{\rho}_\beta = \bar{\bar{\rho}}_\beta\phi_\beta$, and $\mathbf{b}^{(\beta)}$ is replaced by an equivalent, and generally of smaller magnitude, quantity $\mathbf{b}^{(\beta)} + \mathbf{D}_\beta$. To reduce the algebraic complexity of constructing nonlinear constitutive equations it is convenient to simplify (8.5.1) further to mixtures with the constitutive equations of the following form:

$$\Upsilon_\alpha = G_{\alpha\boldsymbol{\kappa}}[\bar{\bar{\rho}}_\beta, \bar{\bar{\theta}}_\beta, \bar{i}_\beta, \phi_\beta, \nabla\phi_\alpha, \mathbf{D}_\alpha, \mathbf{b}^{(\alpha)} + \mathbf{D}_\alpha, \nabla\boldsymbol{\nu}^{(\alpha)}] \tag{8.5.2}$$

where the inertia tensor is isotropic, and the effects of

$$\mathbf{W}_\beta - \mathbf{W}_\gamma, \nabla\bar{\bar{\rho}}_\beta, \nabla\bar{\bar{\theta}}_\beta, \tilde{\mathbf{v}}_\beta - \tilde{\mathbf{v}}_\gamma, \quad \beta = 1, \ldots, \gamma$$
$$\nabla\phi_\beta, \mathbf{D}_\beta, \mathbf{b}^{(\beta)} + \mathbf{D}_\beta, \nabla\boldsymbol{\nu}^{(\beta)}, \quad \beta \neq \alpha$$

can be neglected.

The effects of structure of the multiphase mixture are represented in (8.5.2) by

$$\bar{i}_\beta, \phi_\beta, \nabla\phi_\alpha, \mathbf{b}^{(\alpha)} + \mathbf{D}_\alpha, \nabla\boldsymbol{\nu}^{(\alpha)}$$

and by assuming that these structural effects are small, the following representation of the stress tensor $\bar{\mathbf{T}}_\alpha$, surface traction $\bar{\mathbf{S}}_\alpha$, and intrinsic stress moment $\bar{\boldsymbol{\lambda}}_\alpha$ can be obtained

$$\bar{\mathbf{T}}_\alpha = \mathbf{E}^0_\alpha(\bar{\bar{\rho}}_\beta, \bar{\bar{\theta}}_\beta, \phi_\beta, \bar{i}_\beta, \nabla\phi_\alpha, \mathbf{D}_\alpha, \mathbf{b}^{(\alpha)} + \mathbf{D}_\alpha) + O(\nabla\boldsymbol{\nu}^{(\alpha)})^2 \tag{8.5.3}$$

$$\bar{\mathbf{S}}_\alpha = \mathbf{F}^0_\alpha(\bar{\bar{\rho}}_\beta, \bar{\bar{\theta}}_\beta, \phi_\beta, \bar{i}_\beta, \nabla\phi_\alpha, \mathbf{D}_\alpha, \mathbf{b}^{(\alpha)} + \mathbf{D}_\alpha) + O(\nabla\boldsymbol{\nu}^{(\alpha)})^2 \tag{8.5.4}$$

$$\begin{aligned}\bar{\lambda}_{\alpha ijk} &= K_{\alpha ijk}(\bar{\bar{\rho}}_\beta, \bar{\bar{\theta}}_\beta, \phi_\beta, \bar{i}_\beta, \nabla\phi_\alpha, \mathbf{D}_\alpha, \mathbf{b}^{(\alpha)} + \mathbf{D}_\alpha, \nabla\boldsymbol{\nu}^{(\alpha)}) \\ &= \bar{\lambda}^0_{\alpha ijk}(\bar{\bar{\rho}}_\beta, \bar{\bar{\theta}}_\beta, \phi_\beta, \bar{i}_\beta, \nabla\phi_\alpha = 0, \mathbf{D}_\alpha, \mathbf{b}^{(\alpha)} + \mathbf{D}_\alpha, \nabla\boldsymbol{\nu}^{(\alpha)} = 0) \\ &\quad + \frac{\partial\bar{\lambda}^0_{\alpha ijk}}{\partial\nu^{(\alpha)}_{mn,\ell}}\nu^{(\alpha)}_{mn,\ell} + \frac{\partial\bar{\lambda}^0_{\alpha ijk}}{\partial\phi_{\alpha,\ell}}\phi_{\alpha,\ell} + O((\nabla\boldsymbol{\nu}^{(\alpha)})^3, (\nabla\phi_\alpha)^3)\end{aligned} \tag{8.5.5}$$

Since $\mathbf{K}_\alpha$ is a tensor-valued isotropic function that satisfies

$$\begin{aligned}&\mathbf{K}_\alpha(\bar{\bar{\rho}}_\beta, \bar{\bar{\theta}}_\beta, \phi_\beta, \bar{i}_\beta, \mathbf{Q}\nabla\phi_\alpha, \mathbf{Q}\mathbf{D}_\alpha\mathbf{Q}^T, \mathbf{Q}(\mathbf{b}^{(\alpha)}+\mathbf{D}_\alpha)\mathbf{Q}^T, \mathbf{Q}\nabla\boldsymbol{\nu}^\alpha\mathbf{Q}^T\mathbf{Q}^T)\\ &\quad = \mathbf{Q}\mathbf{K}_\alpha(\bar{\bar{\rho}}_\beta, \bar{\bar{\theta}}_\beta, \phi_\beta, \bar{i}_\beta, \nabla\phi_\alpha, \mathbf{D}_\alpha, \mathbf{b}^{(\alpha)}+\mathbf{D}_\alpha, \nabla\boldsymbol{\nu}^{(\alpha)})\mathbf{Q}^T\mathbf{Q}^T \qquad (8.5.6)\end{aligned}$$

we may choose the necessary condition $\mathbf{Q}=-\mathbf{I}$, giving

$$\begin{aligned}&\mathbf{K}_\alpha(\bar{\bar{\rho}}_\beta, \bar{\bar{\theta}}_\beta, \phi_\beta, \bar{i}_\beta, -\nabla\phi_\alpha, \mathbf{D}_\alpha, \mathbf{b}^{(\alpha)}+\mathbf{D}_\alpha, -\nabla\boldsymbol{\nu}^{(\alpha)})\\ &\quad = -\mathbf{K}_\alpha(\bar{\bar{\rho}}_\beta, \bar{\bar{\theta}}_\beta, \phi_\beta, \bar{i}_\beta, \nabla\phi_\alpha, \mathbf{D}_\alpha, \mathbf{b}^{(\alpha)}+\mathbf{D}_\alpha, \nabla\boldsymbol{\nu}^{(\alpha)})\end{aligned}$$

from where it follows that

$$\mathbf{K}_\alpha(\bar{\bar{\rho}}_\beta, \bar{\bar{\theta}}_\beta, \phi_\beta, \bar{i}_\beta, \nabla\phi_\alpha=\mathbf{0}, \mathbf{D}_\alpha, \mathbf{b}^{(\alpha)}+\mathbf{D}_\alpha, \nabla\boldsymbol{\nu}^\alpha=\mathbf{0})=\mathbf{0} \qquad (8.5.7)$$

In (8.5.5) it is necessary, therefore, to set

$$\bar{\boldsymbol{\lambda}}^0_\alpha(\bar{\bar{\rho}}_\beta, \bar{\bar{\theta}}_\beta, \phi_\beta, \bar{i}_\beta, \nabla\phi_\alpha=\mathbf{0}, \mathbf{D}_\alpha, \mathbf{b}^{(\alpha)}+\mathbf{D}_\alpha, \nabla\boldsymbol{\nu}^{(\alpha)}=\mathbf{0})=\mathbf{0} \qquad (8.5.8)$$

Moreover,

$$\frac{\partial\bar{\lambda}^0_{\alpha ijk}}{\partial\nu^{(\alpha)}_{mn,\ell}} \quad and \quad \frac{\partial\bar{\lambda}^0_{\alpha ijk}}{\partial\phi_{\alpha,\ell}}$$

are isotropic tensors of order six and four, respectively, and upon using equation (A.47) of the appendix, (8.5.5) becomes

$$\begin{aligned}\bar{\lambda}_{\alpha ijk} &= \delta_{ij}d^\alpha_{1,\alpha}\phi_{\alpha,k} + \delta_{ik}d^\alpha_{2,\alpha}\phi_{\alpha,j} + \delta_{jk}d^\alpha_{3,\alpha}\phi_{\alpha,i}\\ &+\delta_{ij}(a^\alpha_{4,\alpha}\nu^{(\alpha)}_{mm,k} + b^\alpha_{4,\alpha}\nu^{(\alpha)}_{k\ell,\ell} + c^\alpha_{4,\alpha}\nu^{(\alpha)}_{\ell k,\ell}) + \delta_{ik}(a^\alpha_{5,\alpha}\nu^{(\alpha)}_{mm,j} + b^\alpha_{5,\alpha}\nu^{(\alpha)}_{\ell j,\ell} + c^\alpha_{5,\alpha}\nu^{(\alpha)}_{j\ell,\ell})\\ &+\delta_{jk}(a^\alpha_{6,\alpha}\nu^{(\alpha)}_{mm,i} + b^\alpha_{6,\alpha}\nu^{(\alpha)}_{i\ell,\ell} + c^\alpha_{6,\alpha}\nu^{(\alpha)}_{\ell i,\ell}) + a^\alpha_{7,\alpha}\nu^{(\alpha)}_{ij,k} + b^\alpha_{7,\alpha}\nu^{(\alpha)}_{ik,j} + c^\alpha_{7,\alpha}\nu^{(\alpha)}_{ji,k}\\ &+a^\alpha_{8,\alpha}\nu^{(\alpha)}_{ki,j} + b^\alpha_{8,\alpha}\nu^{(\alpha)}_{jk,i} + c^\alpha_{8,\alpha}\nu^{(\alpha)}_{kj,i} + O((\nabla\boldsymbol{\nu}^{(\alpha)})^3, (\nabla\phi_\alpha)^3)\end{aligned} \qquad (8.5.9)$$

where the coefficients $a$'s, $b$'s, and $c$'s depend on the invariants of

$$\bar{\bar{\rho}}_\beta, \bar{\bar{\theta}}_\beta, \phi_\beta, \bar{i}_\beta, \mathbf{D}_\alpha, \mathbf{b}^{(\alpha)}+\mathbf{D}_\alpha$$

Using (A.39) and neglecting the second order effect of $\mathbf{b}^{(\alpha)}+\mathbf{D}_\alpha$, this *irreducible set of invariants* is given by

$$\begin{aligned}&\bar{\bar{\rho}}_\beta, \bar{\bar{\theta}}_\beta, \phi_\beta, \bar{i}_\beta, tr\mathbf{D}_\alpha, tr\mathbf{D}^2_\alpha, tr\mathbf{D}^3_\alpha, tr\mathbf{D}_\alpha(\mathbf{b}^{(\alpha)}+\mathbf{D}_\alpha),\\ &\qquad tr\mathbf{D}^2_\alpha(\mathbf{b}^{(\alpha)}+\mathbf{D}_\alpha), tr(\mathbf{b}^{(\alpha)}+\mathbf{D}_\alpha)\end{aligned} \qquad (8.5.10)$$

It should be noted that the coefficients $d^{\alpha}_{1,\alpha}$, $d^{\alpha}_{2,\alpha}$, and $d^{\alpha}_{3,\alpha}$ in (8.5.9) are different from the same coefficients in (8.3.10), since the former coefficients depend on $\mathbf{D}_\alpha$ and $\mathbf{b}^{(\alpha)}+\mathbf{D}_\alpha$, whereas the latter do not.

The stress tensor is uniquely expressable as the sum of symmetric and skew-symmetric parts, *i.e.*

$$\bar{\mathbf{T}}_\alpha = (\bar{\mathbf{T}}_\alpha)_{sym.} + (\bar{\mathbf{T}}_\alpha)_{skew} \tag{8.5.11}$$

where use can be made for each part of the tensor representation theorems of section 4 of the appendix. Thus, be defining

$$\mathbf{M}_\alpha = \nabla\phi_\alpha \times \nabla\phi_\alpha \tag{8.5.12}$$

and using (A.42) and (A.44), gives

$$\begin{aligned}(\bar{\mathbf{T}}_\alpha)_{sym.} &= \beta_0\mathbf{I} + \beta_1\mathbf{D}_\alpha + \beta_2\mathbf{D}_\alpha^2 + \beta_3(\mathbf{b}^{(\alpha)} + \mathbf{D}_\alpha + \mathbf{b}^{(\alpha)T} + \mathbf{D}_\alpha)\\
&+\beta_4[\mathbf{D}_\alpha(\mathbf{b}^{(\alpha)} + \mathbf{D}_\alpha + \mathbf{b}^{(\alpha)T} + \mathbf{D}_\alpha) + (\mathbf{b}^{(\alpha)} + \mathbf{D}_\alpha + \mathbf{b}^{(\alpha)T} + \mathbf{D}_\alpha)\mathbf{D}_\alpha]\\
&+\beta_5[\mathbf{D}_\alpha^2(\mathbf{b}^{(\alpha)} + \mathbf{D}_\alpha + \mathbf{b}^{(\alpha)T} + \mathbf{D}_\alpha) + (\mathbf{b}^{(\alpha)} + \mathbf{D}_\alpha + \mathbf{b}^{(\alpha)T} + \mathbf{D}_\alpha)\mathbf{D}_\alpha^2]\\
&+\beta_6\mathbf{M}_\alpha + \beta_7(\mathbf{M}_\alpha\mathbf{D}_\alpha + \mathbf{D}_\alpha\mathbf{M}_\alpha) + \beta_8(\mathbf{M}_\alpha\mathbf{D}_\alpha^2 + \mathbf{D}_\alpha^2\mathbf{M}_\alpha)\\
&+\beta_9[\mathbf{M}_\alpha(\mathbf{b}^{(\alpha)} + \mathbf{D}_\alpha + \mathbf{b}^{(\alpha)T} + \mathbf{D}_\alpha) + (\mathbf{b}^{(\alpha)} + \mathbf{D}_\alpha + \mathbf{b}^{(\alpha)T} + \mathbf{D}_\alpha)\mathbf{M}_\alpha]\\
&+\beta_{10}[\mathbf{D}_\alpha(\mathbf{b}^{(\alpha)} + \mathbf{D}_\alpha - \mathbf{b}^{(\alpha)T} - \mathbf{D}_\alpha) - (\mathbf{b}^{(\alpha)} + \mathbf{D}_\alpha - \mathbf{b}^{(\alpha)T} - \mathbf{D}_\alpha)\mathbf{D}_\alpha]\\
&+\beta_{11}[\mathbf{D}_\alpha^2(\mathbf{b}^{(\alpha)} + \mathbf{D}_\alpha - \mathbf{b}^{(\alpha)T} - \mathbf{D}_\alpha) - (\mathbf{b}^{(\alpha)} + \mathbf{D}_\alpha - \mathbf{b}^{(\alpha)T} - \mathbf{D}_\alpha)\mathbf{D}_\alpha^2]\\
&+\beta_{12}[\mathbf{M}_\alpha(\mathbf{b}^{(\alpha)} + \mathbf{D}_\alpha - \mathbf{b}^{(\alpha)T} - \mathbf{D}_\alpha) - (\mathbf{b}^{(\alpha)} + \mathbf{D}_\alpha - \mathbf{b}^{(\alpha)T} - \mathbf{D}_\alpha)\mathbf{M}_\alpha]\\
&+O((\nabla\boldsymbol{\nu}^{(\alpha)})^2, (\mathbf{b}^{(\alpha)} + \mathbf{D}_\alpha)^2)\end{aligned} \tag{8.5.13}$$

$$\begin{aligned}(\bar{\mathbf{T}}_\alpha)_{skew} &= \gamma_1(\mathbf{b}^{(\alpha)} + \mathbf{D}_\alpha - \mathbf{b}^{(\alpha)T} - \mathbf{D}_\alpha) + \gamma_2[\mathbf{D}_\alpha(\mathbf{b}^{(\alpha)} + \mathbf{D}_\alpha + \mathbf{b}^{(\alpha)T} + \mathbf{D}_\alpha)\\
&-(\mathbf{b}^{(\alpha)} + \mathbf{D}_\alpha + \mathbf{b}^{(\alpha)T} + \mathbf{D}_\alpha)\mathbf{D}_\alpha] + \gamma_3[\mathbf{D}_\alpha^2(\mathbf{b}^{(\alpha)} + \mathbf{D}_\alpha + \mathbf{b}^{(\alpha)T} + \mathbf{D}_\alpha)\\
&-(\mathbf{b}^{(\alpha)} + \mathbf{D}_\alpha + \mathbf{b}^{(\alpha)T} + \mathbf{D}_\alpha)\mathbf{D}_\alpha^2] + \gamma_4[\mathbf{D}_\alpha(\mathbf{b}^{(\alpha)} + \mathbf{D}_\alpha + \mathbf{b}^{(\alpha)T} + \mathbf{D}_\alpha)\mathbf{D}_\alpha^2\\
&-\mathbf{D}_\alpha^2(\mathbf{b}^{(\alpha)} + \mathbf{D}_\alpha + \mathbf{b}^{(\alpha)T} + \mathbf{D}_\alpha)\mathbf{D}_\alpha] + \gamma_5(\mathbf{M}_\alpha\mathbf{D}_\alpha - \mathbf{D}_\alpha\mathbf{M}_\alpha)\\
&+\gamma_6[\mathbf{M}_\alpha(\mathbf{b}^{(\alpha)} + \mathbf{D}_\alpha + \mathbf{b}^{(\alpha)T} + \mathbf{D}_\alpha) - (\mathbf{b}^{(\alpha)} + \mathbf{D}_\alpha + \mathbf{b}^{(\alpha)T} + \mathbf{D}_\alpha)\mathbf{M}_\alpha]\\
&+\gamma_7(\mathbf{M}_\alpha\mathbf{D}_\alpha^2 - \mathbf{D}_\alpha^2\mathbf{M}_\alpha) + \gamma_8(\mathbf{D}_\alpha\mathbf{M}_\alpha\mathbf{D}_\alpha^2 - \mathbf{D}_\alpha^2\mathbf{M}_\alpha\mathbf{D}_\alpha)\\
&+\gamma_9\{\mathbf{D}_\alpha\mathbf{M}_\alpha(\mathbf{b}^{(\alpha)} + \mathbf{D}_\alpha + \mathbf{b}^{(\alpha)T} + \mathbf{D}_\alpha)\\
&-(\mathbf{b}^{(\alpha)} + \mathbf{D}_\alpha + \mathbf{b}^{(\alpha)T} + \mathbf{D}_\alpha)\mathbf{M}_\alpha\mathbf{D}_\alpha + \mathbf{M}_\alpha[(\mathbf{b}^{(\alpha)} + \mathbf{D}_\alpha + \mathbf{b}^{(\alpha)T} + \mathbf{D}_\alpha)\mathbf{D}_\alpha\\
&-\mathbf{D}_\alpha(\mathbf{b}^{(\alpha)} + \mathbf{D}_\alpha + \mathbf{b}^{(\alpha)T} + \mathbf{D}_\alpha)] - [\mathbf{D}_\alpha(\mathbf{b}^{(\alpha)} + \mathbf{D}_\alpha + \mathbf{b}^{(\alpha)T} + \mathbf{D}_\alpha)\\
&-(\mathbf{b}^{(\alpha)} + \mathbf{D}_\alpha + \mathbf{b}^{(\alpha)T} + \mathbf{D}_\alpha)\mathbf{D}_\alpha]\mathbf{M}_\alpha\} + \gamma_{10}[\mathbf{D}_\alpha(\mathbf{b}^{(\alpha)} + \mathbf{D}_\alpha - \mathbf{b}^{(\alpha)T} - \mathbf{D}_\alpha)\\
&+(\mathbf{b}^{(\alpha)} + \mathbf{D}_\alpha - \mathbf{b}^{(\alpha)T} - \mathbf{D}_\alpha)\mathbf{D}_\alpha] + \gamma_{11}[\mathbf{M}_\alpha(\mathbf{b}^{(\alpha)} + \mathbf{D}_\alpha - \mathbf{b}^{(\alpha)T} - \mathbf{D}_\alpha)\\
&+(\mathbf{b}^{(\alpha)} + \mathbf{D}_\alpha - \mathbf{b}^{(\alpha)T} - \mathbf{D}_\alpha)\mathbf{M}_\alpha] + O((\nabla\boldsymbol{\nu}^{(\alpha)})^2, (\mathbf{b}^{(\alpha)} + \mathbf{D}_\alpha)^2)\end{aligned} \tag{8.5.14}$$

Adding (8.5.13) and (8.5.14) results in a general expression for $\bar{\mathbf{T}}_\alpha$, *i.e.*

$$\begin{aligned}
\bar{\mathbf{T}}_\alpha = \alpha_0\mathbf{I} + \alpha_1\mathbf{D}_\alpha + \alpha_2\mathbf{D}_\alpha^2 + \alpha_3(\mathbf{b}^{(\alpha)} + \mathbf{D}_\alpha) + \alpha_4(\mathbf{b}^{(\alpha)T} + \mathbf{D}_\alpha) \\
+\alpha_5\mathbf{D}_\alpha(\mathbf{b}^{(\alpha)} + \mathbf{D}_\alpha) + \alpha_6(\mathbf{b}^{(\alpha)} + \mathbf{D}_\alpha)\mathbf{D}_\alpha + \alpha_7\mathbf{D}_\alpha(\mathbf{b}^{(\alpha)T} + \mathbf{D}_\alpha) \\
+\alpha_8(\mathbf{b}^{(\alpha)T} + \mathbf{D}_\alpha)\mathbf{D}_\alpha + \alpha_9\mathbf{D}_\alpha^2(\mathbf{b}^{(\alpha)} + \mathbf{D}_\alpha) + \alpha_{10}(\mathbf{b}^{(\alpha)} + \mathbf{D}_\alpha)\mathbf{D}_\alpha^2 \\
+\alpha_{11}\mathbf{D}_\alpha^2(\mathbf{b}^{(\alpha)T} + \mathbf{D}_\alpha) + \alpha_{12}(\mathbf{b}^{(\alpha)T} + \mathbf{D}_\alpha)\mathbf{D}_\alpha^2 \\
+\alpha_{13}[\mathbf{D}_\alpha(\mathbf{b}^{(\alpha)} + \mathbf{D}_\alpha)\mathbf{D}_\alpha^2 + \mathbf{D}_\alpha(\mathbf{b}^{(\alpha)T} + \mathbf{D}_\alpha)\mathbf{D}_\alpha^2 - \mathbf{D}_\alpha^2(\mathbf{b}^{(\alpha)} + \mathbf{D}_\alpha)\mathbf{D}_\alpha \\
-\mathbf{D}_\alpha^2(\mathbf{b}^{(\alpha)T} + \mathbf{D}_\alpha)\mathbf{D}_\alpha] + \alpha_{14}\mathbf{M}_\alpha + \alpha_{15}\mathbf{M}_\alpha\mathbf{D}_\alpha + \alpha_{16}\mathbf{D}_\alpha\mathbf{M}_\alpha + \alpha_{17}\mathbf{M}_\alpha\mathbf{D}_\alpha^2 \\
+\alpha_{18}\mathbf{D}_\alpha^2\mathbf{M}_\alpha + \alpha_{19}\mathbf{M}_\alpha(\mathbf{b}^{(\alpha)} + \mathbf{D}_\alpha) + \alpha_{20}\mathbf{M}_\alpha(\mathbf{b}^{(\alpha)T} + \mathbf{D}_\alpha) \\
+\alpha_{21}(\mathbf{b}^{(\alpha)} + \mathbf{D}_\alpha)\mathbf{M}_\alpha + \alpha_{22}(\mathbf{b}^{(\alpha)T} + \mathbf{D}_\alpha)\mathbf{M}_\alpha + \alpha_{23}(\mathbf{D}_\alpha\mathbf{M}_\alpha\mathbf{D}_\alpha^2 - \mathbf{D}_\alpha^2\mathbf{M}_\alpha\mathbf{D}_\alpha) \\
+\alpha_{24}\{\mathbf{D}_\alpha\mathbf{M}_\alpha(\mathbf{b}^{(\alpha)} + \mathbf{D}_\alpha + \mathbf{b}^{(\alpha)T} + \mathbf{D}_\alpha) - (\mathbf{b}^{(\alpha)} + \mathbf{D}_\alpha + \mathbf{b}^{(\alpha)T} + \mathbf{D}_\alpha)\mathbf{M}_\alpha\mathbf{D}_\alpha \\
+\mathbf{M}_\alpha[(\mathbf{b}^{(\alpha)} + \mathbf{D}_\alpha + \mathbf{b}^{(\alpha)T} + \mathbf{D}_\alpha)\mathbf{D}_\alpha - \mathbf{D}_\alpha(\mathbf{b}^{(\alpha)} + \mathbf{D}_\alpha + \mathbf{b}^{(\alpha)T} + \mathbf{D}_\alpha)] \\
-[\mathbf{D}_\alpha(\mathbf{b}^{(\alpha)} + \mathbf{D}_\alpha + \mathbf{b}^{(\alpha)T} + \mathbf{D}_\alpha) - (\mathbf{b}^{(\alpha)} + \mathbf{D}_\alpha + \mathbf{b}^{(\alpha)T} + \mathbf{D}_\alpha)\mathbf{D}_\alpha]\mathbf{M}_\alpha\} \\
+O((\nabla\boldsymbol{\nu}^{(\alpha)})^2, (\mathbf{b}^{(\alpha)} + \mathbf{D}_\alpha)^2)
\end{aligned} \tag{8.5.15}$$

The coefficients $\beta_i$; $i = 0, \ldots, 12$, $\gamma_j$; $j = 1, \ldots, 11$, and $\alpha_k$; $k = 0, \ldots, 24$ in above equations depend on the invariants of independent constitutive variables in (8.5.2), which by employing equation (A.39) of the appendix are given by

$$\begin{aligned}
\bar{\bar{\rho}}_\beta, \bar{\bar{\theta}}_\beta, \phi_\beta, \bar{i}_\beta, tr\mathbf{M}_\alpha, tr\mathbf{D}_\alpha, tr\mathbf{D}_\alpha^2, tr\mathbf{D}_\alpha^3, tr(\mathbf{b}^{(\alpha)} + \mathbf{D}_\alpha), \\
tr\mathbf{D}_\alpha(\mathbf{b}^{(\alpha)} + \mathbf{D}_\alpha), tr\mathbf{D}_\alpha^2(\mathbf{b}^{(\alpha)} + \mathbf{D}_\alpha), tr\mathbf{M}_\alpha\mathbf{D}_\alpha, tr\mathbf{M}_\alpha\mathbf{D}_\alpha^2, \\
tr\mathbf{M}_\alpha(\mathbf{b}^{(\alpha)} + \mathbf{D}_\alpha), tr\mathbf{M}_\alpha(\mathbf{b}^{(\alpha)} + \mathbf{D}_\alpha + \mathbf{b}^{(\alpha)T} + \mathbf{D}_\alpha)\mathbf{D}_\alpha, \\
tr\mathbf{M}_\alpha(\mathbf{b}^{(\alpha)} + \mathbf{D}_\alpha - \mathbf{b}^{(\alpha)T} - \mathbf{D}_\alpha)\mathbf{D}_\alpha, tr\mathbf{M}_\alpha(\mathbf{b}^{(\alpha)} + \mathbf{D}_\alpha - \mathbf{b}^{(\alpha)T} - \mathbf{D}_\alpha)\mathbf{D}_\alpha^2, \\
\beta = 1, \ldots, \gamma
\end{aligned} \tag{8.5.16}$$

where the terms involving $(\mathbf{b}^{(\alpha)} + \mathbf{D}_\alpha)^2$ have been neglected.

An identical expression to (8.5.15) with $\alpha_k$ replaced by $\eta_k$, $k=1, \ldots, 24$, can also be obtained for $\bar{\mathbf{S}}_\alpha$ and other second order tensor-valued isotropic functions. Similarly, nonlinear constitutive equations for vector-valued isotropic functions can be obtained using equation (A.40), whereas the scalar-valued isotropic functions can be expressed in polynomials of the invariants of independent constitutive variables using (A.39). For multiphase mixtures, the nonlinear terms in constitutive equations should be introduced selectively in order to produce *practical* models of mixtures. The linear forms of $\bar{\mathbf{T}}_\alpha$ and $\bar{\boldsymbol{\lambda}}_\alpha$ developed in previous sections are contained within the nonlinear forms of these equations expressed by (8.5.15) and (8.5.9), *if*, in addition, use is made of (6.12.36).

## 8.6 Concluding Remarks

In this chapter, the linear constitutive equations for multiphase mixtures of fluids with rotational and dilatational effects were considered. The rotational effect considerably complicates the theory of mixtures by introducing three additional balance equations. Further complications in the theory of mixtures can be produced by *not assuming* an isotropy of the inertia tensor and by constructing nonlinear constitutive equations as discussed in section 8.5. The simplest way to introduce nonlinear effects into the constitutive equations is in the phenomenological coefficients in the form of polynomials of irreducible invariants of independent constitutive variables. Models of multiphase mixtures with combined fluidlike and solidlike phases can also be developed by employing the contitutive principles of chapter 6, whereas the higher order theories of structured multiphase mixtures may be constructed by using higher order approximations in equation (3.1.3).

# Bibliography


[1] AHMADI, G. 1980 On mechanics of saturated granular materials. **Int. J. Nonlinear Mech. 15**, 251-262.

[2] AHMADI, G. 1982 A continuum theory for two-phase media. **Acta Mechanica 44**, 299-317.

[3] AHMADI, G. 1985 A generalized continuum theory for multiphase suspension flows. **Int. J. Engng. Sci. 23**, 1-25.

[4] BAGNOLD, R.A. 1956 The flow of cohesionless grains in fluids. **Phil. Trans. R. Soc. London A 249**, 235-297.

[5] BEDFORD, A., & DRUMHELLER, D.S. 1978 A variational theory of immiscible mixtures. **Arch. Rat. Mech. Anal. 68**, 37-51.

[6] BEDFORD, A., & DRUMHELLER, D.S. 1983 Theories of immiscible and structured mixtures. **Int. J. Engng. Sci. 21**, 863-960.

[7] BOEHLER, J.P. 1977 On irreducible representations for isotropic scalar functions. **ZAMM 57**, 323-327.

[8] BOWEN, R. 1976 Theory of Mixtures. **Continuum Physics III** (Edited by A.C. ERINGEN), Academic Press, New York.

[9] BOWEN, R. 1982 Compressible porous media models by use of the theory of mixtures. **Int. J. Engng. Sci. 20**, 697-735.

[10] CALLEN, H.B. 1960 **Thermodynamics**. John Wiley & Sons, New York.

[11] CAPRIZ, G., & PODIO-GUIDUGLI, P. 1981 Materials with spherical structure. **Arch. Rat. Mech. Anal. 75**, 269-279.

[12] CELMINS, A.K.R., & SCHMITT, J.A. 1984 Three dimensional modeling of gas combusting solid two-phase flows. **Multiphase Flow and Heat Transfer III.** Part B (Edited by T.N. VEZIROGLU, & A.E. BERGLES), Elsevier, Netherlands, 681-698.

[13] CHAPMAN, S., & COWLING, T.G. 1970 **The Mathematical Theory of Non-Uniform Gases**. Cambridge Univ. Press, London.

[14] COLEMAN, B.D., & NOLL, W. 1963 The thermodynamics of elastic materials with heat conduction and viscosity. **Arch. Rat. Mech. Anal. 13**, 167-178.

[15] COSSERAT, E., & F. 1909 **Théorie des Corps Déformables**. Hermann, Paris.

[16] COWIN, S.C., & NUNZIATO, J.W. 1981 Waves of dilatancy in a granular material with incompressible granules. **Int. J. Engng. Sci. 19**, 993-1008.

[17] CROSS, J.J. 1973 Mixtures of fluids and isotropic solids. **Archives of Mechanics 25**, 1025-1039.

[18] DEEMER, A.R., & SLATTERY, J.C. 1978 Balance equations and structural models for phase interfaces. **Int. J. Multiphase Flow 4**. 171-193.

[19] DELHAYE, J.M., & ACHARD, J.L. 1977 On the use of averaging operators in two-phase flow modeling. **Symp. Thermal Hydraulic Aspects of Nuclear Reactor Safety 1** (Edited by O.C. JONES, & S.G. BANKOFF), ASME, New York.

[20] DOBRAN, F. 1981 On the consistency conditions of averaging operators of two-phase flow models and on the formulation of magnetohydrodynamic two-phase flow. **Int. J. Engng. Sci. 19**, 1353-1368.

[21] DOBRAN, F. 1982a Theory of Multiphase Mixtures. SIT Technical Report ME-RT-81015.

[22] DOBRAN, F. 1982b A two-phase fluid model based on the linearized constitutive equations. **Advances in Two-Phase Flow and Heat Transfer 1**, Martinus Nijhoff Publishers, Netherlands, 41-49.

[23] DOBRAN, F. 1983 An acceleration wave model for the speed of propagation of shock waves in a bubbly two-phase flow. **ASME/JSME Thermal Engng. Conf., Vol. 1** (Edited by Y. MORI, & W. YANG), ASME, New York.

[24] DOBRAN, F. 1984a On the formulation of conservation, balance and constitutive equations for multiphase flows. **Multiphase Flow and Heat Transfer III**. Part A (Edited by T.N. VEZIROGLU, & A.E. BERGLES), Elsevier, Netherlands, 23-39.

[25] DOBRAN, F. 1984b Constitutive equations for multiphase mixtures of fluids. **Int. J. Multiphase Flow 10**, 273-305.

[26] DOBRAN, F. 1985a Theory of multiphase mixtures. **Int. J. Multiphase Flow 11**, 1-30.

[27] DOBRAN, F. 1985b On the foundations of theory of multiphase mixtures. Seminar delivered at Ballistic Research Lab., Aberdeen Proving Grounds, Sept. 18, 1985.

[28] DREW, D.A. 1971 Averaged field equations for two-phase media. **Studies in Appl. Math. L**, 133-166.

[29] DREW, D.A., & SEGEL, L.A. 1971 Averaged equations for two-phase flows. **Studies in Appl. Math. L.**, 205-231.

[30] DRUMHELLER, D.S., & BEDFORD, A. 1980 A thermomechanical theory of reacting immiscible mixtures. **Arch. Rat. Mech. Anal. 73**, 257-284.

[31] DRUMHELLER, D.S., KIPP, M.E., & BEDFORD, A. 1982 Transient wave propagation in bubbly liquids. **J. Fluid Mech. 119**, 347-365.

[32] EDELEN, D.G.B., & MCLENNAN, D.A. 1973 Material indifference: a principle of convenience. **Int. J. Engng. Sci. 11**, 813-817.

[33] ERICKSEN, J.L. 1960 Anisotropic fluids. **Arch. Rat. Mech. Anal. 4**, 231-237.

[34] ERINGEN, A.C. 1964 Simple microfluids. **Int. J. Engng. Sci. 2**, 205-217.

[35] ERINGEN, A.C., & SUHUBI, E.S. 1964 Nonlinear theory of simple micro-elastic solids I. **Int. J. Engng. Sci. 2**, 189-203.

[36] ERINGEN, A.C. 1967 Theory of micropolar fluids. **J. Math. Anal. 16**, 1-18.

[37] ERINGEN, A.C. 1975 **Continuum Physics II**. Academic Press, New York.

[38] ERINGEN, A.C. 1980 **Mechanics of Continua**. R.E. Krieger Publ. Company.

[39] GOODMAN, M.A., & COWIN, S.C. 1971 Two problems in the gravity flow of granular materials. **J. Fluid Mech. 45**, 321-339.

[40] GOODMAN, M.A., & COWIN, S.C. 1972 A continuum theory for granular materials. **Arch. Rat. Mech. Anal. 44**, 249-266.

[41] GRAD, H. 1952 Statistical mechanics, thermodynamics, and fluid dynamics of systems with an arbitrary number of integrals. **Comm. Pure Appl. Math. 5**, 455-494.

[42] GURTIN, M.E., & GUIDUGLI, P.P. 1973 The thermodynamics of constrained materials. **Arch. Rat. Mech. Anal. 51**, 192-208.

[43] HANES, D.M., & INMAN, D.L. 1985a Observations of rapidly flowing granular-fluid materials. **J. Fluid Mech. 150**, 357-380.

[44] HANES, D.M., & INMAN, D.L. 1985b Experimental evaluation of a dynamic yield criterion for granular fluid flows. **J. Geophys. Res. 90**, 3670-3674.

[45] HASSANIZADEH, M., & GRAY, W.G. 1979 General conservation equations for multi-phase systems: 1. Averaging procedure. **Adv. Water Resources 2**, 131-144.

[46] HECKL, M., & MÜLLER, I. 1983 Frame indifference, entropy, entropy flux, and wave speeds in mixtures of gases. **Acta Mechanica 50**, 71-95.

[47] HUTTER, K., SZIDAROVZSKY, F., & YAKOWITZ, S. 1986 Plane steady shear flow of a cohesionless granular material down an inclined plane: A model for flow avalanches. Part I: Theory. **Acta Mechanica 63**, 87-112.

[48] HUTTER, K., SZIDAROVZSKY, F., & YAKOWITZ, S. 1986 Plane Steady shear flow of a cohesionless granular material down an inclined plane: A model for flow avalanches. Part II: Numerical results. **Acta Mechanica 65**, 239-261.

[49] ISHII, M. 1975 **Thermo-Fluid Dynamic Theory of Two-Phase Flows.** Eyrolles, Paris.

[50] JENKINS, J.T., & SAVAGE, S.B. 1983 A theory for the rapid flow of identical, smooth, nearly elastic, spherical particles. **J. Fluid Mech. 130**, 187-202.

[51] JENKINS, J.T., & MANCINI, F. 1987 Balance laws and constitutive relations for plane flows of a dense, binary mixture of smooth, nearly elastic, circular discs. **J. Appl. Mech. 54**, 27-34.

[52] JOHNSON, P.C., & JACKSON, R. 1987 Frictional-collisional constitutive relations for granular materials, with application to plane shearing. **J. Fluid Mech. 176**, 67-93.

[53] LUN, C.K.K., & SAVAGE, S.B. 1987 A simple kinetic theory for granular flow of rough, inelastic spherical particles. **J. Appl. Mech. 54**, 47-53.

[54] LUN, C.K.K., SAVAGE, S.B., JEFFREY, D.J., & CHEPURNITY, N. 1984 Kinetic theories for granular flow: inelastic particles in Couette flow and slightly inelastic particles in a general flow field. **J. Fluid Mech. 140**, 223-256.

[55] MURDOCH, A.I. 1985 A corpuscular approach to continuum mechanics: Basic considerations. **Arch. Rat. Mech. Anal. 88**, 291-321.

[56] MÜLLER, I. 1968 A thermodynamic theory of mixtures of fluids. **Arch. Rat. Mech. Anal. 28**, 1-39.

[57] MÜLLER, I. 1972 On the frame dependence of stress and heat flux. **Arch. Rat. Mech. Anal. 45**, 241-250.

[58] NIGMATULIN, R.I. 1979 Spatial averaging in the mechanics of heterogeneous and dispersed systems. **Int. J. Multiphase Flow 5**, 353-385.

[59] NITSCHE, L.C., & BRENNER, H. 1989 Eulerian kinematics of flow through spatially periodic models of porous media. **Arch. Rat. Mech. Anal. 107**, 225-292.

[60] NOLL, W. 1958 A mathematical theory of the mechanical behavior of continuous media. **Arch. Rat. Mech. Anal. 2**, 197-226.

[61] NOLL, W. 1973 Lectures on the foundations of continuum mechanics and thermodynamics. **Arch. Rat. Mech. Anal. 52**, 62-92.

[62] NUNZIATO, J.W., & WALSH, E.K. 1980 On ideal multiphase mixtures with chemical reactions and diffusion. **Arch. Rat. Mech. Anal. 73**, 285-311.

[63] PASSMAN, S.L. 1977 Mixtures of granular materials. **Int. J. Engng. Sci. 15**, 117-129.

[64] PASSMAN, S.L., & NUNZIATO, J.W. 1981 Shearing flow of a fluid saturated granular material. In **Studies in Appl. Mech. 5**, Part A (Edited by A.P.S. SELVADURAI), Elsevier, Amsterdam, 343-353.

[65] PASSMAN, S.L., NUNZIATO, J.W., & WALSH, E.K. 1984 A theory of multiphase mixtures. In **Rational Thermodynamics**, C. TRUESDELL, Springer Verlag, New York.

[66] SAVAGE, S.B. 1979 Gravity flow of cohesionless granular materials in chutes and channels. **J. Fluid Mechanics 92**, 53-96.

[67] SAVAGE, S.B. 1984 The mechanics of rapid granular flows. **Adv. Appl. Mech. 24**, 289-366.

[68] SAVAGE, S.B., & JEFFREY, D.J. 1981 The stress tensor in granular flow at high shear rates. **J. Fluid Mech. 110**, 255-272.

[69] SAVAGE, S.B., & MCKEOWN, S. 1983 Shear stresses developed during rapid shear of concentrated suspensions of large spherical particles between concentric cylinders. **J. Fluid Mech. 127**, 453-472.

[70] SAVAGE, S.B., & SAYED, M. 1984 Stresses developed by very cohesionless granular materials sheared in an annular shear cell. **J. Fluid Mech. 142**, 391-430.

[71] SMITH, G.F. 1971 On isotropic functions of symmetric tensors, skew-symmetric tensors and vectors. **Int. J. Engng. Sci. 9**, 899-916.

[72] SOKOLOVSKI, V.V. 1965 **Statics of Granular Media**. Pergamon.

[73] SPENCER, A.J.M. 1971 Theory of Invariants. In **Continuum Physics I** (Edited by A.C. ERINGEN), Academic Press, New York.

[74] SPEZIALE, C.G. 1986 The effect of the reference frame on the thermophysical properties of an ideal gas. **Int. J. Thermophysics 7**, 99-110.

[75] SPEZIALE, C.G. 1988 The Einstein equivalence principle, intrinsic spin and the invariance of constitutive equations in continuum mechanics. **Int. J. Engng. Sci. 26**, 211-220.

[76] TOUPIN, R.A. 1964 Theories of elasticity with couple-stress. **Arch. Rat. Mech. Anal. 17**, 85-112.

[77] TRAPP, J.A. 1976 On the relationship between continuum mixture theory and integral averaged equations for immiscible fluids. **Int. J. Engng. Sci. 14**, 991-998.

[78] TRUESDELL, C. 1976 Correction of two errors in the kinetic theory of gases which have been used to cast unfounded doubt upon a principle of material frame-indifference. **Meccanica 11**, 196-199.

[79] TRUESDELL, C. 1977 **A First Course in Rational Continuum Mechanics**. Academic Press, New York.

[80] TRUESDELL, C. 1984 **Rational Thermodynamics**. 2nd ed., Springer Verlag, New York.

[81] TRUESDELL, C., & NOLL, W. 1965 The nonlinear field theories of mechanics. **Handbuch der Physik Band III/1**, Springer Verlag, Berlin.

[82] TRUESDELL, C., & TOUPIN, R.A. 1960 The classical field theories. **Handbuch der Physic Band III/1**, Springer, Berlin.

[83] TWISS, R.J., & ERINGEN, A.C. 1971 Theory of mixtures for micromorphic materials I. **Int. J. Engng. Sci. 9**, 1019-1044.

[84] VAN WIJNGAARDEN, L. 1972 One dimensional flow of liquids containing small gas bubbles. **Annual Review of Fluid Mechanics** 4, 369-376.

[85] WANG, C.C. 1971 Corrigendum to representations for isotropic functions. **Arch. Rat. Mech. Anal. 43**, 392-395.

[86] WANG, C.C. 1975 On the concept of frame-indifference in continuum mechanics and in the kinetic theory of gases. **Arch. Rat. Mech. Anal. 58**, 381-393.

APPENDIX

# THEOREMS OF ALGEBRA, GEOMETRY, AND CALCULUS

The purpose of this appendix is to summarize some basic results from algebra, geometry, and calculus for the purpose of making the monograph more readily accessible to those less familiar with the basic mathematical concepts used in the book. The symbols below have the following meaning:

| | | |
|---|---|---|
| $\forall$ | *meaning* | *"for every"* |
| $A \rightarrow B$ | *meaning* | *"mapping of A into B"* |
| $\rightarrow$ | *meaning* | *"implies"* |
| $x$ | *meaning* | *"cartesian product of sets"* |
| $\times$ | *meaning* | *"tensor product"* |
| $\epsilon$ | *meaning* | *"belongs to"* |
| $\cup$ | *meaning* | *"union"* |
| $\subset$ | *meaning* | *"is included"* |

## 1 Concepts from Abstract Algebra

**Necessity and Sufficiency.** A statement that $P$ is true *if and only if* a condition $Q$ holds is equivalent to the statement that a necessary and sufficient condition for $P$ to be true is that $Q$ holds. The proof must proceed as follows:

*Sufficiency* (if) ($Q$ is sufficient for $P$) - Assume $Q$ holds; then show that this implies that $P$ is true.

*Necessity* (only if) ($Q$ is necessary for $P$) - Assume $P$ holds; then show that $Q$ follows as a consequence.

**Functions.** A *function* $f$ from a set $A$ into a set $B$ is an ordered triple of sets $(f, A, B)$ denoted by $f$: $A \rightarrow B$ ($f$ maps $A$ into $B$), where:

1. $f \subseteq AxB$ ($f$ is a *proper subset* of $AxB$).

2. $\forall\ x \epsilon A$ there exists a $y \epsilon B$ such that $(x, y) \epsilon f$.

3. $\forall\ x \epsilon A$ and $y_1, y_2\ \epsilon B$, if $(x, y_1) \epsilon f$ *and* $(x, y_2) \epsilon f$, then $y_1 = y_2$.

When $(x,y) \in f$, we write

$$y = f(x) \tag{A.1}$$

1. *Surjective (Onto) Functions.* A function $f : A \to B$ is *surjective*, or from $A$ *onto* $B$, if and only if every $b \in B$ is the image of some element of $A$.

2. *Injective (One-to-One) Functions.* A function $f : A \to B$ is *injective*, or *one-to-one*, if for every $b$ belonging to the range of $f$ there is exactly one $a \in A$ such that $b = f(a)$.

3. *Bijective (One-to-One and Onto) Functions.* A function $f : A \to B$ is *bijective*, or *one-to-one and onto*, if and only if it is both injective and surjective; *i.e.*, if and only if every $b \in B$ is the unique image of some $a \in A$.

A function $f : A \to B$ is invertible if and only if it is bijective.

**Homomorphism.** Let $@ = \{A, *\}$ and $\# = \{B, o\}$ denote two systems, @ consisting of a set $A$ and a binary operation $*$ defined on $A$, and # consisting of a set $B$ and a binary operation $o$ defined on $B$. A *homomorphism* of @ into # is a mapping $H : @ \to \#$ such that for each $a, b \in A$

$$H(a * b) = H(a) \; o \; H(b) \tag{A.2}$$

Note that there need not be a one-to-one correspondence between the elements of @ and their images in #.

**Isomorphism.** Let @ and # be two systems as above. The systems @ and # are *isomorphic* if and only if the following hold:

1. There exists a bijective map $F : A \to B$.
2. The operations are preserved by the mapping $F$ in the sense that if $a, b \in A$, then $F(a * b) = F(a) \; o \; F(b)$.

The mapping is then referred to as an *isomorphism* or an *isomorphic mapping* of @ into #. Isomorphism is a special case of homomorphism in which the mapping is bijective.

**Group.** Group is a system consisting of a set $G$ and a binary operation $*$ on $G$ such that the following axioms must hold:

1. *Closure.* If $a, b \in G$, then $a * b \in G$, *i.e.* a group is *closed* under the operation $*$.
2. *Associative Law.* For every $a, b, c \in G$, $a * (b * c) = (a * b) * c$.
3. *Identity Element.* There exists an element $e \in G$ such that $a * e = e * a = a$ for every $a \in G$.
4. *Inverse.* For each $a \in G$, there exists an inverse element $a^{-1} \in G$ such that $a * a^{-1} = a^{-1} * a = e$.

If $\mathfrak{G} = \{G, *\}$ is a group, any group $\mathfrak{G}_1 = \{G_1, *\}$ such that $G_1 \subset G$ and $e \in G_1$ is a *subgroup* of $\mathfrak{G}$, denoted by $\mathfrak{G}_1 \subset \mathfrak{G}$.

## 2 Concepts from Linear Vector Spaces

The real linear vector space under consideration in the book is Euclidean with a cartesian coordinate system. The elements of this space are scalars usually indicated by the light-faced italics $a, b, \phi, A, B, \Phi, \ldots$, vectors indicated by bold-faced miniscule letters $\mathbf{a}$,$\mathbf{b}$, $\boldsymbol{\xi}, \ldots$, and tensors indicated by bold-faced letters $\mathbf{A}$,$\mathbf{B}$,$\mathbf{T}$,$\boldsymbol{\Phi}, \ldots$. The standard basis of the space is defined as follows:

$$\mathbf{e}_1 = (1,0,0), \quad \mathbf{e}_2 = (0,1,0), \quad \mathbf{e}_3 = (0,0,1) \tag{A.3}$$

such that if $\mathbf{x}$ is any vector, then

$$\mathbf{x} = x_k \mathbf{e}_k \tag{A.4}$$

where $x_k$ are unique scalars or *components* of vector $\mathbf{x}$. The subscript $k$ is the *tensor index* and repeated implies a summation from 1 to 3. The tensorial indices always occur as subscripts and are denoted by the italic light-faced miniscules $i, j, k, \ldots$ The basis $\mathbf{e}_1$, $\mathbf{e}_2$, $\mathbf{e}_3$ is *orthonormal*, since

$$\begin{aligned} \mathbf{e}_q \cdot \mathbf{e}_\ell = \delta_{q\ell} &= 1 \;\; if \;\; q = \ell \\ &= 0 \;\; if \;\; q \neq \ell \end{aligned} \tag{A.5}$$

where $\delta_{q\ell}$ is the Kronecker delta, with $\delta_{qq} = 3$. The magnitude $|\mathbf{x}|$ of the vector $\mathbf{x}$ is defined with an *inner product*, denoted by a dot, *i.e.*

$$|\mathbf{x}|^2 = \mathbf{x} \cdot \mathbf{x} \tag{A.6}$$

The two vectors $\mathbf{x}$ and $\mathbf{y}$ are *orthogonal* if

$$\mathbf{x} \cdot \mathbf{y} = 0 \tag{A.7}$$

The second order tensor $\mathbf{A}$ is a linear transformation; a mapping $\mathbf{A}$ of the vector space $\mathfrak{S}$ into the vector space $\mathfrak{S}'$, *i.e.*

$$\mathbf{A}(a\mathbf{x} + b\mathbf{y}) = a\,\mathbf{A}(\mathbf{x}) + b\,\mathbf{A}(\mathbf{y}) = a\,\mathbf{A}\mathbf{x} + b\,\mathbf{A}\mathbf{y} \tag{A.8}$$

for all vectors $\mathbf{x}$ and $\mathbf{y}$ and scalars $a$ and $b$. The linear transformation $\mathbf{A}$ can have an *inverse* $\mathbf{A}^{-1}$. If $dim\mathfrak{S} = dim\mathfrak{S}'$, then any of the following statements is a necessary and sufficient condition that $\mathbf{A}$ has an inverse: $\mathbf{A}$ is one-to-one, $\mathbf{A}$ maps $\mathfrak{S}$ onto $\mathfrak{S}'$, the nullspace of $\mathbf{A}$ contains only vector $\mathbf{0}$.

If $\mathbf{A}^{-1}$ exists, it is a linear mapping, and $(\mathbf{A}^{-1})^{-1} = \mathbf{A}$. The unit linear transformation $\mathbf{I}$ is a *unit tensor*, satisfying for all $\mathbf{x}$

$$\mathbf{I}\mathbf{x} = \mathbf{x} \tag{A.9}$$

With $\mathbf{e}_1$, $\mathbf{e}_2$, $\mathbf{e}_3$ defining a basis of the vector space, the condition

$$\mathbf{A}\mathbf{e}_i = A_{ij}\mathbf{e}_j \tag{A.10}$$

defines the *components* $A_{ij}$ of $\mathbf{A}$ relative to the basis. The components of $\mathbf{A}$ form a matrix $\| A_{ij} \|$, with the row index $i$ and column index $j$. The *determinant* of the tensor is the determinant of the matrix $\| A_{ij} \|$, and $\mathbf{A}$ is invertible if $det\mathbf{A} \neq 0$. If $\mathbf{A}$ and $\mathbf{B}$ are tensors, their composition $\mathbf{AB}$ is the *product* of $\mathbf{A}$ and $\mathbf{B}$. The components of $\mathbf{AB}$ are $A_{ij}B_{jk}$, and

$$det(\mathbf{AB}) = det\mathbf{A}\, det\mathbf{B} = det(\mathbf{BA}) \tag{A.11}$$

If $det\mathbf{A} = \pm 1$, then $\mathbf{A}$ is *unimodular.*

The *tensor product* of vectors $\mathbf{a}$ and $\mathbf{b}$ is the tensor $\mathbf{a} \times \mathbf{b}$, such that

$$(\mathbf{a} \times \mathbf{b})\mathbf{u} = (\mathbf{u}\cdot\mathbf{b})\mathbf{a} \quad \forall \quad \mathbf{u} \in \$ \tag{A.12}$$

The set of tensors $\mathbf{e}_i \times \mathbf{e}_j$ forms a basis for the space of tensors, *i.e.*

$$\begin{aligned} \mathbf{A} &= A_{ij}\, \mathbf{e}_i \times \mathbf{e}_j \\ \mathbf{I} &= \delta_{ij}\, \mathbf{e}_i \times \mathbf{e}_j \end{aligned} \tag{A.13}$$

Similarly, the products $\mathbf{e}_i \times \mathbf{e}_j \times \mathbf{e}_k$ form a basis for a third order tensor $\mathbf{U}$,

$$\mathbf{U} = U_{ijk}\, \mathbf{e}_i \times \mathbf{e}_j \times \mathbf{e}_k \tag{A.14}$$

with components $U_{ijk}$.

The *transpose* $\mathbf{A}^T$ of the tensor $\mathbf{A}$ satisfies

$$(A^T)_{ij} = A_{ji} \tag{A.15}$$

and

$$\begin{gathered} (\mathbf{A} + \mathbf{B})^T = \mathbf{A}^T + \mathbf{B}^T, \quad (\mathbf{AB})^T = \mathbf{B}^T\mathbf{A}^T, \quad (\mathbf{A}^T)^T = \mathbf{A} \\ (\mathbf{a} \times \mathbf{b})^T = \mathbf{b} \times \mathbf{a} \end{gathered} \tag{A.16}$$

and if $\mathbf{A}$ is invertible, then so is $\mathbf{A}^T$, and

$$(\mathbf{A}^T)^{-1} = (\mathbf{A}^{-1})^T \tag{A.17}$$

A tensor $\mathbf{S}$ is *symmetric*, or $\mathbf{W}$ *skew-symmetric* if

$$\mathbf{S} = \mathbf{S}^T, \quad S_{ij} = S_{ji}; \qquad \mathbf{W} = -\mathbf{W}^T, \quad W_{ij} = -W_{ji} \tag{A.18}$$

and any second order tensor **A** has a unique representation as a sum of symmetric and skew-symmetric tensors,

$$\mathbf{A} = \mathbf{S} + \mathbf{W} \tag{A.19}$$

where

$$\mathbf{S} = \frac{1}{2}(\mathbf{A} + \mathbf{A}^T), \quad \mathbf{W} = \frac{1}{2}(\mathbf{A} - \mathbf{A}^T) \tag{A.20}$$

or in terms of brackets $(\ldots)$ indicating symmetrization and $[\ldots]$ indicating skew-symmetrization

$$A_{ij} = A_{(ij)} + A_{[ij]}; \quad A_{(ij)} = S_{ij}, \quad A_{[ij]} = W_{ij} \tag{A.21}$$

The symmetry and skew-symmetry can also be defined for a third-order tensor **U**, *i.e.*

$$U_{ijk} = U_{(ijk)} + U_{[ijk]} + \frac{2}{3}(U_{[ij]k} + U_{[kj]i} + U_{(ij)k} - U_{k(ij)}) \tag{A.22}$$

where

$$\begin{aligned} U_{(ijk)} &= \frac{1}{6}(U_{ijk} + U_{jki} + U_{kij} + U_{jik} + U_{kji} + U_{ikj}) \\ U_{[ijk]} &= \frac{1}{6}(U_{ijk} + U_{jki} + U_{kij} - U_{jik} - U_{kji} - U_{ikj}) \\ U_{(ij)k} &= \frac{1}{2}(U_{ijk} + U_{jik}) \\ U_{[ij]k} &= \frac{1}{2}(U_{ijk} - U_{jik}) \end{aligned} \tag{A.23}$$

The third order *alternating tensor* is defined as

$$\epsilon_{ijk}\, \mathbf{e}_i \times \mathbf{e}_j \times \mathbf{e}_k \tag{A.24}$$

$$\epsilon_{ijk} = \left| \begin{array}{c} +1 \text{ if } (i,j,k) \text{ is an even permutation of } (1,2,3) \\ -1 \text{ if } (i,j,k) \text{ is an odd permutation of } (1,2,3) \\ 0 \text{ if two or more indices } i,j,k \text{ are equal} \end{array} \right|$$

In particular, the following results are useful

$$\begin{aligned} \epsilon_{pqs}\epsilon_{snr} &= \delta_{np}\delta_{rq} - \delta_{nq}\delta_{rp} \\ \epsilon_{pqs}\epsilon_{sqr} &= \delta_{qp}\delta_{rq} - \delta_{qq}\delta_{rp} = \delta_{pr} - 3\delta_{rp} = -2\delta_{pr} \end{aligned} \tag{A.25}$$

The *contraction* operation of a tensor lowers its index by two. For example, the contraction of a second order tensor **A** leads to a tensor $A_{ii}$ of order zero, or the *trace* of **A**, $tr\mathbf{A}$, *i.e.*

$$A_{ii} = tr\mathbf{A}, \quad \mathbf{a}\cdot\mathbf{b} = tr(\mathbf{a} \times \mathbf{b}) \tag{A.26}$$

The *cross product* $\mathbf{w}=\mathbf{a}\wedge\mathbf{b}$ of vectors $\mathbf{a}$ and $\mathbf{b}$ is a pseudovector

$$w_i = \epsilon_{ijk} a_j a_k \tag{A.27}$$

and with any second order skew-symmetric tensor $\mathbf{W}$ can be associated a vector $\mathbf{w}$, such that

$$w_i = \frac{1}{2}\epsilon_{ijk} W_{jk}, \quad W_{ij} = \epsilon_{ijk} w_k \tag{A.28}$$

or in matrix form

$$\| W_{ij} \| = \begin{vmatrix} 0 & W_{12} & W_{13} \\ -W_{12} & 0 & W_{23} \\ -W_{13} & -W_{23} & 0 \end{vmatrix} = \begin{vmatrix} 0 & w_3 & -w_2 \\ -w_3 & 0 & w_1 \\ w_2 & -w_1 & 0 \end{vmatrix} \tag{A.29}$$

A mapping $\mathbf{Q}$ of an inner-product space is *orthogonal* if it preserves the inner-product:

$$\mathbf{Q}(\mathbf{a})\mathbf{Q}(\mathbf{b}) = \mathbf{a}\cdot\mathbf{b} \tag{A.30}$$

This condition is satisfied *if and only if* $\mathbf{Q}$ is an invertible tensor such that

$$\mathbf{Q}^{-1} = \mathbf{Q}^T \tag{A.31}$$

and hence

$$det\mathbf{Q} = \pm 1 \tag{A.32}$$

If $det\mathbf{Q}=1$, the orthogonal tensor is *proper* or a *rotation*.

Two topological spaces $(X,\mathcal{X})$ and $(Y,\mathcal{Y})$, where $X$ and $Y$ are sets and $\mathcal{X}$ and $\mathcal{Y}$ are topologies, are *homeomorphic* (or topologically equivalent) if and only if there exists a map $H: X \to Y$ such that

1. $H$ is bijective.
2. $H$ is continuous.
3. $H^{-1}$ is continuous.

The map is then a *homeomorphism* from $(X,Y)$ to $(\mathcal{X},\mathcal{Y})$.

The distance between the points $\mathbf{a}$ and $\mathbf{b}$ is the magnitude $|\mathbf{b}-\mathbf{a}|$ or metric that can define a topology of the Euclidean space, with standard procedures defining then the continuity, convergence, limits, compactness, *etc.* When a tensor $\mathbf{A}$ maps a subspace into itself, that subspace is *invariant* under $\mathbf{A}$. Every tensor $\mathbf{A}$ has invariant subspaces: the whole vector space, the subspace $\{\mathbf{0}\}$, the range of $\mathbf{A}$, and the nullspace of $\mathbf{A}$. Moreover, if $\lambda$ is any scalar the nullspace of $\mathbf{A}-\lambda\mathbf{I}$ is also an invariant subspace of $\mathbf{A}$. It is called the *proper space* of **A** corresponding to $\lambda$, and its dimension is the *multiplicity* of $\lambda$ for **A**. The *characteristic equation* of **A** is

$$\lambda^3 - I_1\lambda^2 + I_2\lambda - I_3 = 0 \tag{A.33}$$

where the *principal invariants* $I_k$ of **A** are uniquely determined by the complex numbers $a_1, a_2, a_3$ such that

$$\lambda^3 - I_1\lambda^2 + I_2\lambda - I_3 = (\lambda - a_1)(\lambda - a_2)(\lambda - a_3)$$
$$I_1 = a_1 + a_2 + a_3, \quad I_2 = a_2a_3 + a_3a_1 + a_1a_2, \quad I_3 = a_1a_2a_3 \tag{A.34}$$

If **A** is symmetric then $a_1$, $a_2$, and $a_3$ are basic invariants in the sense that any invariant of **A** can be expressed in terms of them, *i.e.*

$$I_1 = tr\mathbf{A}, \quad I_2 = \frac{1}{2}[(tr\mathbf{A})^2 - tr\mathbf{A}^2], \quad I_3 = det\mathbf{A} \tag{A.35}$$

Another set of invariants of **A** are $tr\mathbf{A}$, $tr\mathbf{A}^2$, and $tr\mathbf{A}^3$. The invariants, by definition, are independent of the choice of the coordinate system and play a central role in the theory of constitutive equations as further discussed below in section 4.

## 3 Basic Identities from Calculus

Let $f$ and $g$ be scalars, **f** and **g** vectors, **A** a second order tensor, and all differentiable. Then,

$$\nabla(fg) = f\nabla g + g\nabla f, \quad (fg)_{,i} = fg_{,i} + gf_{,i}$$
$$\nabla(\mathbf{f}\cdot\mathbf{g}) = (\nabla\mathbf{f})^T\mathbf{g} + (\nabla\mathbf{g})^T\mathbf{f}, \quad (f_ig_i)_{,j} = f_{i,j}\,g_i + f_i\,g_{i,j}$$
$$\nabla(f\mathbf{g}) = f\nabla\mathbf{g} + \mathbf{g}\times\nabla f, \quad (fg_i)_{,j} = fg_{i,j} + g_if_{,j}$$
$$div(\mathbf{f}) = \nabla\cdot\mathbf{f} = tr(\nabla\mathbf{f}) = f_{i,i}$$
$$\Delta f = div(\nabla f) = tr(\nabla^2 f), \quad f_{,ii} = (f_{,i})_{,i}$$
$$\nabla\cdot f\mathbf{g} = div(f\mathbf{g}) = \mathbf{g}\cdot\nabla f + f\nabla\cdot\mathbf{g}, \quad (fg_i)_{,i} = g_if_{,i} + fg_{i,i}$$
$$\nabla\cdot(\mathbf{A}\mathbf{g}) = (\nabla\cdot\mathbf{A}^T)\cdot\mathbf{g} + tr(\mathbf{A}\nabla\mathbf{g}), \quad (A_{ij}g_j)_{,i} = A_{ij,i}\,g_j + A_{ij}\,g_{j,i}$$
$$\nabla\cdot(\nabla\mathbf{g})^T = \nabla(\nabla\cdot\mathbf{g}), \quad g_{i,ji} = g_{i,ij}$$
$$\nabla\cdot(\nabla\mathbf{g} \pm (\nabla\mathbf{g})^T) = \Delta\mathbf{g} \pm \nabla(\nabla\cdot\mathbf{g}), \quad (g_{i,j} \pm g_{j,i})_{,j} = g_{i,jj} \pm g_{j,ji} \tag{A.36}$$

## 4 Results from the Theory of Invariants and Tensor Representation Theorems

The invariance of constitutive equations under orthogonal transformation **Q** is discussed in chapters 4 and 6. A function

$$f(\mathbf{v}_1, \ldots, \mathbf{v}_P, \mathbf{A}_1, \ldots, \mathbf{A}_N, \mathbf{W}_1, \ldots, \mathbf{W}_M) \tag{A.37}$$

of $P$ vectors $\mathbf{v}$, $N$ symmetric second order tensors $\mathbf{A}$, and $M$ skew-symmetric second order tensors $\mathbf{W}$ is an *absolute invariant* of the vectors and tensors under the transformation $\mathbf{Q}$ if

$$\begin{aligned} f(\mathbf{Q}\mathbf{v}_1,\ldots,\,\mathbf{Q}\mathbf{v}_P,\,\mathbf{Q}\mathbf{A}_1\mathbf{Q}^T,\ldots,\,\mathbf{Q}\mathbf{A}_N\mathbf{Q}^T,\,\mathbf{Q}\mathbf{W}_1\mathbf{Q}^T,\ldots,\,\mathbf{Q}\mathbf{W}_M\mathbf{Q}^T) \\ = f(\mathbf{v}_1,\ldots,\,\mathbf{v}_P,\,\mathbf{A}_1,\ldots,\,\mathbf{A}_N,\,\mathbf{W}_1,\ldots,\,\mathbf{W}_M) \end{aligned} \tag{A.38}$$

The central problem in the theory of invariants is to determine a set of basic invariants from which all other can be generated, given a set of variables (vectors and tensors) and a group of transformations. When it is sufficient to consider a function of polynomial invariants, the polynomial invariant is *reducible* if it can be expressed as a polynomial in other invariants. A set of polynomial invariants with the property that any polynomial invariant can be expressed as a polynomial in members of the given set is an *integrity basis*. The main problem is to determine *minimal* integrity bases and polynomial relations between invariants which do not permit any one invariant to be expressed as a polynomial in the remainder.

In this section, a summary of representations of isotropic scalar, vector, and tensor functions will be presented from the works of WANG (1971), SMITH (1971), and corrections introduced by BOEHLER (1977). For polynomial invariants, the article by SPENCER (1971) may also be consulted.

Let $f$, $\mathbf{g}$, $\mathbf{H}$, and $\mathbf{K}$ be scalar, vector, symmetric second order tensor, and skew-symmetric second order tensor isotropic functions, respectively, of vectors and tensors as in (A.37). The *irreducible set of invariants* of $P$ vectors $\mathbf{v}$, $N$ tensors $\mathbf{A}$, and $M$ tensors $\mathbf{W}$ are as follows:

$$\begin{gathered} \mathbf{v}_m\cdot\mathbf{v}_m,\ \mathbf{v}_n\cdot\mathbf{v}_m,\ tr\mathbf{A}_i,\ tr\mathbf{A}_i^2,\ tr\mathbf{A}_i^3,\ tr\mathbf{A}_i\mathbf{A}_j, \\ tr\mathbf{A}_i^2\mathbf{A}_j,\ tr\mathbf{A}_i\mathbf{A}_j^2,\ tr\mathbf{A}_i^2\mathbf{A}_j^2,\ tr\mathbf{A}_i\mathbf{A}_j\mathbf{A}_k,\ tr\mathbf{W}_p^2,\ tr\mathbf{W}_p\mathbf{W}_q, \\ tr\mathbf{W}_p\mathbf{W}_q\mathbf{W}_r,\ \mathbf{v}_m\cdot\mathbf{A}_i\mathbf{v}_m,\ \mathbf{v}_m\cdot\mathbf{A}_i^2\mathbf{v}_m,\ \mathbf{v}_m\cdot\mathbf{A}_i\mathbf{A}_j\mathbf{v}_m,\ \mathbf{v}_m\cdot\mathbf{A}_i\mathbf{v}_n,\ \mathbf{v}_m\cdot\mathbf{A}_i^2\mathbf{v}_n, \\ \mathbf{v}_m\cdot(\mathbf{A}_i\mathbf{A}_j-\mathbf{A}_j\mathbf{A}_i)\mathbf{v}_n,\ \mathbf{v}_m\cdot\mathbf{W}_p^2\mathbf{v}_m,\ \mathbf{v}_m\cdot\mathbf{W}_p\mathbf{W}_q\mathbf{v}_m,\ \mathbf{v}_m\cdot\mathbf{W}_p^2\mathbf{W}_q\mathbf{v}_m, \\ \mathbf{v}_m\cdot\mathbf{W}_p\mathbf{W}_q^2\mathbf{v}_m,\ \mathbf{v}_m\cdot\mathbf{W}_p\mathbf{v}_n,\ \mathbf{v}_m\cdot\mathbf{W}_p^2\mathbf{v}_n,\ \mathbf{v}_m\cdot(\mathbf{W}_p\mathbf{W}_q-\mathbf{W}_q\mathbf{W}_p)\mathbf{v}_n, \\ tr\mathbf{A}_i\mathbf{W}_p^2,\ tr\mathbf{A}_i^2\mathbf{W}_p^2,\ tr\mathbf{A}_i^2\mathbf{W}_p^2\mathbf{A}_i\mathbf{W}_p,\ tr\mathbf{A}_i\mathbf{W}_p\mathbf{W}_q,\ tr\mathbf{A}_i\mathbf{W}_p\mathbf{W}_q^2, \\ tr\mathbf{A}_i\mathbf{W}_p^2\mathbf{W}_q,\ tr\mathbf{A}_i\mathbf{A}_j\mathbf{W}_p,\ tr\mathbf{A}_i\mathbf{W}_p^2\mathbf{A}_j\mathbf{W}_p,\ tr\mathbf{A}_i\mathbf{A}_j^2\mathbf{W}_p,\ tr\mathbf{A}_i^2\mathbf{A}_j\mathbf{W}_p, \\ \mathbf{v}_m\cdot\mathbf{A}_i\mathbf{W}_p\mathbf{v}_m,\ \mathbf{v}_m\cdot\mathbf{W}_p\mathbf{A}_i\mathbf{W}_p^2\mathbf{v}_m,\ \mathbf{v}_m\cdot\mathbf{A}_i^2\mathbf{W}_p\mathbf{v}_m,\ \mathbf{v}_m\cdot(\mathbf{A}_i\mathbf{W}_p-\mathbf{W}_p\mathbf{A}_i)\mathbf{v}_n \end{gathered} \tag{A.39}$$

where $i,j,k=1,\ldots,N$; $i<j<k$, where $p,q,r=1,\ldots,M$; $p<q<r$, and where $m,n=1,\ldots,P$; $m<n$.

The representation of a *vector-valued isotropic function* $\mathbf{g}$ is given by

$$\mathbf{g}(\mathbf{v}_m,\,\mathbf{A}_i,\,\mathbf{W}_p)=\sum_r \alpha_r\,\mathbf{g}_r(\mathbf{v}_m,\,\mathbf{A}_i,\,\mathbf{W}_p) \tag{A.40}$$

where $\alpha_r$ are scalar-valued isotropic functions of the invariants (A.39), whereas $\mathbf{g}_r$ are vector-valued functions given by

$$\begin{gathered}\mathbf{v}_m,\ \mathbf{A}_i\mathbf{v}_m,\ \mathbf{A}_i^2\mathbf{v}_m,\ (\mathbf{A}_i\mathbf{A}_j-\mathbf{A}_j\mathbf{A}_i)\mathbf{v}_m,\ \mathbf{W}_p\mathbf{v}_m,\\ \mathbf{W}_p^2\mathbf{v}_m,\ (\mathbf{W}_p\mathbf{W}_q-\mathbf{W}_q\mathbf{W}_p)\mathbf{v}_m,\ (\mathbf{A}_i\mathbf{W}_p-\mathbf{W}_p\mathbf{A}_i)\mathbf{v}_m\end{gathered}\tag{A.41}$$

where $i,j=1,\dots,N;\ i<j$, where $p,q=1,\dots,M;\ p<q$, and where $m=1,\dots,P$.

A *symmetric tensor-valued isotropic function* $\mathbf{H}$ has the following representation

$$\mathbf{H}(\mathbf{v}_m,\ \mathbf{A}_i,\ \mathbf{W}_p)=\sum_r\beta_r\,\mathbf{H}_r(\mathbf{v}_m,\ \mathbf{A}_i,\ \mathbf{W}_p)\tag{A.42}$$

where $\beta_r$ are scalar-valued isotropic functions of the invariants (A.39), and where $\mathbf{H}_r$ are symmetric tensor-valued isotropic functions given by

$$\begin{gathered}\mathbf{I},\ \mathbf{A}_i,\ \mathbf{A}_i^2,\ \mathbf{A}_i\mathbf{A}_j+\mathbf{A}_j\mathbf{A}_i,\ \mathbf{A}_i^2\mathbf{A}_j+\mathbf{A}_j\mathbf{A}_i^2,\ \mathbf{A}_i\mathbf{A}_j^2+\mathbf{A}_j^2\mathbf{A}_i,\\ \mathbf{v}_m\times\mathbf{v}_m,\ \mathbf{v}_m\times\mathbf{v}_n+\mathbf{v}_n\times\mathbf{v}_m,\ \mathbf{W}_p^2,\ \mathbf{W}_p\mathbf{W}_q+\mathbf{W}_q\mathbf{W}_p,\ \mathbf{W}_p\mathbf{W}_q^2-\mathbf{W}_q^2\mathbf{W}_p,\\ \mathbf{W}_p^2\mathbf{W}_q-\mathbf{W}_q\mathbf{W}_p^2,\ \mathbf{v}_m\times\mathbf{A}_i\mathbf{v}_m+\mathbf{A}_i\mathbf{v}_m\times\mathbf{v}_m,\ \mathbf{v}_m\times\mathbf{A}_i^2\mathbf{v}_m+\mathbf{A}_i^2\mathbf{v}_m\times\mathbf{v}_m,\\ \mathbf{A}_i(\mathbf{v}_m\times\mathbf{v}_n-\mathbf{v}_n\times\mathbf{v}_m)-(\mathbf{v}_m\times\mathbf{v}_n-\mathbf{v}_n\times\mathbf{v}_m)\mathbf{A}_i,\ \mathbf{A}_i\mathbf{W}_p-\mathbf{W}_p\mathbf{A}_i,\\ \mathbf{W}_p\mathbf{A}_i\mathbf{W}_p,\ \mathbf{A}_i^2\mathbf{W}_p-\mathbf{W}_p\mathbf{A}_i^2,\ \mathbf{W}_p\mathbf{A}_i\mathbf{W}_p^2-\mathbf{W}_p^2\mathbf{A}_i\mathbf{W}_p,\ \mathbf{W}_p\mathbf{v}_m\times\mathbf{W}_p\mathbf{v}_m,\\ \mathbf{v}_m\times\mathbf{W}_p\mathbf{v}_m+\mathbf{W}_p\mathbf{v}_m\times\mathbf{v}_m,\ \mathbf{W}_p\mathbf{v}_m\times\mathbf{W}_p^2\mathbf{v}_m+\mathbf{W}_p^2\mathbf{v}_m\times\mathbf{W}_p\mathbf{v}_m,\\ \mathbf{W}_p(\mathbf{v}_m\times\mathbf{v}_n-\mathbf{v}_n\times\mathbf{v}_m)+(\mathbf{v}_m\times\mathbf{v}_n-\mathbf{v}_n\times\mathbf{v}_m)\mathbf{W}_p\end{gathered}\tag{A.43}$$

where $i,j=1,\dots,N;\ i<j$, where $p,q=1,\dots,M;\ p<q$, and where $m,n=1,\dots,P;\ m<n$.

A *skew-symmetric tensor valued isotropic function* $\mathbf{K}$ has the representation as follows

$$\mathbf{K}(\mathbf{v}_m,\ \mathbf{A}_i,\ \mathbf{W}_p)=\sum_r\gamma_r\,\mathbf{K}_r(\mathbf{v}_m,\ \mathbf{A}_i,\ \mathbf{W}_p)\tag{A.44}$$

where $\gamma_r$ are scalar-valued isotropic functions of the invariants (A.39), whereas $\mathbf{K}_r$ are skew-symmetric tensor-valued isotropic functions given by

$$\begin{gathered}\mathbf{W}_p,\ \mathbf{W}_p\mathbf{W}_q-\mathbf{W}_q\mathbf{W}_p,\ \mathbf{A}_i\mathbf{A}_j-\mathbf{A}_j\mathbf{A}_i,\\ \mathbf{A}_i^2\mathbf{A}_j-\mathbf{A}_j\mathbf{A}_i^2,\ \mathbf{A}_i^2\mathbf{A}_j^2-\mathbf{A}_j^2\mathbf{A}_i^2,\ \mathbf{A}_i\mathbf{A}_j\mathbf{A}_i^2-\mathbf{A}_i^2\mathbf{A}_j\mathbf{A}_i,\ \mathbf{A}_j\mathbf{A}_i\mathbf{A}_j^2-\mathbf{A}_j^2\mathbf{A}_i\mathbf{A}_j,\\ \mathbf{A}_i\mathbf{A}_j\mathbf{A}_k+\mathbf{A}_j\mathbf{A}_k\mathbf{A}_i+\mathbf{A}_k\mathbf{A}_i\mathbf{A}_j-\mathbf{A}_j\mathbf{A}_i\mathbf{A}_k-\mathbf{A}_i\mathbf{A}_k\mathbf{A}_j-\mathbf{A}_k\mathbf{A}_j\mathbf{A}_i,\\ \mathbf{v}_m\times\mathbf{v}_n-\mathbf{v}_n\times\mathbf{v}_m,\ \mathbf{v}_m\times\mathbf{A}_i\mathbf{v}_m-\mathbf{A}_i\mathbf{v}_m\times\mathbf{v}_m,\ \mathbf{v}_m\times\mathbf{A}_i^2\mathbf{v}_m-\mathbf{A}_i^2\mathbf{v}_m\times\mathbf{v}_m,\end{gathered}$$

$$\begin{gathered}\mathbf{A}_i\mathbf{v}_m \times \mathbf{A}_i^2\mathbf{v}_m - \mathbf{A}_i^2\mathbf{v}_m \times \mathbf{A}_i\mathbf{v}_m,\ \mathbf{A}_i\mathbf{v}_m \times \mathbf{A}_j\mathbf{v}_m \\ -\mathbf{A}_j\mathbf{v}_m \times \mathbf{A}_i\mathbf{v}_m + \mathbf{v}_m \times (\mathbf{A}_i\mathbf{A}_j - \mathbf{A}_j\mathbf{A}_i)\mathbf{v}_m - (\mathbf{A}_i\mathbf{A}_j - \mathbf{A}_j\mathbf{A}_i)\mathbf{v}_m \times \mathbf{v}_m, \\ \mathbf{A}_i(\mathbf{v}_m \times \mathbf{v}_n - \mathbf{v}_n \times \mathbf{v}_m) + (\mathbf{v}_m \times \mathbf{v}_n - \mathbf{v}_n \times \mathbf{v}_m)\mathbf{A}_i,\ \mathbf{A}_i\mathbf{W}_p + \mathbf{W}_p\mathbf{A}_i, \\ \mathbf{A}_i\mathbf{W}_p^2 - \mathbf{W}_p^2\mathbf{A}_i,\ \mathbf{v}_m \times \mathbf{W}_p\mathbf{v}_m - \mathbf{W}_p\mathbf{v}_m \times \mathbf{v}_m,\ \mathbf{v}_m \times \mathbf{W}_p^2\mathbf{v}_m - \mathbf{W}_p^2\mathbf{v}_m \times \mathbf{v}_m, \\ \mathbf{W}_p(\mathbf{v}_m \times \mathbf{v}_n - \mathbf{v}_n \times \mathbf{v}_m) - (\mathbf{v}_m \times \mathbf{v}_n - \mathbf{v}_n \times \mathbf{v}_m)\mathbf{W}_p\end{gathered} \tag{A.45}$$

where $i,j,k=1,\ldots,N$; $i<j<k$, where $p,q=1,\ldots,M$; $p<q$, and where $m,n=1,\ldots,P$; $m<n$.

The representation of a general second order tensor-valued isotropic function is obtained by adding (A.42) and (A.44). *Form-invariant functionals* may also be constructed in terms of polynomial invariants (SPENCER, 1971).

An occurring problem in the studies of linearized constitutive equations is to obtain representations for *isotropic tensors.* An isotropic tensor of order $\mu$ has *components* which are unchanged by an orthogonal transformation of rectangular cartesian coordinates (SPENCER,1971), *i.e.*

$$\alpha^*_{i_1 i_2 \ldots i_\mu} = \alpha_{i_1 i_2 \ldots i_\mu} \tag{A.46}$$

$\mu$ is *even:* $\alpha_{i_1 i_2 \ldots i_\mu}$ is the sum of the terms of the type

$$\delta_{i_\alpha i_\beta}\, \delta_{i_\gamma i_\delta} \ldots \delta_{i_\sigma i_\tau}$$

$\mu$ is *odd:* $\alpha_{i_1 i_2 \ldots i_\mu}$ is a sum of the terms of the type

$$\delta_{i_\alpha i_\beta}\, \delta_{i_\gamma i_\delta} \ldots \delta_{i_\lambda i_\nu}\, \epsilon_{i_\rho i_\sigma i_\tau}$$

where in each case $\alpha,\beta,\gamma,\delta,\ldots,\rho,\sigma,\tau$ is a permutation of $1,2,\ldots,\mu$. As can be seen, the isotropic tensors of even order are invariant under improper orthogonal transformations, whereas those of odd order change sign under such transformations. For the use in chapters 7 and 8, it can be shown from the above that

$$\begin{gathered}\alpha_{ij} = a\,\delta_{ij}, \qquad \alpha_{ijk} = a\,\epsilon_{ijk} \\ \alpha_{ijpq} = a\,\delta_{ij}\delta_{pq} + b\,\delta_{ip}\delta_{jq} + c\,\delta_{iq}\delta_{jp} \\ \alpha_{ijkmn\ell} = a_1\,\delta_{ij}\delta_{k\ell}\delta_{mn} + a_2\,\delta_{ij}\delta_{km}\delta_{\ell n} + a_3\,\delta_{ij}\delta_{\ell m}\delta_{kn} + a_4\,\delta_{ik}\delta_{j\ell}\delta_{mn} \\ +a_5\,\delta_{ik}\delta_{\ell m}\delta_{jn} + a_6\,\delta_{ik}\delta_{jm}\delta_{\ell n} + a_7\,\delta_{jk}\delta_{i\ell}\delta_{mn} + a_8\,\delta_{jk}\delta_{im}\delta_{\ell n} \\ +a_9\,\delta_{jk}\delta_{in}\delta_{m\ell} + a_{10}\,\delta_{im}\delta_{jn}\delta_{k\ell} + a_{11}\,\delta_{im}\delta_{j\ell}\delta_{kn} + a_{12}\,\delta_{in}\delta_{jm}\delta_{k\ell} \\ +a_{13}\,\delta_{in}\delta_{j\ell}\delta_{km} + a_{14}\,\delta_{i\ell}\delta_{jm}\delta_{kn} + a_{15}\,\delta_{i\ell}\delta_{jn}\delta_{km}\end{gathered} \tag{A.47}$$

# Index

## Lecture Notes in Mathematics

Vol. 1236: Stochastic Partial Differential Equations and Applications. Proceedings, 1985. Edited by G. Da Prato and L. Tubaro. V, 257 pages. 1987.

Vol. 1237: Rational Approximation and its Applications in Mathematics and Physics. Proceedings, 1985. Edited by J. Gilewicz, M. Pindor and W. Siemaszko. XII, 350 pages. 1987.

Vol. 1250: Stochastic Processes – Mathematics and Physics II. Proceedings 1985. Edited by S. Albeverio, Ph. Blanchard and L. Streit. VI, 359 pages. 1987.

Vol. 1251: Differential Geometric Methods in Mathematical Physics. Proceedings, 1985. Edited by P.L. García and A. Pérez-Rendón. VII, 300 pages. 1987.

Vol. 1255: Differential Geometry and Differential Equations. Proceedings, 1985. Edited by C. Gu, M. Berger and R.L. Bryant. XII, 243 pages. 1987.

Vol. 1256: Pseudo-Differential Operators. Proceedings, 1986. Edited by H.O. Cordes, B. Gramsch and H. Widom. X, 479 pages. 1987.

Vol. 1258: J. Weidmann, Spectral Theory of Ordinary Differential Operators. VI, 303 pages. 1987.

Vol. 1260: N.H. Pavel, Nonlinear Evolution Operators and Semigroups. VI, 285 pages. 1987.

Vol. 1263: V.L. Hansen (Ed.), Differential Geometry. Proceedings, 1985. XI, 288 pages. 1987.

Vol. 1265: W. Van Assche, Asymptotics for Orthogonal Polynomials. VI, 201 pages. 1987.

Vol. 1267: J. Lindenstrauss, V.D. Milman (Eds.), Geometrical Aspects of Functional Analysis. Seminar. VII, 212 pages. 1987.

Vol. 1269: M. Shiota, Nash Manifolds. VI, 223 pages. 1987.

Vol. 1270: C. Carasso, P.-A. Raviart, D. Serre (Eds.), Nonlinear Hyperbolic Problems. Proceedings, 1986. XV, 341 pages. 1987.

Vol. 1272: M.S. Livšic, L.L. Waksman, Commuting Nonselfadjoint Operators in Hilbert Space. III, 115 pages. 1987.

Vol. 1273: G.-M. Greuel, G. Trautmann (Eds.), Singularities, Representation of Algebras, and Vector Bundles. Proceedings, 1985. XIV, 383 pages. 1987.

Vol. 1275: C.A. Berenstein (Ed.), Complex Analysis I. Proceedings, 1985–86. XV, 331 pages. 1987.

Vol. 1276: C.A. Berenstein (Ed.), Complex Analysis II. Proceedings, 1985–86. IX, 320 pages. 1987.

Vol. 1277: C.A. Berenstein (Ed.), Complex Analysis III. Proceedings, 1985–86. X, 350 pages. 1987.

Vol. 1283: S. Mardešić, J. Segal (Eds.), Geometric Topology and Shape Theory. Proceedings, 1986. V, 261 pages. 1987.

Vol. 1285: I.W. Knowles, Y. Saitō (Eds.), Differential Equations and Mathematical Physics. Proceedings, 1986. XVI, 499 pages. 1987.

Vol. 1287: E.B. Saff (Ed.), Approximation Theory, Tampa. Proceedings, 1985–1986. V, 228 pages. 1987.

Vol. 1288: Yu. L. Rodin, Generalized Analytic Functions on Riemann Surfaces. V, 128 pages, 1987.

Vol. 1294: M. Queffélec, Substitution Dynamical Systems – Spectral Analysis. XIII, 240 pages. 1987.

Vol. 1299: S. Watanabe, Yu.V. Prokhorov (Eds.), Probability Theory and Mathematical Statistics. Proceedings, 1986. VIII, 589 pages. 1988.

Vol. 1300: G.B. Seligman, Constructions of Lie Algebras and their Modules. VI, 190 pages. 1988.

Vol. 1302: M. Cwikel, J. Peetre, Y. Sagher, H. Wallin (Eds.), Function Spaces and Applications. Proceedings, 1986. VI, 445 pages. 1988.

Vol. 1303: L. Accardi, W. von Waldenfels (Eds.), Quantum Probability and Applications III. Proceedings, 1987. VI, 373 pages. 1988.

## Lecture Notes in Physics

Vol. 347: Ph. Blanchard, J.-P. Dias, J. Stubbe (Eds.), New Methods and Results in Non-linear Field Equations. Proceedings. 1987. VII, 133 pages. 1989.

Vol. 348: A. Bohm, M. Gadella, Dirac Kets, Gamow Vectors and Gel'fand Triplets. VII, 119 pages. 1989.

Vol. 349: Y. A. Kubyshin, J. M. Mourão, G. Rudolph, I. P. Volobujev, Dimensional Reduction of Gauge Theories, Spontaneous Compactification and Model Building. X, 80 pages. 1989.

Vol. 350: G. Tenorio-Tagle, M. Moles, J. Melnick (Eds.), Structure and Dynamics of the Interstellar Medium. Proceedings, 1989. XXI, 537 pages. 1989.

Vol. 351: A. Dervieux, B. Larrouturou (Eds.), Numerical Combustion. Proceedings, 1989. VII, 481 pages. 1989.

Vol. 352: K. B. Wolf (Ed.), Lie Methods in Optics II. Proceedings, 1988. XII, 197 pages. 1989.

Vol. 353: M. Barthes, J. Léon (Eds.), Nonlinear Coherent Structures. Proceedings, 1989. X, 280 pages. 1990.

Vol. 354: J. De Coninck, F. Dunlop (Eds.), Wetting Phenomena. Proceedings, 1988. VII, 112 pages. 1990.

Vol. 355: R. Lima, L. Streit, R. Vilela Mendes (Eds.), Dynamics and Stochastic Processes. Proceedings, 1988. V, 298 pages. 1990.

Vol. 356: R. Wehrse (Ed.), Accuracy of Element Abundances from Stellar Atmospheres. Proceedings, 1988. V, 109 pages. 1990.

Vol. 357: J. L. Lumley (Ed.), Whither Turbulence? Turbulence at the Crossroads. Proceedings, 1989. V, 535 pages. 1990.

Vol. 358: J. Kovalevsky, Astrométrie moderne. VI, 294 pages. 1990.

Vol. 359: K. Kirchgässner (Ed.), Problems Involving Change of Type. Proceedings, 1988. XII, 207 pages. 1990.

Vol. 360: Y. Mellier, B. Fort, G. Soucail (Eds.), Gravitational Lensing. Proceedings 1989. XV, 315 pages. 1990.

Vol. 361: P. Gaigg, W. Kummer, M. Schweda (Eds.), Physical and Nonstandard Ganges. Proceedings, 1989. IX, 310 pages. 1990.

Vol. 362: N.E. Kassim, K.W. Weiler (Eds.), Low Frequency Astrophysics from Space. Proceedings. XII, 280 pages. 1990.

Vol. 363: V. Ruždjak, E. Tandberg-Hanssen (Eds.), Dynamics of Quiescent Prominences. Proceedings, 1989. XI, 304 pages. 1990.

Vol. 364: T. T. S. Kuo, E. Osnes, Folded-Diagram Theory of the Effective Interaction in Nuclei, Atoms and Molecules. V, 175 pages. 1990.

Vol. 365: M. Schumacher, G. Tamas (Eds.), Perspectives of Photon Interactions with Hadrons and Nuclei. Proceedings. IX, 251 pages. 1990.

Vol. 366: M.-J. Goupil, J.-P. Zahn (Eds.), Rotation and Mixing in Stellar Interiors. Proceedings, 1989. XIII, 183 pages. 1990.

Vol. 367: Y. Osaki, H. Shibahashi (Eds.), Progress of Seismology of the Sun and Stars. Proceedings, 1989. XIII, 467 pages. 1990.

Vol. 368: L. Garrido (Ed.), Statistical Mechanics of Neural Networks. Proceedings. VI, 477 pages. 1990.

Vol. 370: H.-D. Doebner, J.-D. Hennig (Eds.), Quantum Groups. Proceedings, 1989. X, 434 pages. 1990.

Vol. 372: F. Dobran, Theory of Structured Multiphase Mixtures. IX, 223 pages. 1991.